Topics on Chaotic Systems

Selected Papers from CHAOS 2008 International Conference

Topics on Chaotic Systems

Selected Papers from CHAOS 2008 International Conference

Chania, Crete, Greece 3 – 6 June 2008

editors

Christos H Skiadas
Technical University of Crete, Greece

Ioannis Dimotikalis
Technological Educational Institute of Crete, Greece

Charilaos Skiadas
Hanover College, Indiana, USA

World Scientific

NEW JERSEY • LONDON • SINGAPORE • BEIJING • SHANGHAI • HONG KONG • TAIPEI • CHENNAI

Published by

World Scientific Publishing Co. Pte. Ltd.
5 Toh Tuck Link, Singapore 596224
USA office: 27 Warren Street, Suite 401-402, Hackensack, NJ 07601
UK office: 57 Shelton Street, Covent Garden, London WC2H 9HE

British Library Cataloguing-in-Publication Data
A catalogue record for this book is available from the British Library.

TOPICS ON CHAOTIC SYSTEMS
Selected Papers from CHAOS 2008 International Conference

ISBN-13 978-981-4271-33-2
ISBN-10 981-4271-33-0

Printed in Singapore by B & Jo Enterprise Pte Ltd

PREFACE

This collection includes the main part of the best papers presented in the International Conference (CHAOS2008) on Chaotic Modeling, Simulation and Applications, Chania, Crete, Greece, June 3–6, 2008. The task was to bring together various groups working in the area of Nonlinear Systems and Dynamics, Chaotic Theory and Application for exchanging views and reporting research findings. The topics included are relevant to the study of nonlinear systems and dynamics in an interdisciplinary research, and include many very interesting applications. This book provides a valuable collection of new ideas, methods, and techniques in the field of Nonlinear Dynamics, Chaos, Fractals and their applications in General Science and Engineering Sciences.

The book focuses on many fields such as Chaos and Dynamical Systems, Nonlinear Systems, Fractals, Chaotic Attractors, Mechanics, Hydro-Fluid Dynamics, Chaotic Advection, Chaos in Meteorology and Cosmology, Bifurcation, Hamiltonian and Quantum Chaos, Plasma Physics, Chaos in Biology and Genetics, Chaos in Medicine and Physiology, Chaotic Control, Time Series Analysis and Forecasting Chaotic Systems, Chaos in Economy and Markets, Traffic Flow, and Chaotic Simulations. These contributions present new ideas and methods for solving problems by analyzing the relevant data. Also, the use of recent advances in different fields are emphasized, especially on chaotic simulation methods and techniques.

We would like to acknowledge the valuable support and hospitality provided by the Mediterranean Agronomic Institute, Chania, Greece. Sincere thanks must also go to those whose contributions have been essential in organizing the Conference and creating these Proceedings. Finally, We would like to thank Anthi Katsirikou, Mary Karadima, Aggeliki Oikonomou and George Matalliotakis for their valuable support.

February 10, 2009

Christos H. Skiadas, Technical University of Crete, Greece
Ioannis Dimotikalis, Technological Educational Institute of Crete, Greece
Charilaos Skiadas, Hanover College, Indiana, USA

INTERNATIONAL ADVISORY COMMITTEE

for the Series of International Conferences on Chaotic Modeling, Simulation and Applications

C. H. Skiadas (chair)	– Technical University of Crete, Chania, Greece
H. Adeli (hon. chair)	– The Ohio State University, USA
N. Akhmediev	– Australian National University, Australia
M. Amabili	– Universita di Parma, Parma, Italy
J. Awrejcewicz	– Technical University of Lodz, Poland
T. Bountis	– University of Patras, Greece
Y. S. Boutalis	– Democritus University of Thrace, Greece
C. Chandre	– Centre de Physique Theorique, Marseille, France
M. Christodoulou	– Technical University of Crete, Chania, Crete, Greece
P. Commendatore	– Università di Napoli 'Federico II', Italy
B. Dacorogna	– Ecole Polytechnique Fdrale de Lausanne, Suisse
D. Dhar	– Tata Institute of Fundamental Research, India
J. Dimotikalis	– Technological Educational Institute, Crete, Greece
B. Epureanu	– University of Michigan, Ann Arbor, MI, USA
G. Fagiolo	– Sant'Anna School of Advanced Studies, Pisa, Italy
V. Grigoras	– University of Iasi, Romania
A. Harb	– University of Central Florida, USA
J.-H. He	– Donghua University, Shanghai, China
L. Hong	– Xi'an Jiaotong University, Xi'an, Shaanxi, China
T. Kapitaniak	– Technical University of Lodz, Lodz, Poland
G. P. Kapoor	– Indian Institute of Technology Kanpur, Kanpur, India
W. Li	– Northwestern Polytechnical University, China
V. Lucarini	– University of Bologna, Italy

PLENARY AND KEYNOTE TALKS

Hojjat Adeli
The Ohio State University, USA
Chaos in the Brain: Chaos-Wavelet Models for EEG-Based Diagnosis of Neurological Disorders

Julien Clinton Sprott
University of Wisconsin, Madison, USA
Chaotic Dynamics on Large Networks

Sergey V. Prants
Pacific Oceanological Institute of the RAS Vladivostok, RUSSIA
Chaotic Advection in the Ocean

Tomasz Kapitaniak
Institute of Applied Mechanics, Technical University of Lodz, POLAND
Coin Tossing is Predictable

Nail Akhmediev
Inst. of Adv. Studies, Australian National Univ., Canberra, AUSTRALIA
Chaotic Solitons in Dissipative Systems

Bogdan I. Epureanu
Dept. of Mech. Eng., Univ. of Michigan, Ann Arbor, USA
Creating and Exploiting Nonlinear and Chaotic Dynamics for System Interrogation in Sensing and Damage Detection

Valerio Lucarini
Department of Physics, University of Bologna, Bologna, ITALY
Response Theory for Equilibrium and Non-Equilibrium Statistical Mechanics: Causality and Generalized Kramers-Kronig Relations

Alexander G. Ramm
Math. Dept., Kansas State Univ, Manhattan, Kansas, USA
Creating Materials with a Desired Refraction Coefficient

CONTENTS

The Effects of Machine Components on Bifurcation and Chaos as Applied to Multimachine System

M. M. Alomari* and B. S. Rodanski

University of Technology, Sydney (UTS)
P.O. Box 123, Broadway NSW 2007, Australia
*Email: *majdi.m.alomari@eng.uts.edu.au*

The second system of the IEEE second benchmark model of Subsynchronous Resonance (SSR) is considered. The system can be mathematically modeled as a set of first order nonlinear ordinary differential equations with the compensation factor ($\mu = X_c/X_L$) as a bifurcation (control) parameter. So, bifurcation theory can be applied to nonlinear dynamical systems, which can be written as $dx/dt = F(x;\mu)$. The effects of machine components, i.e. damper winding, automatic voltage regulator (AVR), and power system stabilizer (PSS) on SSR in power system are studied. The results show that these components affect the locations, number and type of the Hopf bifurcations.

Keywords: Hopf bifurcation; chaos; subsynchronous resonance; damper windings; PSS.

1. Introduction

In power systems series compensation is considered as a powerful technique based on economic and technical considerations for increasing effectively the power transfer capability as well as improving the stability of these systems. However, this introduces problems as well as with the benefits, namely the electromechanical interaction between electrical resonant circuits of the transmission system and the torsional natural frequencies of the turbine-generator rotor. This phenomenon is called subsynchronous resonance (SSR), and it can cause shaft fatigue and possible damage or failure.

The phenomenon of subsynchronous resonance occurs mainly in series capacitor-compensated transmission systems. SSR has been studied extensively since 1970, when a major transmission network in southern California experienced shaft failure to its turbine-generator unit with series compensation. The subsynchronous torques on the rotor is a matter of concern because the turbine-generator shaft itself has natural modes of oscillation that are typical of any spring mass system. It happens that the shaft oscillatory modes are at subsynchronous frequencies. Should the induced subsynchronous torques coincide with one of the shaft natural modes of oscillation, the shaft will oscillate at this natural frequency, sometimes with high amplitude.

Three types of SSR can identify the interaction of the system and the generator under the subsynchronous resonance. They have been called torsional interaction effect, induction generator effect, and transient torque effect. In this research, we focus on the torsional interaction effect, which results from the interaction of the electrical subsynchronous mode with the torsional mode. Several methods have been used in SSR study. The most common of these methods are eigenvalue analysis, frequency scanning, and time-domain analysis. The eigenvalue analysis is used in this research. It is a very valuable technique because it provides both the frequencies of oscillation and the damping at each frequency.

Recently, power system dynamics has been studied using the nonlinear dynamics point of view, which utilizes the bifurcation theory. Actually, power systems have rich bifurcation phenomena. Bifurcation is used to indicate a qualitative change in the features of a system, such as the number and types of solution upon a small variation in the parameters of a system. It has been revealed that there are different types of bifurcation in power system models. In general, the power system model can be represented by a system of nonlinear algebraic and ordinary differential equations.

The bifurcation theorem was used by Zhu *et al.* [1] to demonstrate the existence of a Hopf bifurcation in a single machine infinite busbar (SMIB) power system, in which the dynamics of the damper windings and the AVR are neglected. Nayfeh *et al.* [2] applied the bifurcation theory to a practical series capacitor compensated single machine power system, the BOARDMAN turbine-generator system. Harb *et al.* [3] applied a bifurcation analysis together with the method of multiple scales and Floquet theory to the CHOLLA #4 turbine-generator system. Tomim *et al.* [4] proposed an index that identifies Hopf bifurcation points in power systems susceptible to subsynchronous resonance. Dobson and Barocio [5] analyzed general perturbations of a weak resonance and found two distinct behaviors, including interactions near strong resonances in which the eigenvalues quickly change direction.

In this paper, we focus on the torsional interaction effect, which results from the interaction of the electrical subsynchronous mode with the torsional mode. The second system of the IEEE second benchmark model is considered. We use bifurcation theory and chaos to investigate the complex dynamics of the considered system. The type of the Hopf bifurcation is determined by numerical integration of the system, with specific amount of initial disturbances, slightly before and after the bifurcation value. On further increase of the compensation factor, the system experiences chaos via torus attractor. Chaos is a bounded steady-state behavior that is not an equilibrium solution or a periodic solution or a quasiperiodic solution.[6]

2. System Description

The system considered is the two different machine infinite bus system, shown in Figure 1(a). The two machines have a common torsional mode connected to a single series compensated transmission line. The model and the parameters are provided in the second system of the IEEE second benchmark model.

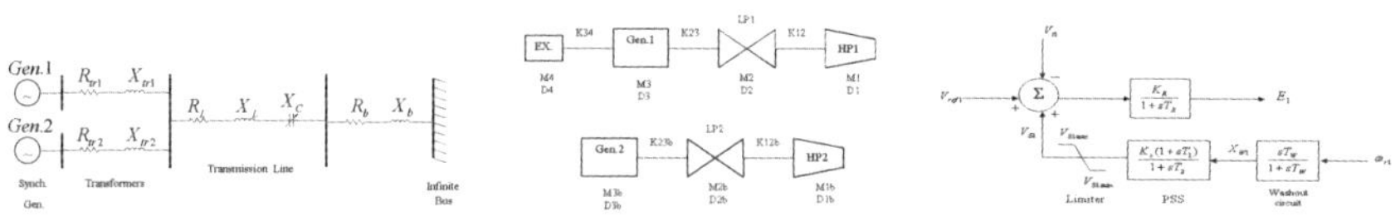

(a) Electrical system (Two different machine infinite bus system)

(b) Electro-mechanical systems for the first and second units

(c) Block diagram of the use of AVR and PSS to the first generator

Fig. 1.

The electro-mechanical systems for the first and second units are shown in Figure 1(b). The first unit consists of exciter (EX), generator (Gen.1), low-pressure (LP1) and high-pressure (HP1) turbine sections. And the second unit consists of generator (Gen.2), low-pressure (LP2) and high-pressure (HP2) turbine sections. Every section has its own angular momentum constant M and damping coefficient D, and every pair of successive masses have their own shaft stiffness constant K, as shown in Figure 1(b). The data for electrical and mechanical system are provided in [7]. Replacement of these generators with a single equivalent generator will change the resonance characteristics and therefore is not justified. Consequently, each generator is represented in its own rotor frame of reference and suitable a transformation is made.

3. Mathematical Model

The mathematical model of the electrical and mechanical system will be presented in this section. Actually, the electrical system includes the dynamic nonlinear mathematical model of a synchronous generator and that of the transmission line. The generator model considered in this study includes five equations, d-axis stator winding, q-axis stator winding, d-axis rotor field winding, q-axis rotor damper winding and d-axis rotor damper winding equations. Each mass of the mechanical system can be modeled by a second order ordinary differential equation (swing equation), which is presented in state space model as two first order ordinary differential equations.

Using the direct and quadrature d-q axes and Park's transformation, we can write the complete mathematical model that describes the dynamics of the system as follows. For the first generator:

$$\begin{aligned}
&-(X_{tr1}+X_L+X_b+X_{d1})\frac{di_{d1}}{dt}-(X_L+X_b)\cos\delta_{r12}\frac{di_{d2}}{dt}\\
&\quad-(X_L+X_b)\sin\delta_{r12}\frac{di_{q2}}{dt}+X_{md1}\frac{di_{f1}}{dt}+X_{md1}\frac{di_{D1}}{dt}=\\
&\omega_b\Big\{(R_{tr1}+R_L+R_b+R_{a1})i_{d1}\\
&\quad+\left[(R_L+R_b)\cos\delta_{r12}+\left(\frac{\omega_{r2}}{\omega_b}+1\right)(X_L+X_b)\sin\delta_{r12}\right]i_{d2}\\
&\quad-\left[\frac{\omega_{r1}}{\omega_b}(X_{tr1}+X_L+X_b)+(X_L+X_b)+\omega_{r1}X_{q1}\right]i_{q1}\\
&\quad+\left[(R_L+R_b)\sin\delta_{r12}-\left(\frac{\omega_{r2}}{\omega_b}+1\right)(X_L+X_b)\cos\delta_{r12}\right]i_{q2}+X_{mq1}\omega_{r1}i_{Q1}\\
&\quad+e_{cd}\sin\delta_{r1}-e_{cq}\cos\delta_{r1}+V_{0D}\sin\delta_{r1}-V_{0Q}\cos\delta_{r1}\Big\}
\end{aligned} \tag{1}$$

$$\begin{aligned}
&-(X_{tr1}+X_L+X_b+X_{q1})\frac{di_{q1}}{dt}-(X_L+X_b)\cos\delta_{r12}\frac{di_{q2}}{dt}\\
&\quad+(X_L+X_b)\sin\delta_{r12}\frac{di_{d2}}{dt}+X_{mq1}\frac{di_{Q1}}{dt}=\\
&\omega_b\Big\{(R_{tr1}+R_L+R_b+R_{a1})i_{q1}\\
&\quad+\left[(R_L+R_b)\cos\delta_{r12}+\left(\frac{\omega_{r2}}{\omega_b}+1\right)(X_L+X_b)\sin\delta_{r12}\right]i_{q2}\\
&\quad+\left[\frac{\omega_{r1}}{\omega_b}(X_{tr1}+X_L+X_b)+(X_L+X_b)+\omega_{r1}X_{d1}\right]i_{d1}\\
&\quad+\left[-(R_L+R_b)\sin\delta_{r12}+\left(\frac{\omega_{r2}}{\omega_b}+1\right)(X_L+X_b)\cos\delta_{r12}\right]i_{d2}-X_{md1}\omega_{r1}i_{f1}\\
&\quad-X_{md1}\omega_{r1}i_{D1}+e_{cd}\cos\delta_{r1}+e_{cq}\sin\delta_{r1}+V_{0D}\cos\delta_{r1}+V_{0Q}\sin\delta_{r1}\Big\}
\end{aligned} \tag{2}$$

$$-X_{md1}\frac{di_{d1}}{dt}+X_{f1}\frac{di_{f1}}{dt}+X_{md1}\frac{di_{D1}}{dt}=\omega_b\left[-R_{f1}i_{f1}+\frac{R_{f1}E_{fd1}}{X_{md1}}\right] \tag{3}$$

$$-X_{mq1}\frac{di_{q1}}{dt}+X_{Q1}\frac{di_{Q1}}{dt}=-\omega_b R_{Q1}i_{Q1} \tag{4}$$

$$-X_{md1}\frac{di_{d1}}{dt}+X_{md1}\frac{di_{f1}}{dt}+X_{D1}\frac{di_{D1}}{dt}=-\omega_b R_{D1}i_{D1} \tag{5}$$

where: $v_{f1} = \frac{R_{f1}E_{fd1}}{X_{md1}}$, and for $i, j = 1, 2$ and $i \neq j$, $\sin\delta_{rij} = \sin(\delta_{ri} - \delta_{rj})$, $\cos\delta_{rij} = \cos(\delta_{ri} - \delta_{rj})$.

Similarly, for the second generator the generator model includes five equations as follows:

$$\begin{aligned}
&-(X_{tr2} + X_L + X_b + X_{d2})\frac{di_{d2}}{dt} - (X_L + X_b)\cos\delta_{r21}\frac{di_{d1}}{dt} \\
&\quad - (X_L + X_b)\sin\delta_{r21}\frac{di_{q1}}{dt} + X_{md2}\frac{di_{f2}}{dt} + X_{md2}\frac{di_{D2}}{dt} = \\
&\omega_b\Big\{(R_{tr2} + R_L + R_b + R_{a2})i_{d2} \\
&\quad + \left[(R_L + R_b)\cos\delta_{r21} + \left(\frac{\omega_{r1}}{\omega_b} + 1\right)(X_L + X_b)\sin\delta_{r21}\right]i_{d1} \\
&\quad - \left[\frac{\omega_{r2}}{\omega_b}(X_{tr2} + X_L + X_b) + (X_L + X_b) + \omega_{r2}X_{q2}\right]i_{q2} \\
&\quad + \left[(R_L + R_b)\sin\delta_{r21} - (\frac{\omega_{r1}}{\omega_b} + 1)(X_L + X_b)\cos\delta_{r21}\right]i_{q1} + X_{mq2}\omega_{r2}i_{Q2} \\
&\quad + e_{cd}\sin\delta_{r2} - e_{cq}\cos\delta_{r2} + V_{0D}\sin\delta_{r2} - V_{0Q}\cos\delta_{r2}\Big\}
\end{aligned} \tag{6}$$

$$\begin{aligned}
&-(X_{tr2} + X_L + X_b + X_{q2})\frac{di_{q2}}{dt} - (X_L + X_b)\cos\delta_{r21}\frac{di_{q1}}{dt} \\
&\quad + (X_L + X_b)\sin\delta_{r21}\frac{di_{d1}}{dt} + X_{mq2}\frac{di_{Q2}}{dt} = \\
&\omega_b\Big\{(R_{tr2} + R_L + R_b + R_{a2})i_{q2} \\
&\quad + \left[(R_L + R_b)\cos\delta_{r21} + \left(\frac{\omega_{r1}}{\omega_b} + 1\right)(X_L + X_b)\sin\delta_{r21}\right]i_{q1} \\
&\quad + \left[\frac{\omega_{r2}}{\omega_b}(X_{tr2} + X_L + X_b) + (X_L + X_b) + \omega_{r2}X_{d2}\right]i_{d2} \\
&\quad + \left[-(R_L + R_b)\sin\delta_{r21} + (\frac{\omega_{r1}}{\omega_b} + 1)(X_L + X_b)\cos\delta_{r21}\right]i_{d1} - X_{md2}\omega_{r2}i_{f2} \\
&\quad - X_{md2}\omega_{r2}i_{D2} + e_{cd}\cos\delta_{r2} + e_{cq}\sin\delta_{r2} + V_{0D}\cos\delta_{r2} + V_{0Q}\sin\delta_{r2}\Big\}
\end{aligned} \tag{7}$$

$$-X_{md2}\frac{di_{d2}}{dt} + X_{f2}\frac{di_{f2}}{dt} + X_{md2}\frac{di_{D2}}{dt} = \omega_b\left[-R_{f2}i_{f2} + \frac{R_{f2}E_2}{X_{md2}}\right] \tag{8}$$

$$-X_{mq2}\frac{di_{q2}}{dt} + X_{Q2}\frac{di_{Q2}}{dt} = -\omega_b R_{Q2}i_{Q2} \tag{9}$$

$$-X_{md2}\frac{di_{d2}}{dt} + X_{md2}\frac{di_{f2}}{dt} + X_{D2}\frac{di_{D2}}{dt} = -\omega_b R_{D2}i_{D2} \tag{10}$$

where $v_{f2} = \frac{R_{f2}E_2}{X_{md2}}$.

Voltage drop across Xc:

$$\frac{de_{cd}}{dt} = \omega_b[X_c I_{LD} + e_{cq}] \tag{11}$$

$$\frac{de_{cq}}{dt} = \omega_b[X_c I_{LQ} - e_{cd}] \tag{12}$$

where:

$$I_{LD} = I_{q1}\cos\delta_1 + I_{q2}\cos\delta_2 + I_{d1}\sin\delta_1 + I_{d2}\sin\delta_2$$
$$I_{LQ} = I_{q1}\sin\delta_1 + I_{q2}\sin\delta_2 - I_{d1}\cos\delta_1 - I_{d2}\cos\delta_2$$

For the mechanical system, we have the following equations. High-pressure turbine #1:

$$\frac{d\omega_1}{dt} = \frac{1}{M_1}\left[-D_1(\omega_1 - 1) - K_{12}(\theta_1 - \theta_2)\right] \tag{13}$$

$$\frac{d\theta_1}{dt} = \omega_b(\omega_1 - 1) \tag{14}$$

Low-pressure turbine #1:

$$\frac{d\omega_2}{dt} = \frac{1}{M_2}\left[-D_2(\omega_2 - 1) + K_{12}(\theta_1 - \theta_2) - K_{23}(\theta_2 - \delta_{r1})\right] \tag{15}$$

$$\frac{d\theta_2}{dt} = \omega_b(\omega_2 - 1) \tag{16}$$

Similarly, for the second high- and low-pressure turbines. High-pressure turbine # 2:

$$\frac{d\omega_{1b}}{dt} = \frac{1}{M_{1b}}\left[-D_{1b}(\omega_{1b} - 1) - K_{12b}(\theta_{1b} - \theta_{2b})\right] \tag{17}$$

$$\frac{d\theta_{1b}}{dt} = \omega_b(\omega_{1b} - 1) \tag{18}$$

Low-pressure turbine # 2:

$$\frac{d\omega_{2b}}{dt} = \frac{1}{M_{2b}}\left[-D_{2b}(\omega_{2b} - 1) + K_{12b}(\theta_{1b} - \theta_{2b}) - K_{23b}(\theta_{2b} - \delta_{r2})\right] \tag{19}$$

$$\frac{d\theta_{2b}}{dt} = \omega_b(\omega_{2b} - 1) \tag{20}$$

Generator #1:

$$\frac{d\omega_{r1}}{dt} = \frac{1}{M_3}\left[T_{m1} - T_{e1} + K_{23}(\theta_2 - \delta_{r1}) - K_{34}(\delta_{r1} - \theta_4) - D_3(\omega_{r1} - 1)\right] \tag{21}$$

$$\frac{d\delta_{r1}}{dt} = \omega_b(\omega_r - 1) \tag{22}$$

where:

$$\begin{aligned} T_{e1} &= i_{q1}\Psi_{d1} - i_{d1}\Psi_{q1} \\ &= (X_{q1} - X_{d1})i_{d1}i_{q1} + X_{md1}i_{f1}i_{q1} - X_{mq1}i_{Q1}i_{d1} + X_{md1}i_{D1}i_{q1} \end{aligned}$$

Generator #2:

$$\frac{d\omega_{r2}}{dt} = \frac{1}{M_{3b}}\left[T_{m2} - T_{e2} + K_{23b}(\theta_{2b} - \delta_{r2}) - D_{3b}(\omega_{r2} - 1)\right] \tag{23}$$

$$\frac{d\delta_{r2}}{dt} = \omega_b(\omega_{r2} - 1) \tag{24}$$

where:

$$\begin{aligned}T_{e2} &= i_{q2}\Psi_{d2} - i_{d2}\Psi_{q2}\\ &= (X_{q2} - X_{d2})i_{d2}i_{q2} + X_{md2}i_{f2}i_{q2} - X_{mq2}i_{Q2}i_{d2} + X_{md2}i_{D2}i_{q2}\end{aligned}$$

Exciter:

$$\frac{d\omega_4}{dt} = \frac{1}{M_4}\left[-D_4(\omega_4 - 1) + K_{34}(\delta_{r1} - \theta_4)\right] \tag{25}$$

$$\frac{d\theta_4}{dt} = \omega_b(\omega_4 - 1) \tag{26}$$

The mathematical models of AVR and PSS (shown in Figure 1(c)) can be written as follows:

$$\frac{d\omega_{r1}}{dt} - \frac{dX_{W1}}{dt} = \frac{1}{T_W}X_{W1} \tag{27}$$

$$K_sT_1\frac{dX_{W1}}{dt} - T_2\frac{dV_{S1}}{dt} = V_{s1} - K_sX_{W1} \tag{28}$$

$$T_R\frac{dE_1}{dt} = K_R(V_{ref1} + V_{S1} - V_{t1}) - E_1 \tag{29}$$

where $V_{t1} = \sqrt{V_{d1}^2 + V_{q1}^2}$. Neglecting stator transients yields:

$$\begin{aligned}V_{d1} &= -R_{a1}i_{d1} + X_{q1}i_{q1}\\ V_{q1} &= -R_{a1}i_{q1} - X_{d1}i_{d1} + X_{md1}i_{fd1}\end{aligned}$$

Consequently,

$$V_{t1} = \sqrt{(-R_{a1}i_{d1} + X_{q1}i_{q1})^2 + (-R_{a1}i_{q1} - X_{d1}i_{d1} + X_{md1}i_{fd1})^2} \tag{30}$$

Therefore, these systems can be mathematically modeled as a set of first order nonlinear ordinary differential equations with the compensation factor ($\mu = X_c/X_L$) as a bifurcation (control) parameter. So, bifurcation theory can be applied to nonlinear dynamical systems, which can be written in the form $dx/dt = F(x;\mu)$.

4. System Response Without Damper Windings, AVR and PSS

To study the case of neglecting the dynamics of d- and q-axes damper windings, we set $i_{D1} = i_{Q1} = 0$ and $i_{D2} = i_{Q2} = 0$ in (4, 5, 9 and 10). Then, the dynamics of the system are governed by (1-3, 6-8, and 11-26). At any value of the compensation factor μ, the operating points (equilibrium solutions) are obtained by setting the derivatives of the 22 state variables in the system equal to zero. The stability of the equilibrium solution is studied by examination of the eigenvalues of the Jacobian matrix evaluated at the operating point. There are 20 complex conjugate eigenvalues and two real eigenvalues. The operating point stability regions for the case of neglect of the dynamics of both d- and q-axes damper windings are shown in Figure 2(a). It can be observed that the power system has stable operating point in the region $0 < \mu < H \approx 0.420692$, unstable operating point in the region $H \approx 0.420692 < \mu \leq 1$ and a Hopf bifurcation point at $\mu = H \approx 0.420692$.

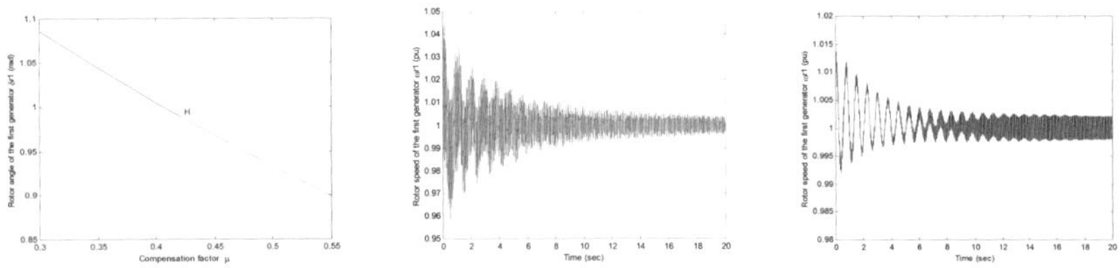

(a) Electrical system (bifurcation diagram showing variation of the first generator rotor angle δ_{r1} with the compensation factor μ)

(b) Rotor speed of the first generator at $\mu = 0.410600$ with 5% initial disturbance in rotor speed of generator

(c) Rotor speed of the first generator at $\mu = 0.420760$ with 2% initial disturbance in rotor speed of generator

Fig. 2.

Figure 2(b) shows the response of the system with 5% initial disturbance on the speed of the generator at $\mu = 0.410600$, which is less than H. It can be observed that the system is stable. While Figure 2(c) shows the system response after a 2% initial disturbance in generator rotor speed at $\mu = 0.420760$, which is greater than H. It can be observed that, the system routes to a periodic solution giving rise to oscillations. Hence, the type of this Hopf bifurcation is supercritical. So, the periodic solution emanating at the bifurcation point is stable.

Figure 3 shows the two-dimensional projections and time histories of the system at different compensation factors. It can be observed that, at $\mu = 0.420692$ the system oscillates just at one frequency with a small amplitude limit cycle. While it oscillates with more than one incommensurate frequency slightly after the bifurcation point at $\mu = 0.421583$ and obtain a period quasiperiodic attractor. As the compensation factors increase further, up to $\mu = 0.423580$, the system oscillates at infinite number of incommensurate frequencies. The motion is still bounded. This is however called chaos.

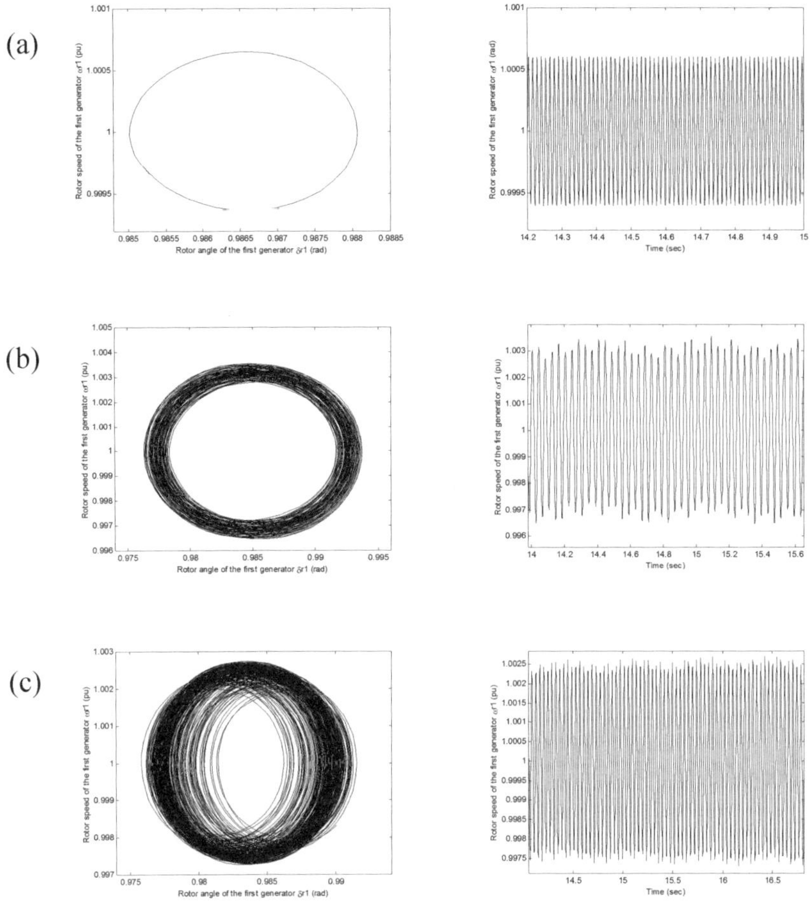

Fig. 3. Two-dimensional projection of the phases portrait onto $\omega_{rn} - \delta_{rn}$ plane (left) and the time histories of the corresponding rotor speed of generator (right). The solution at (a) limit cycle ($\mu = 0.420692$), (b) torus-attractor ($\mu = 0.421583$), and (c) chaotic attractor ($\mu = 0.423580$). (For the case of no damper windings)

5. Addition of Damper Windings

In this section, we study the effect of adding damper windings on both axes of the first generator. We evaluate the influence of these damper windings on the Hopf bifurcations. Although damper windings have some benefits in transient stability, such as damping off speed oscillations, they may induce subsynchronous resonance.[2] We will show that damper windings reduce the compensation level at which subsynchronous resonance occurs.

In this case we have 24 ordinary nonlinear differential equations, (1-8, 11-26). As in the previous section, the stability of the equilibrium solution is studied by examination of the eigenvalues of the Jacobian matrix evaluated at the operating point. There are 22 complex conjugate eigenvalues and four real eigenvalues.

The operating point stability regions in the $\delta_{r1} - \mu$ plane together with two Hopf bifurcation points are depicted in Figure 4. We observe that the

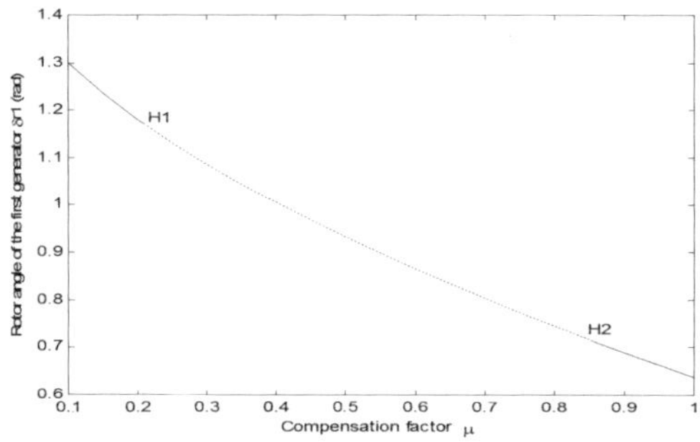

Fig. 4. Bifurcation diagram showing variation of the first generator rotor angle δ_{r1} with the compensation factor μ (for the case of adding both damper windings to the first generator)

power system has a stable operating point to the left of H_1≈0.207621 and to the right of H_2≈0.857447, and has an unstable operating point between H_1 and H_2. The operating point loses stability at a Hopf bifurcation point where $\mu = H_1$, the type of this Hopf bifurcation is supercritical. While it regains stability at a reverse Hopf bifurcation for $\mu = H_2$. In this case a pair of complex conjugate eigenvalues will transversally cross from left half and right half of the complex plane, and then back to left half. According to the Hopf bifurcation theory, a Hopf bifurcation of a fixed point of a continuous-time system leads to a periodic solution. So, a limit cycle is born at $\mu = H_1 \approx 0.206721$.

6. Addition of Damper Windings, AVR and PSS

In this section we investigate the influence of adding damper windings, automatic voltage regulator (AVR) and power system stabilizer (PSS) to the first generator. The automatic voltage regulator (AVR) controls the excitation voltage of the synchronous generator with the terminal voltage of synchronous generator as an input signal, while the output signal is the d-axis field voltage. Power system stabilizer (PSS) uses auxiliary (supplementary) stabilizing signals to control the excitation system so as to improve power system dynamic performance. Commonly used input signals to PSS are shaft speed, terminal frequency or power. Power system dynamic performance can be improved by the damping of system oscillations. This is a very effective method of enhancing small-signal stability performance. In this study, the input signal to the PSS is the generator rotor speed. Figure 1(c) shows the block diagram of the use of AVR together with the PSS.[8] The first block on the right side that is used to wash out the compensation effect after a time lag TW, with a typical values between 4 to 20 sec.

In this study, the mathematical model of the system is extended to 27 differential equations by including (27)-(30). The internal generated voltage E_1 of the first synchronous generator is no more a fixed value i.e. E_1 in addition to X_{W1} and V_{S1} are the new state variables. The output of the PSS is one of the inputs of the AVR. The other inputs are V_{ref1} and the terminal voltage of the synchronous generator V_{t1}. The terminal voltage is not a state variable, but it can be expressed in terms of other state variables, (30). As in the previous cases, the stability of the equilibrium solution is studied by examination of the eigenvalues of the Jacobian matrix evaluated at the operating point. This study is carried out on a heavily loaded synchronous generator with $V_{t1} = 1.14$, $V_{ref1} = 1.152$ and $T_m = 0.91$ pu at a compensation factor $\mu = 0.1$.

In addition to the parameters given before, the parameters for the AVR and PSS are:

$K_R = 200$, $T_R = 0.025$ sec., $T_W = 10$ sec., $K_S = 12$, $T_1 = 0.048$ sec., $T_2 = 0.032$ sec.

In this case, a new stable oscillatory mode with a frequency of 3.65 Hz and a stable real eigenvalue were added. The operating point stability regions in the δ_{r1} plane together with two Hopf bifurcation points are depicted in Figure 5.

We observe that the power system has a stable operating point to the left of $H_1 \approx 0.198377$ and to the right of $H_2 \approx 0.824135$, and has an unstable operating point between H_1 and H_2. The operating point loses stability at a Hopf bifurcation point, namely $\mu = H_1$. It regains stability at a reverse Hopf bifurcation, namely $\mu = H_2$. In this case a pair of complex conjugate eigenvalues will transversally cross from left half to right half of the complex plane, and then back to the left half.

To determine whether the limit cycles created due to the Hopf bifurca-

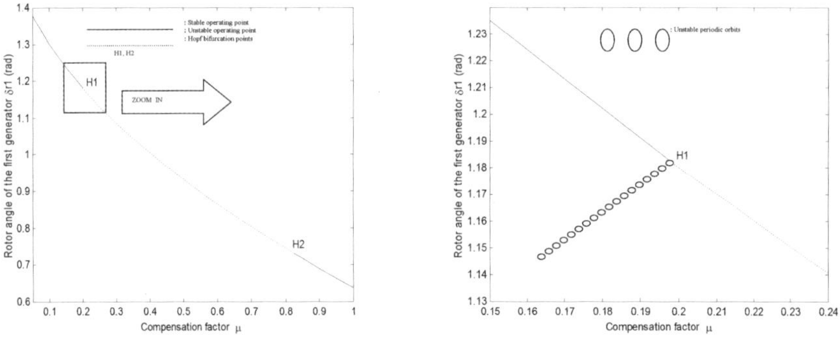

Fig. 5. Bifurcation diagram showing variation of the first generator rotor angle δ_{r1} with the compensation factor μ (for the case of including damper windings, AVR and PSS to the first generator)

tion are stable or unstable, we obtain the time response of the system by numerical integration with small disturbance slightly before H_1. Figure 6 shows the response of the system with 7% initial disturbance on the speed of the generator at μ =0.182265, which is less than H_1. It can be observed that the system is unstable. Therefore, the type of this Hopf bifurcation is subcritical. So, the periodic solution emanating at the bifurcation point is unstable.

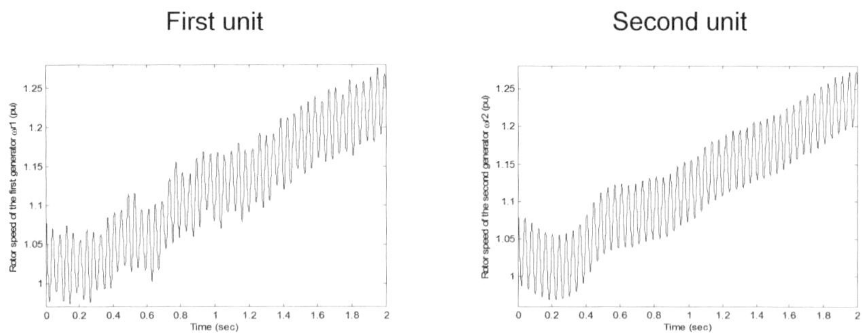

Fig. 6. Rotor speed of the generator at $\mu = 0.182265$ with 7% initial disturbance in rotor speed of generator (for the case of including damper windings, AVR and PSS to the first generator)

7. Conclusions

Bifurcation theory is applied to the second system of the IEEE second benchmark model of SSR to investigate the complex dynamics of the system. The case of neglecting the dynamics of both damper windings is considered. The results show that as the compensation factor ($\mu = X_c/X_L$) increases the operating point loses stability via Hopf bifurcation point. The type of the Hopf bifurcation can be determined by numerical method based on the response of the perturbed system. Also, the influence of adding damper windings along the q- and d-axes on subsynchronous resonance has been investigated. The results show that these damper windings affect the locations, number and type of the Hopf bifurcations. Adding damper windings shift the Hopf bifurcation to smaller value of μ and hence shrink the stable region. In other words, the damper windings destabilize the system by reducing the compensation level at which subsynchronous resonance occurs. In the case of adding damper windings, AVR and PSS the Hopf bifurcation slightly shifted to the left. Also, the results showed that the type of this Hopf bifurcation is subcritical. Consequently, we can conclude that the change of Hopf bifurcation caused by AVR and PSS is mainly qualitative with some quantitative change.

References

1. W. Zhu, R. R. Mohler, R. Spce, W. A. Mittelstadt and D. Maratukulman, *IEEE Trans. on Power Systems* **11**(3), 1579 (1996).
2. A. H. Nayfeh, A. M. Harb, C.-M. Chin, A. M. Hamdan and L. Mili, *Int. J. Bifurcation and Chaos* **8**(1), 157 (1998).
3. A. M. Harb, A. H. Nayfeh and L. Mili, Bifurcation control for mitigating subsynchronous oscillations in power systems, in *14th PSCC, Sevilla, 24-28 June*, (2002).
4. M. A. Tomim, A. C. Z. de Souza, P. P. C. Mendes and G. Lambert-Torres, Identification of hopf bifurcation in power systems susceptible to subsynchronous resonance, in *IEEE Bologna Power Tech Conference*, (Bologna, Italy, 2003).
5. I. Dobson and E. Barocio, Perturbations of weakly resonant power system electromechanical modes, in *IEEE Bologna Power Tech Conference*, (Bologna, Italy, 2003).
6. A. H. Nayfeh and B. Balachandran, *Applied Nonlinear Dynamics* (Wiley, New York, 1995).
7. IEEE SSR Working Group, *IEEE Trans. On Power Apparatus and Systems* **PAS-104**, 1057 (May 1985).
8. K. R. Padiyar, M. Geetha and K. Rao, A novel power flow controller for controlled series compensation, in *IEE, AC and DC Power Transmission*, (Conference Publication No. 423, 29 April-3 May 1996).

Measurement and Simulation of Random Matrix Statistics in Aluminum Mesoscopic Cavities

O. Antoniuk* and R. Sprik

Van der Waals-Zeeman Institute, University of Amsterdam
Valckenierstraat 65, 1018 XE Amsterdam, The Netherlands
*Email: *oantoniu@science.uva.nl*

We investigated spectral fluctuations in mesoscopic aluminum cavities for acoustic waves with and without of reflection plane symmetry. Results of measurements and simulations of the cavity responses are compared to the predictions of a random Hamiltonian model with time reversal symmetry.

Keywords: Acoustic waves; mesoscopic cavity; spectral fluctuation; time-reversal symmetry.

1. Introduction

We study the propagation of acoustical waves in aluminum cavities as a model system for chaotic classical waves. Responses of the cavities to short excitation pulses were recorded for different positions of the source and receiver. Sequences of resonance frequencies were obtained from the spectral density of the recorded responses. These sequences were used to calculate the nearest neighbor resonance spacing distributions (NNSD) and the spectral rigidity (SR), a parameter[1,2] characterizing fluctuation of number of resonances per frequency band. Exact model representations of the cavities were simulated with the program Wave3000.[3] So the NNSD and the SR were also calculated from simulated acoustic wave dynamics. Results are compared to predictions[1,2] of Random Matrix Theory (RMT) for random Hamiltonian with time reversal symmetry. These predictions[1,2] are valid for the eigenvalues of an ensemble of random symmetric matrices with real Gaussian distributed elements, or Gaussian Orthogonal Ensemble (GOE).

2. The Samples

The samples used in measurement and simulation are shown in Fig. 1. The shapes of the samples are cubic with a side length of the cube $d = 2$ cm. A cylindrical volume of aluminum is removed from each cube (see Fig. 1). In addition cavity Nr. 2 has a prism of aluminum removed from the side corner. Sound velocities for the aluminium used can be estimated as 6.4 km/s and 3.2 km/s for longitudinal and transverse waves respectively.[4] Therefore the corresponding wavelengths ranges of the frequency band 200..450 kHz used in the measurement are approximately $1.6d..0.7d$ in wavelength of the longitudinal wave and $0.8d..0.35d$ in wavelength of the transversal wave.

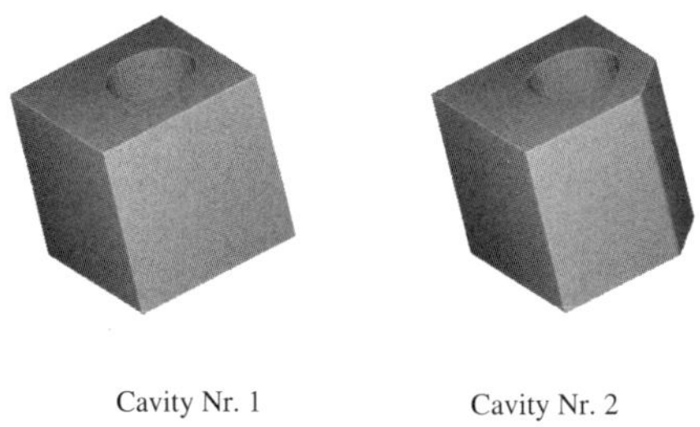

Fig. 1. Samples: cavity with reflection symmetry (Nr. 1) and cavity without reflection symmetry (Nr. 2).

3. Measurements and Simulation

The schematic of experimental setup is shown in Fig. 2. Sending the short exciting pulse (one period of 400 kHz sine signal) and receiving the cavity response is done by piezoelectric sensors, connected to the sample. The amplified response signal is recorded by the digitizing oscilloscope. The recorded 5 ms long responses with a time step of 0.1 μs are used further for the calculation of the spectral density and determining the sequence of cavity resonances detected at the current position of source and receiver.

The same procedure is repeated for different positions of source and receiver. Spectral statistics (NNSD and SR) is calculated for each sequence of resonances and averaged over different sequences corresponding to different positions of source and receiver.

We also performed simulations of the cavity responses using the pro-

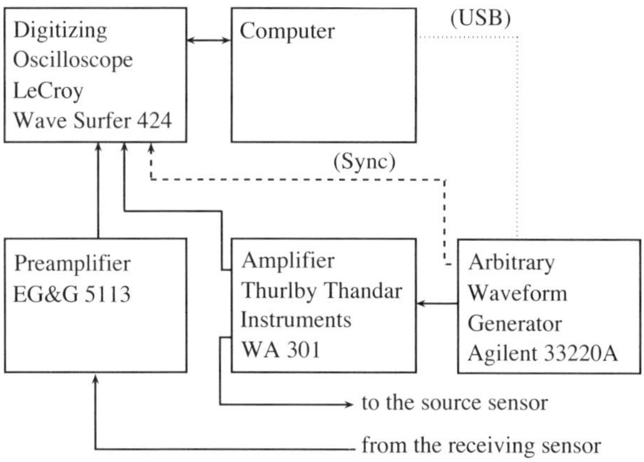

Fig. 2. Schematic of the experimental setup.

gram Wave3000[3] that has been developed by Kaufman et al. to calculate ultrasonic responses in complex solid materials using an adaptive finite difference method. Fig. 3(a) and Fig. 3(b) show examples of measured and simulated responses. The details of the cavity geometries, the coupling to the environment, and the exact response of the transducers are too complex to expect a detailed match between experiment and simulation. In particular the reverberation time is much shorter in the experimental results. However, the statistcal properties of the spectra should be the same.

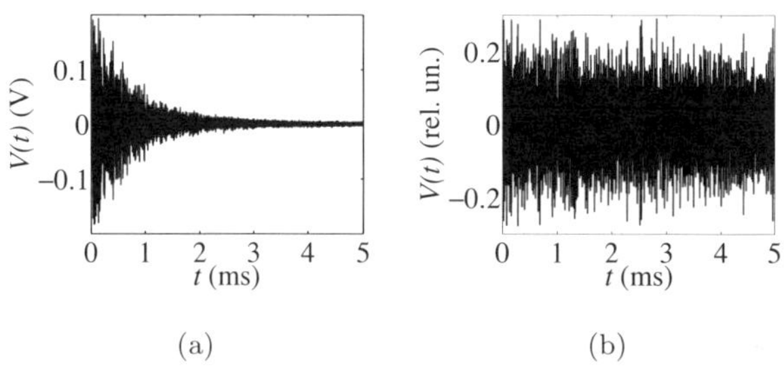

Fig. 3. Measured (a) and simulated (b) cavity responses at certain positions of the source and receiver.

4. Discussion of Results

4.1. *Staircase function determined from the experiment*

Staircase function $N(f)$ gives the number of resonances in the frequency band below frequency f. It represents the sequence of cavity resonances detected at a certain fixed position of source and receiver. Typical staircase functions obtained from the measurement are shown in Fig. 4. To compensate for slow increase of density of resonances with frequency an unfolding procedure[5] has been applied to the staircase functions using a fit of the staircase functions with a smooth cubic polynomial $p(f)$ (Fig. 4).

4.2. *NNSD and SR determined from the experiment*

The NNSD from data measured in experiment for cavity Nr. 2 is shown in Fig. 5. The dotted line shows exponential distribution (as for the case of spacings between "randomly chosen" resonant frequencies in the studied band, Poisson distribution). The dashed line shows the distribution corresponding to the GOE model. We include in our analysis the effect of missing resonance frequencies in the experimental detection. The dash dotted line shows the same GOE distribution when 25% of the resonances in the sequence are not detected. Points with error bars show the NNSD from the experimental data. s_{avg} is the average spacing between the neighboring resonances.

The SR (Δ) determined from data measured in experiment is shown in Fig. 6. Dashed lines show the fluctuation of the number of resonances per frequency band in case of "randomly chosen frequencies" (straight line) and fluctuation corresponding to GOE model[1,2] (logarithmically saturating curve). $2L$ is the size of frequency interval $f-L..f+L$ over which fluctuation of staircase function is averaged as e.g. in,[1] p.113.

The points of the SR dependencies for both cavities are in much better agreement with the GOE prediction than with the straight line. The SR for cavity Nr. 1 (with reflection plane) is found to be slightly larger than for cavity Nr. 2. The difference may be caused by a larger amount of lost resonances in cavity Nr. 1, as well as by the coexistence of two independent families of resonances (symmetric and antisymmetric) due to presence of reflection plane symmetry for cavity Nr. 1.

The fluctuation of NNSD and SR values shown by error bars in Fig. 5 and Fig. 6 appears due to different positions of source and receiver in the measurement.

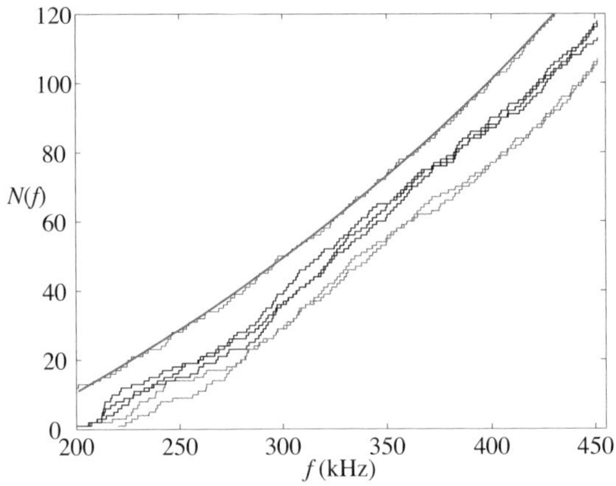

Fig. 4. Staircase functions obtained from the cavity responses measured in experiments. Staircase function and cubic polynomial fitting it (top curves).

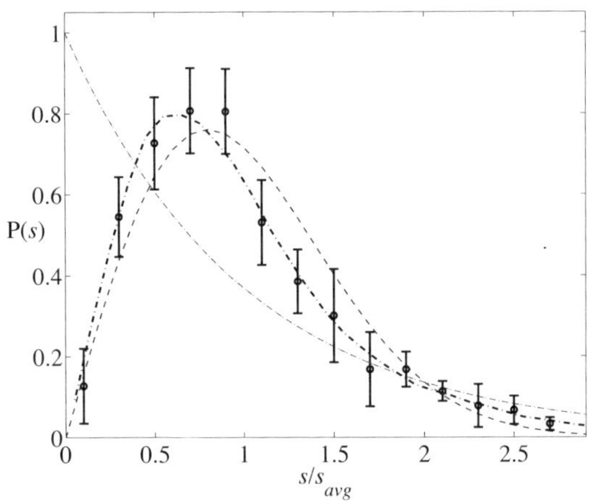

Fig. 5. NNSD for cavity Nr. 2 from responses measured in experiments.

4.3. *NNSD and SR determined from simulation*

The NNSD calculated from simulated responses of cavity Nr. 2 is shown in Fig. 7. The simulation does not show the effect of missing resonances, something inherently present in the experiment, Fig. 5. However the NNSD determined from simulation shows greater repelling of resonant frequencies

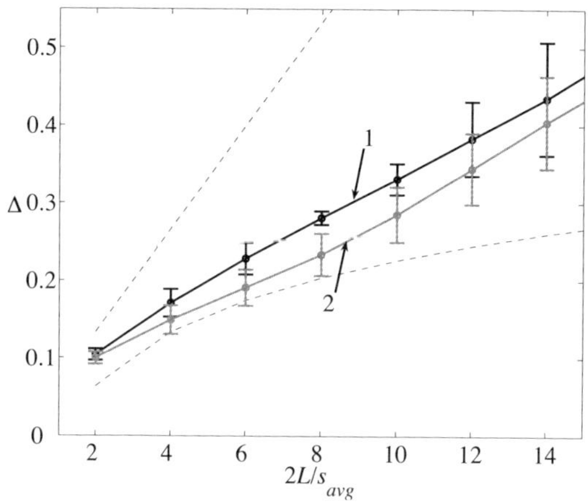

Fig. 6. Spectral rigidity from spectral density of responses measured in experiments for both cavities Nr. 1 and 2 (shown by arrows in the picture).

than in the GOE model distribution. We see that the NNSD determined from simulation has higher peak value while both the NNSD from simulation and the GOE model distribution are normalized to the unit area under the curve.

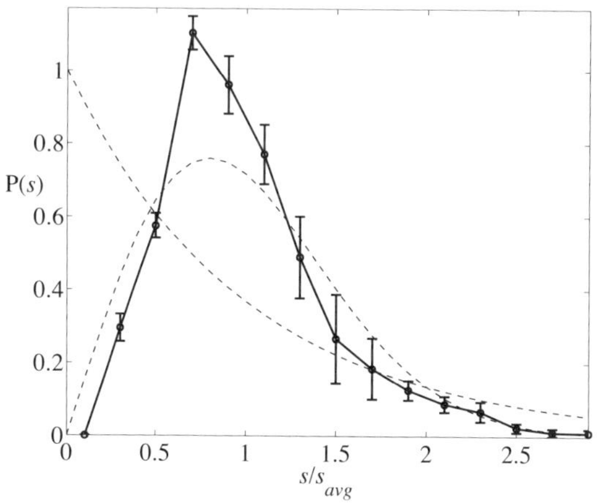

Fig. 7. NNSD calculated for cavity Nr. 2 from responces obtained in Wave3000 simulation program for different positions of source and receiver.

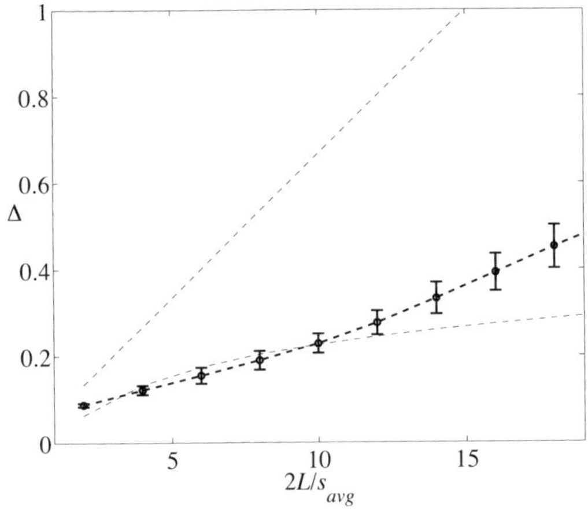

Fig. 8. Spectral rigidity for cavity Nr. 2. From responses obtained in Wave3000 simulation program.

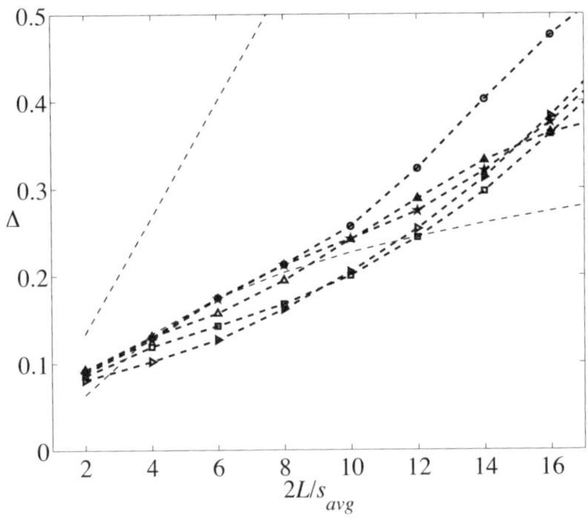

Fig. 9. Spectral rigidity for cavity Nr. 2. From response obtained in Wave3000 simulation program. Different curves in this picture show results for different positions of source and receiver.

The SR (Δ) calculated from the simulated responses for cavity Nr. 2 is shown in Fig. 8. It is also in agreement with the GOE model. However unlike the dependence determined from experiment, the simulation data gives lower values of SR than predicted by the model. This is better shown by Fig. 9. The SR calculated from simulation for each particular position of source and receiver is shown in Fig. 9. We see that these dependencies either exactly coincide with GOE model for frequency bands $2L < 10s_{\text{avg}}$ or take lower value.

5. Conclusions

The NNSD obtained from the measured responses agrees with the model of random Hamiltonian preserving the time-reversal symmetry. Agreement was achieved when accounting for a 25% of resonances missing (not detected). The SR obtained from the measured responses behaves according to the GOE model when averaged over frequency bands smaller than 10 average resonance frequency spacings s_{avg}. The simulated acoustic wave dynamics shows greater amount of resonances and noticeably lower spectral fluctuations. This is consistent with the longer reverberation times in the calculations. The NNSD determined from simulation data is better peaked around average value and the SR averaged over frequency bands smaller than $10s_{\text{avg}}$ has lower value than predicted by the GOE model.

Acknowledgements

This work is part of the research programme of the Stichting voor Fundamenteel Onderzoek de Materie (FOM), which is financially supported by the Nederlandse Organisatie voor Wetenschappelijk Onderzoek (NWO).

References

1. H.-J. Stöckmann, *Quantum Chaos* (Cambridge University Press, Cambridge, 1999).
2. T. Guhr, A. Müller-Groeling, H. A. Weidenmüller, Phys. Rep. **299**, 189-425 (1998).
3. Software (Wave3000 program) used for simulation of acoustic responses, http://www.cyberlogic.org/index.html
4. H. H. Demarest, J. Acoust. Soc. Am. **49**, 768-775 (1971).
5. F. Haake, *Quantum Signatures of Chaos*, 2-nd edition, pp. 58-60 (Springer Series in Synergetics, 2001).

On a Novel Dry Friction Modeling: Differential Equations Computation and Lyapunov Exponent Estimation

J. Awrejcewicz, D. Grzelczyk and Yu. Pyryev

Technical University of Lodz
Lodz, Poland

A novel dry friction modeling and its impact on differential equations computation and Lyapunov exponent estimation are studied in this paper. A brief review of some existing standard friction laws is presented and a novel continuous friction model taking into account some elements of the mentioned classical friction models is proposed. Its advantages are illustrated and discussed with the use of classical differential equation of motion of a 1-DOF system with dry friction and harmonic excitation. The behavior of the mentioned system is monitored via standard tools, i.e. time series, phase portraits, bifurcation diagrams as well as the Lyapunov exponents.

Keywords: Dry friction; discontinuous systems; stick-slip vibrations.

1. Introduction

Computing Lyapunov exponents is one of the most known identification tools in nonlinear dynamics. Although Lyapunov's theory for estimation of chaotic dynamics was used by Oseledec as early as forty years ago and there are numerous works in the scientific literature dedicated to chaotic vibrations, in general there are not many effective methods available for Lyapunov exponents estimation. The classical Lyapunov exponents definition can be used with success only in systems governed by differential equations with smooth right-hand sides. In cases of discontinuous systems arising due to physical discontinuities (dry friction, impact or backlash in mechanical systems) other numerical/analytical approaches may be also applied, i.e. phase portraits, bifurcation diagrams, the Melnikov-type methods, Fourier spectra, as well as various methods experimentally and numerically oriented.

Our investigations in this paper will be concerned with systems governed by differential equations with non-smooth right-hand sides due to dry friction phenomena. Although dry friction belongs to one of the known phenomena exhibited by mechanical systems, its proper mathematical modeling is not an easy task. Friction force between sliding surfaces is a complex process and depends on several parameters. An extensive literature review on dry friction models can be found in [1–5]. Even in the last decade stick-slip vibrations were the aim of research of many authors.[6–8] Contrary to those works, we study a mechanical system exhibiting stick-slip vibrations with constant and non-constant foundation velocity.

2. The Novel Friction Model

Several dry friction formulations have been proposed based on the classical (discontinuous) Coulomb model. One of the simplest friction laws is the classical Amontons-Coulomb law. The constitutive relation for friction force F_{fr} in this case is known as the signum model with static friction point and it describes dry friction phenomena in the correct and accurate way. During numerical simulation an exact value of zero is rather rarely computed (so the mentioned signum model from a numerical point of view is equivalent to the classical Coulomb model). In addition, the dependence of friction force on the relative velocity based on the signum model is a non-continuous function for argument equal to zero, and the standard numerical procedures devoted for solving differential equations cannot be used. For this reason, the friction curve is therefore often approximated by a continuous or smooth function, but in each case of these smoothing methods for relative velocity equal to zero one gets always friction force equal to zero. In one of the recent papers[8] devoted to mathematical modeling of dry friction, the so called switch model is proposed and used. Switch model (from a mathematical point of view) is governed by three nonlinear ordinary differential equations (stick, slip and transition from stick to slip). In [8] it has been shown that from a computational point of view the smoothing methods are more expensive than the switch model based methods. In what follows we present (in non-dimensional form) an alternative continuous friction model taking into account some elements of the mentioned friction models. First we divide the space $F_{\mathrm{ex}} - v_r$ into the following four regions:

$$
\begin{aligned}
&V_1 : |v_r| > \varepsilon \\
&V_2 : \big[(0 \le v_r \le \varepsilon) \cap (F_{\mathrm{ex}} > \quad F_s)\big] \cup \big[(-\varepsilon \le v_r \le 0) \cap (F_{\mathrm{ex}} < -F_s)\big] \\
&V_3 : \big[(0 < v_r \le \varepsilon) \cap (F_{\mathrm{ex}} < -F_s)\big] \cup \big[(-\varepsilon \le v_r < 0) \cap (F_{\mathrm{ex}} > \quad F_s)\big] \\
&V_4 : \big(|v_r| \le \varepsilon\big) \cap \big(|F_{\mathrm{ex}}| \le F_s\big)
\end{aligned}
$$

where v_r is the relative velocity of sliding surfaces, F_{ex} is externally applied force (all forces acting in the system excluding friction force), F_s denotes maximum static friction force and ε is the steepness parameter. The proposed continuous friction model has the form (see Fig. 1)

$$F_{\text{fr}}(v_r, F_{\text{ex}}) = \begin{cases} F(|v_r|)\,\operatorname{sgn} v_r & \text{for } V_1 \\ F_s \operatorname{sgn} F_{\text{ex}} & \text{for } V_2 \\ (2A_3 - 1)\,F_s \operatorname{sgn} v_r & \text{for } V_3 \\ A_3\,(-F_{\text{ex}} + F_s \operatorname{sgn} v_r) + F_{\text{ex}} & \text{for } V_4 \end{cases}$$

where

$$A_3 = \frac{v_r^2}{\varepsilon^2}\left(3 - 2\frac{|v_r|}{\varepsilon}\right).$$

In this work we take

$$F(|v_r|) = \frac{F_s}{1 + \delta(|v_r| - \varepsilon)}.$$

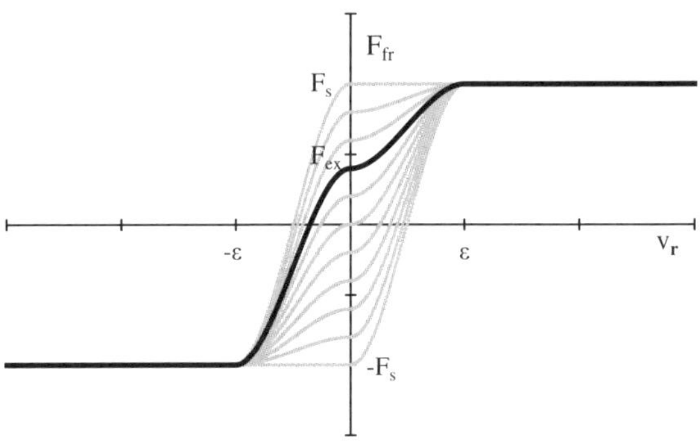

Fig. 1. The friction force as a function of relative velocity in stick phase for several fixed external forces F_{ex}

3. The Analyzed System

Advantages of the proposed friction model and its impact on differential equations computation are illustrated and discussed with the use of a classical second order differential equation of motion of 1-DOF system with dry friction and harmonic excitation of the following non-dimensional form:

$$\frac{d^2x}{d\tau^2} + x = F_{\text{fr}}(v_r, F_{\text{ex}}) + F_0 \cos\omega_0\tau,$$

where x is the non-dimensional displacement, τ is the non-dimensional time, $F_{\rm fr}(v_r, F_{\rm ex})$ is the non-dimensional friction force, F_0 is the non-dimensional amplitude of force of harmonic excitation, whereas ω_0 is the non-dimensional harmonic excitation angular velocity. Although there are numerous works in the scientific literature dedicated to stick-slip vibrations, a rigid body lying on a belt which moves with non-constant velocity, and the associated problems with Lyapunov exponents estimation, are less investigated. In this work the mechanical model with both constant and non-constant belt (foundation) velocity is studied. In our case $F_{\rm ex} = x - F_0 \cos \omega_0 \tau$ and $v_r = v - v_{dr}$, where $v = \dot{x}$ and $v_{\rm dr}$ is constant or non-constant foundation velocity. In the first case we take $v_{\rm dr} = v_0 = {\rm const}$, whereas in the second we apply $v_{\rm dr} = v_0 \cos \omega \tau$.

4. Computer Simulation Results

The methods commonly used to compute the Lyapunov exponents require smooth vector fields as a necessary condition. Our friction model is a continuous function of relative velocity and can be used during analysis of the systems, whereas the Lyapunov exponents are computed by the standard procedures.[9] Note, that while computing Lyapunov exponents, besides the equation of motion, additional systems of equations with respect to perturbations are solved. Finally, all equations are solved via the Runge-Kutta-Fehlberg (RKF 45) method with the varied time step h ($h_{\rm min} = 10^{-5}$, $h_{\rm max} = 10^{-1}$), and with the Runge-Kutta-Fehlberg tolerance of $\eta = 10^{-6}$ and steepness parameter $\varepsilon = 10^{-3}$ (the Gramm-Schmidt orthonormalization technique is applied[9]). The behavior of the system is monitored via monitoring of time series, phase portraits, bifurcation diagram as well as the Lyapunov exponents. The initial non-dimensional parameters of our model are: $F_s = 1$, $\delta = 3$ and $v_0 = 0.2$.

Let us consider first dynamics of the system without harmonic excitation, i.e. $F_0 = 0$ and $\omega = 0$, i.e. when $v_{\rm dr} = v_0 = {\rm const}$. For this case, the phase portraits obtained via signum model approach, smoothing method (with hyperbolic tangent approximation), the switch model technique and our continuous friction modeling are shown in Fig. 2.

It is visible that in the sticking phase some differences are observed. The differences occur as a result of another approximating friction force application in a near-zero relative velocity neighborhood. Contrary to the other model results, our results (from numerical point of view) are better. In addition, we have obtained almost exact results (high precision numerical computations) even for larger (10 times) parameter ε. It allows to obtain

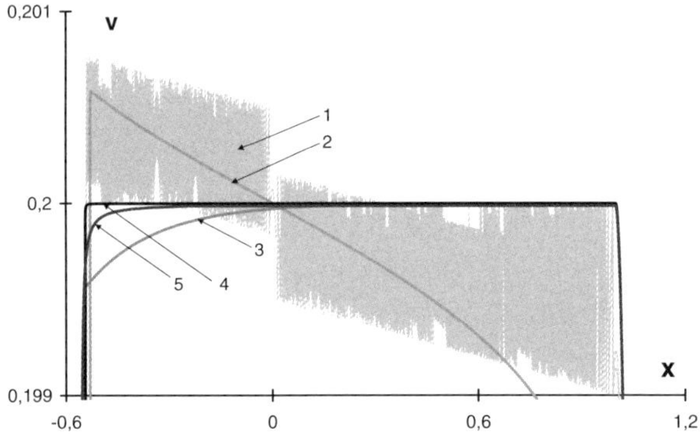

Fig. 2. Phase portraits for $v_{dr} = v_0 =$ const. for different friction models: signum model (curve 1); smoothing method (curve 2); switch model (curve 3); our friction model (curve 4); our friction model for 10 times bigger steepness parameter ε (curve 5)

the same accuracy as in the switch model, but for larger both time steps and steepness parameter ε.

Fig. 3 shows phase portraits obtained with the two compared friction models (the standard approximation using a signum function modeled by the hyperbolic tangent function in the stick phase and our friction model). For the first friction model the computation took 3353 integration points to obtain the orbit with the non-dimensional period time 12. Small time steps are not necessary near the transitions, but during the whole stick phase, as Fig. 3(a) shows. For our friction model the computation took only 104 integrations points to obtain the same orbit (Fig. 3(b)).

In our calculations small time steps are taken only near the transitions between stick and slip phases, and the time step in the stick phase is bounded by the maximum time step $h_{\max}$. Consequently, we show that in this case the smoothing approximation of classical signum function is clearly more expensive than the proposed model. The differential in the first friction model is extremely large for relative velocity equal to zero, whereas in our model it is equal to zero and this is an advantage for numerical computations.

Our proposed friction model is a continuous function of the relative velocity and therefore it can be used during the analysis of the systems, where the Lyapunov exponents are computed by the standard procedures.

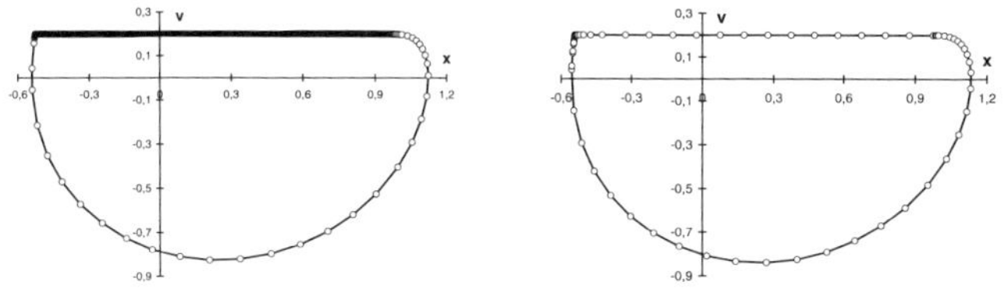

(a) The smoothing hyperbolic tangent approximation (b) The proposed model

Fig. 3. Points of trajectories of motion in the system's phase space for different friction models

In our model (for $F_0 = 0.5$, $\omega_0 = 2$) the dependences of the largest Lyapunov exponent λ with $v_{\rm dr}$ and F_0 as control parameters and displacement x on the vertical axis are reported in Fig. 4 using the proposed continuous friction model.

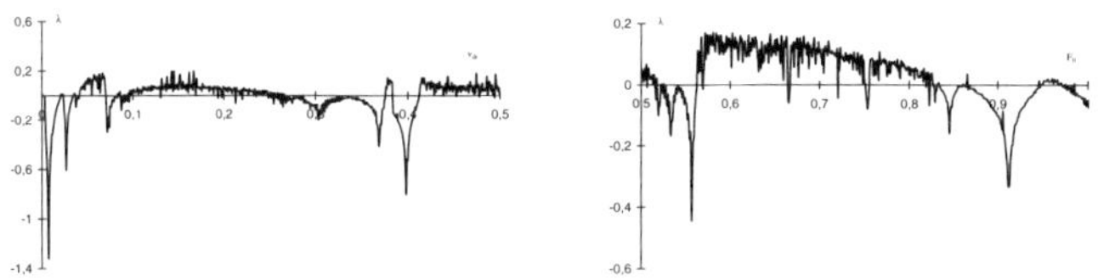

(a) With $v_{\rm dr}$ as control parameter (b) With F_0 as control parameter

Fig. 4. Dependences of the largest Lyapunov exponent λ on control parameters

Both periodic and non-periodic solutions are also detected with the use of a bifurcation diagram, in the velocity-displacement ($v_{\rm dr}$-x) plane, in Fig. 5.

Consider now the dynamics of the system without harmonic excitation, but for $\omega \neq 0$, i.e. with the non-constant velocity $v_{dr} = v_0 \cos \omega \tau$, where $v_0 = 0.1$. The studied mechanical system possesses various interesting solutions including the stick-slip vibrations, as it is further shown. Figs. 6 and 7 present different behaviors of the analyzed mechanical system using our proposed model.

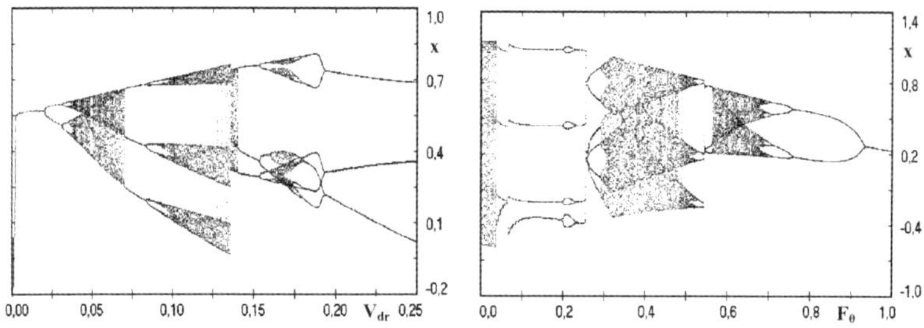

(a) With v_{dr} as control parameter (b) With F_0 as control parameter

Fig. 5. Bifurcation diagrams of the analyzed system with v_{dr} (to the left) and F_0 (to the right) as the control parameters and x on the vertical axis

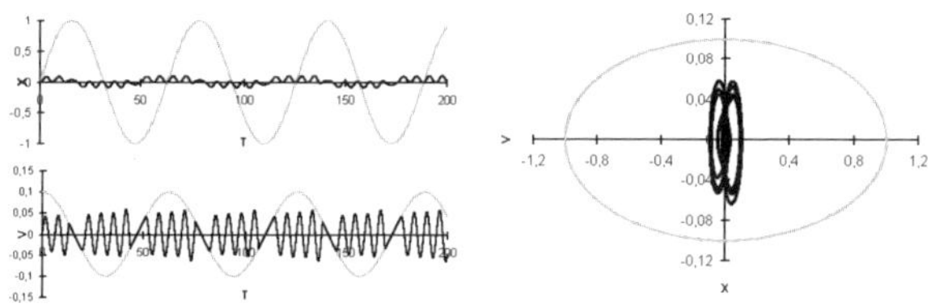

(a) $F_s = 0.05$

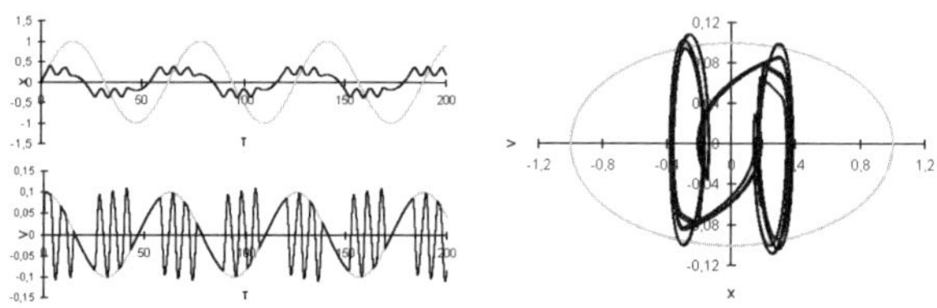

(b) $F_s = 0.3$

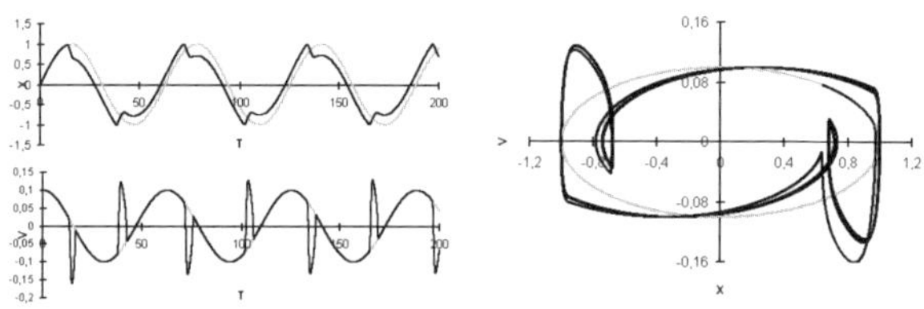

(c) $F_s = 0.95$

Fig. 6. Time series and phase portraits of the analyzed system for $\omega = 0.1$ and for various values of the static friction force F_s

5. Conclusions

A novel friction model, which takes into account some elements of existing standard friction models, is proposed. Its impact on differential equations

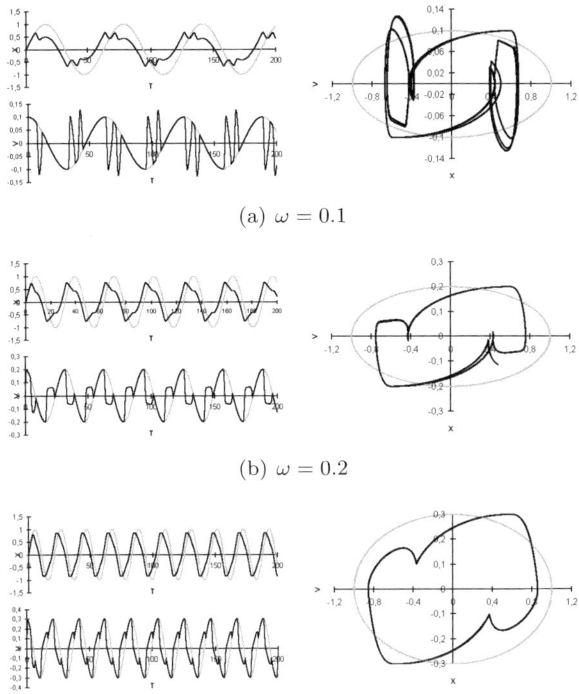

(a) $\omega = 0.1$

(b) $\omega = 0.2$

(c) $\omega = 0.3$

Fig. 7. Time series and phase portraits of the analyzed system for $F_s = 0.6$ and for various value of the angular velocity ω

computation and Lyapunov exponents estimation are studied. We show that our model is suitable in particular for stick-slip vibration simulations. It has been observed that our continuous friction model yields engineering accepted results and possesses advantages in comparison to other friction models. The obtained results have been compared with those given by switch model application, and they indicate better numerical efficiency of our proposed continuous model. In addition, the obtained results exhibit advantages of the proposed algorithm in comparison to the algorithm using smoothing approaches. Continuous friction model is validated and it gives correct results and almost exact solutions (high precision numerical computations), even for extremely large numerical steepness parameter ε. It allows obtaining the same accuracy as in the switch model but faster and for larger steepness parameter ε. For these reasons our calculations are less expensive from a computational point of view. One of the important advantages of our novel model is also associated with the direct application of the standard numerical procedures devoted to either solving nonlinear differential equations or to computation and estimation of the Lyapunov exponents.

References

1. S. Andersson, A. Soderberg and S. Bjorklund, *Tribology International* **40**, 580 (2007).
2. J. Awrejcewicz and C.-H. Lamarque, *Bifurcation and chaos in nonsmooth mechanical system* (Vol. 45 Series A. , World Scientific, New Jersey, London, Singapore, 2003).
3. J. Awrejcewicz and P. Olejnik, *Applied Mechanics Reviews* **58**(6), 389 (2005).
4. R. A. Ibrahim, *ASME Applied Mechanics Reviews* **47**(7), 209 (1994).
5. J. A. C. Martins, J. T. Oden and F. M. F. Simoes, *International Journal of Engineering Science* **28**(1), 29 (1990).
6. U. Galvanetto and S. Bishop, *MECCANICA* **34**, 337 (1999).
7. N. Hinrichs, M. Oestreich and K. Popp, *Chaos, Solitons and Fractals* **8**(4), 535 (1997).
8. R. Leine, D. H. van Campen, A. de Kraker and L. van den Steen, *Nonlinear Dynamics* **16**(1), 41 (1998).
9. S. P. Kuznetsov, *Dynamical chaos, in Russian* (Fizmatlit, Moskva, 2001).

Magic Angle Chaotic Precession

B. Binder

Quanics, Weildorferstr. 22, 88682, Salem, Germany
Email: binder@quanics.com

This paper explores the properties of a precessing rotor or a coupled system of precessing rotors (gyroscopes), where a special chaotic behavior in the precession angle can be found if the change of rotor angular velocity is linearly coupled by (an)holonomy to the precession angular velocity and angle. The linear coupling provides for rolling cone paths and allows spinning up and controlling the rotor simply by forcing precession at special quantum magic precession angles. The geometric phase induced by the curved path of the rotor or external curvature and part of the coupling increases with precession angle. This leads to bifurcations in coupling strength resulting in chaotic precession. As an alternative to the $SO(3)$ matrix or quaternion representation the treatment of the three coupled rotations is here based on Euler's dynamical equations. First, the classical Magic Angle Precession (MAP) dynamics is realized by a geometric or mechanical condition (type I, transcendental solutions), where it can be experimentally demonstrated how MAP can "slave" angular degrees of freedom allowing the external control of high-frequent spin by slow oscillations. MAP is found in a commercial fitness device and is conceptually approached via Chua's electric circuit. Second, the quantum-gravitational MAP (type II, rational solutions) with discrete precession angles is analyzed on a deeper level requiring intrinsic curvature/relativistic effects adjusting holonomy to quantum numbers. Third, a macroscopic network of MAP elements is presented as a discrete-time recurrent neural network synchronizing to one common MAP I/II dynamics under special pairing and symmetry conditions (type III). In all three cases MAP can be treated as a time-discrete chaotic system with singularities given by the cosine map with several possible links to interesting applications on all scales.

Keywords: Chaotic precession; spin-orbit coupling; gyroscope; Magic Angle; Euler's equations; $SO(3)$; bifurcation; neural net; Chua; Berry; Hannay.

1. Introduction

Rotated spin systems like gyroscopes, precessing stars, rotating solids, superconductors, electromagnetic particles, or others are very interesting from many different points of view. Adding a few extra coupling or damping terms, which is typical for a real system, can lead to chaotic precession (like a dissipation induced instability or bifurcation[1]), where it is rather expensive to actively control chaotic motion to obtain the sufficient stability conditions at the equilibrium points even for relatively simple gyroscope systems.[2] Regarding this challenge and the broad spectrum of possible physical applications we focus on special conditions supporting fast auto-tuning and self-stabilization capabilities. In [3–5] we have already identified a rather basic chaotic mechanism given by the cosine map, where the precession angle of rotated rotors can iterate towards an optimum. This situation arises if the precession angle induced by an external or internal coupling becomes linearly related by a connection or (an)holonomy to the spin frequency shift of precession. Recently we were able to show[5] that the basic mechanism can be illustrated and experimented with well known commercial gyroscopes useful as a hand fitness device,[6] but there is almost no published literature about its chaotic precession properties.[7] In this paper we will work out the dynamical equations of MAP. Comparing the classical transcendental recursion (type I) to the rational recursion (type II), there is some number theoretical evidence that type II requires relativistic degrees of freedom to realize quantum boundary conditions like quantum spin and precession numbers. Introducing the MAP spin network (type III) and some symmetry conditions the network behaves as one MAP element. The emerging differential or difference equations of precession provide for attracting and repelling "charge" singularities including broken symmetry, which has similarities to Chua's Circuit. This helps to evaluate both, the classical and quantum-gravitational context of MAP.

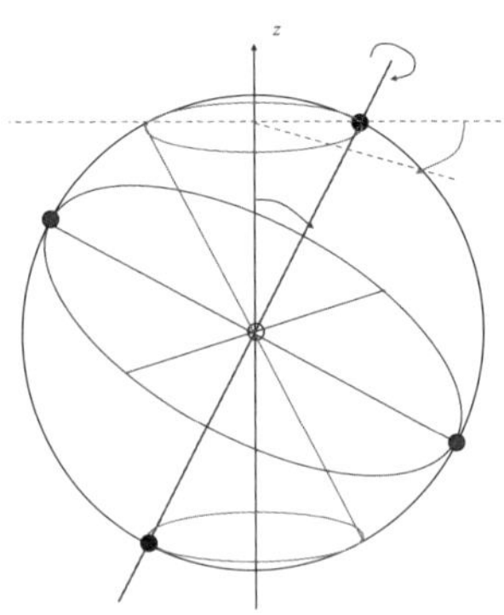

2. Transcendental Spin-Orbit Recursion (MAP Type I)

It makes sense to separate MAP into irrational transcendental and rational solutions since rational quantum angles require some curvature assistance.

2.1. *Kinematics, frequencies, and frequency shift in flat space*

As an alternative to the $SO(3)$ matrix[3] or quaternion representation, we start with Euler's dynamical equations[8,9] based on angular relations. In Euler's dynamical equations the angular velocity vector is expressed by three Euler angles describing the motion of the advancing frame with respect to the inertial frame. With mass points rotating and precessing on the sphere in flat space (which is a classical connection) the three Euler angles are (see Fig. 1):

(1) θ, first Euler angle describing the precession tilt angle,
(2) φ, second Euler angle describing the spin of the rotor axis,
(3) ψ, third Euler angle describing the precession of the axis.

In Cartesian coordinates the rotor will precess around the z-axis. Assume that the rotor axis of a spherical rotating rotor precesses at frequency $\omega_\psi = d\psi/dt$. The temporal change of precession tilt called nutation is given by $\omega_\theta = d\theta/dt$. As a characteristics of precession the rotor rotating at frequency $\omega_\varphi = d\varphi/dt$ will experience a frequency shift ω_Δ , which is part of the invariant component ω_r given by the 3rd dynamical equation

$$\omega_r = \omega_\varphi + \omega_\Delta = \omega_\varphi + \omega_\psi \cos\theta, \tag{1}$$

(see [9] with notation $\omega_r = r$, $\omega_{p,q}^2 = p^2 + q^2$). With eq.(1) and according to [9] the kinetic energy T of the spherical symmetry is given by

$$T = \tfrac{1}{2}\omega_{p,q}^2 + \tfrac{1}{2}\omega_r^2 = \tfrac{1}{2}(\omega_\theta^2 + \omega_\psi^2 + \omega_\varphi^2) + \omega_\Delta(\omega_r - \omega_\Delta). \tag{2}$$

From one Euler's equation, the other two follow by symmetry,[9] where

$$\omega_{p,q}^2 = \omega_\theta^2 + \omega_\psi^2 \sin^2\theta \text{ and } \omega_\theta^2 + \omega_\psi^2 = \omega_{p,q}^2 + \omega_\Delta^2. \tag{3}$$

2.2. *Geometric phase from (an)holonomy*

Geometric phases naturally arise in the dynamics of coupled rigid bodies,[1,10,11] (i.e. rolling on each other) or in curved space-time. Therefore, we will introduce $\bar{\theta}$ as a geometric phase or the Hannay anholonomy of the $U(1)$ bundle along the curve, where the 'parallel transported' spin vector (rotor axis) will come back after every loop or cyclic evolution with extra rotations equal to the curvature enclosed by the path.[1,11] On the unit sphere with unit Gauss curvature the j extra rotations are given by the spherical area Ω the precession axis sweeps, it is a curve integral within a precession period T providing for the difference between the total and dynamical phase

$$\Omega = T\omega_{\bar{\theta}} = 2\pi\frac{\omega_{\bar{\theta}}}{\omega_{\psi}} = 2\pi\left(1-\cos\theta\right) = 2\bar{\theta} = \frac{2\pi j}{N},\ T = \frac{2\pi}{\omega_{\psi}}. \tag{4}$$

j and N are numbers characterizing the geometric phase shift induced by external accelerations, fields, or relativistic effects (by motion or curvature). Dividing the precession period into M invariant units, the 3rd dynamical invariant ω_r is M-times the precession frequency

$$\omega_r = \omega_{\psi} M. \tag{5}$$

The special case $M = 1$ corresponds to $\omega_r = \omega_{\psi}$, $|\theta| = \pi/2$, and $j = 1$, with $\omega_r = N\omega_1$ and $\omega_{\bar{\theta}} = \omega_1 = \omega_{\varphi}$, see Figs. 1 and 2a.

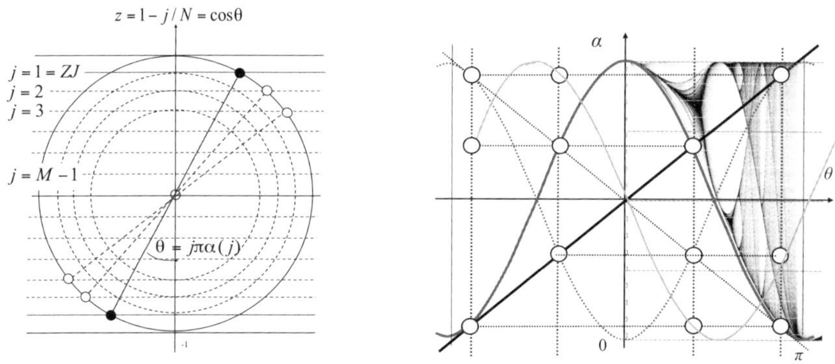

(a) Geometric phases and precession with $|j/N| > 0$. Here $M = 7$

(b) First bifurcation singularities (yellow) at $\theta\tan\theta = 1$, $\theta = \pm 0.8603336\ldots$

Fig. 1.

2.3. *The coupling parameter* α

According to eqs. (1)-(5) the dynamical phase or total phase minus geometric phase can be found in the z-axis with $z = \cos\theta$ from the tilt/precession angle θ in terms of j and N, see Fig. 2a, resulting in a relative frequency shift given by the transcendental number (the rational case will be MAP type II)

$$\frac{\omega_\Delta}{\omega_\psi} = \cos\theta = \frac{N-j}{N}, \quad 0 < j < N \tag{6}$$

According to eqs. (4) and (6) the coupling and corresponding frequency shift can be interpreted as a geometric phase changing the precession frequency, where the product $\omega_\Delta \omega_{\bar{\theta}}$ is linearly related to the nutation frequency ω_θ

$$\omega_\Delta = \omega_\psi - \omega_{\bar{\theta}} = \omega_r - \omega_\varphi, \quad \omega_\theta \omega_\psi = \omega_\Delta \omega_{\bar{\theta}} \tag{7}$$

The coupling parameter describing the shift and dynamical phase evolution within one precession loop can be defined according to eqs. (4-7) by

$$\alpha = \frac{\cos\theta}{M} = \frac{\omega_\Delta}{\omega_r} = \frac{\omega_\psi - \omega_{\bar{\theta}}}{\omega_r} = \frac{1}{M} - \frac{\omega_{\bar{\theta}}}{\omega_r} = \frac{N-j}{NM}. \tag{8}$$

Eq. (8) can characterize the fine structure of spin-orbit coupling.[3] With eq. (2) the coupling constant $|\alpha| > 0$ leads to a shift $|\omega_\Delta| > 0$, $\alpha = 0$ corresponds either to $\omega_\psi = 0$ or $|\theta| = \pi/2$, see the red curve in Fig. 2b. Assuming positive frequencies and $\omega_1 = \omega_{\bar{\theta}}(j = 1)$ we get with eqs. (1)-(8) $\omega_r > \omega_\varphi > \omega_\psi > \omega_\Delta > \omega_{\bar{\theta}} > \omega_\theta$, where

$$\omega_1 = \frac{\omega_r}{MN} = \frac{\omega_\varphi}{N(M-1)+j} = \frac{\omega_\psi}{N} = \frac{\omega_\Delta}{N-j} = \frac{\omega_{\bar{\theta}}}{j} = \frac{N\omega_\theta}{j(N-j)}. \tag{9}$$

2.4. *The geometric feedback condition leading to chaotic precession*

As a criterion for precessional coupling, the precession angle subject to dissipation will nutate according to eq. (7). The amplitude of nutation will decrease during the dissipation process. The corresponding angular contraction (Jacobian flow) and response in T will be given by one rotator loop out of M

$$-\omega_1 \frac{\theta}{\pi} = \frac{1}{M}\frac{d\theta}{dt} = \frac{\omega_\theta}{M} = \omega_{\bar{\theta}} \frac{\cos\theta}{M}. \tag{10}$$

Since the geometric phase $\bar{\theta}_t$ is evaluated within one discrete precession loop period, the precession angle for a new precession period T_{t+1} at time $t+1$ is determined by the dynamics and precession angle of the previous period T_t at time t providing with eq. (10) for a time-discrete coupling $\alpha_t(\theta_t)$ and angle

$$\theta_{t+1} = \pi j_t \alpha_t(\theta_t) = \frac{\pi}{M} \frac{j_t}{N_t} (N_t - j_t) = \pi j_t \frac{\omega_{\Delta,t}}{\omega_{r,t}}. \tag{11}$$

Because of the recursive relation between geometric phase contribution $\bar{\theta}_t$ and the precession angle in eq. (11) we can expect chaotic precession. With a dimensionless z-variable according to Fig. 2a and the z-Phase $\tilde{\theta}_t$

$$z_{t+1} \propto \theta_t(z_t), \quad z_t = 1 - \frac{j_t}{N_t} = 1 - \frac{\bar{\theta}_t}{\pi} = \frac{\tilde{\theta}_t}{\pi}, \tag{12}$$

the number j of geometric extra loops in eq. (11) become in the stable case the fixed points $j = N(1 - M\alpha)$ of a logistic map, whereas $\tilde{\theta}_{t+1}$ or the precession angle can be iteratively computed from a time-discrete cosine chaotic map showing bifurcations (see Figs. 2b, 3b) with transcendental solution

$$\tilde{\theta}_{t+1} = \pi \cos \tilde{\theta}_t, \quad \theta_{t+1} M = j\pi \cos \theta_t. \tag{13}$$

With the precession angle linearly related to the spinning phase/frequency offset induced by precession we have the MAP feedback situation recursively adjusting or auto-tuning the precession angle and frequency to the external acceleration source eventually leading to coupling anomalies.[12]

2.5. *Approaching the dynamics of MAP with Chua's Circuit*

To model the chaotic precession/nutation dynamics in eqs. (10)-(13) we can refer to a well known and rather simple system of coupled differential equations, where the angular /phase space occupation density and flow of the attractor can be simulated and visualized, see Fig. 3a (the time discrete simulation uses the difference equation version)

$$\begin{aligned} dz &= \omega_\psi M\,(y - z)\,dt + \omega_\psi f(y - z)dt, \\ dy &= \omega_\psi x dt - \omega_\psi M\,(y - z)\,dt, \\ dx &= -\omega_\psi y dt, \\ &\text{with } \pi x = \tilde{\theta},\ \pi j \alpha = \theta = j(y - z),\ f(y - z) = \pm\pi \cos[j(y - z)]. \end{aligned} \tag{14}$$

This is a form of Chua's electronic circuit, a well known and real-world example showing chaotic dynamics [13]. MAP can be approached and illustrated if we take the precession angle as the voltage term (showing Coulomb type "charge" and dipole effects) and the rotor spin as the electric current (showing magnetic monopole[8] and dipole effects). Both systems have 3 degrees of freedom (two voltage y, z, one current x) and 3 energy storage elements (two capacitors and one inductivity as the rotor angular momentum setting the timescale $\omega_r = \omega_\psi M$). In both cases a linear oscillator (precession in MAP) is coupled to a nonlinear element (anholonomy in MAP). The nonlinear element $f(\theta/j)$ responsible for chaos and bifurcation is driven by precession $\theta = j(y - z)$, where the geometric coupling current is delivered by the conductivity term $M\,(y - z)$ providing for j missing or extra loops (monopoles with charge M). Both systems show vortex singularities at points where the linear coupling current equals the nonlinear current. Asymmetric pairs of attracting vortex singularities show $|j/M| \ \geq 0.839801$, see Figs. 3 and 4a. With eqs. (10)-(13) we get near the vortex singularities ($dx/dt = dy/dt = 0$) a spiraling in plane $j\tilde{\theta} = M\theta$, the MAP condition can be found in eq.(14) from $dz/dt = d\bar{\theta}/dt = 0$, nutation can be assigned to $d\theta/dt \neq 0$. The first bifurcation is the second contact or intersection (singularity) of the nonlinear $\cos\theta$ and the linear coupling term proportional to θ , see Fig. 2b, the tangent to the cosine function (slope is given by the sine function) at $\theta \tan\theta = 1$ with $\theta = 0.5\pi \cdot 0.8603336...$ and $j/M \sim 0.839801\ldots$.

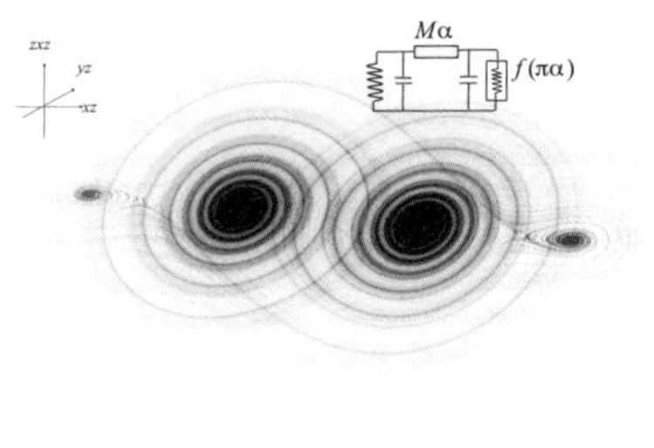

(a) Precession/nutation phase space with singularities. For $M = 12$ small connected vortex singularities with "charge pairs" (blue/red) appear for $j > 10$, here $j = 12$, see (14). The yellow conductive/coupling term provides for dissipation.

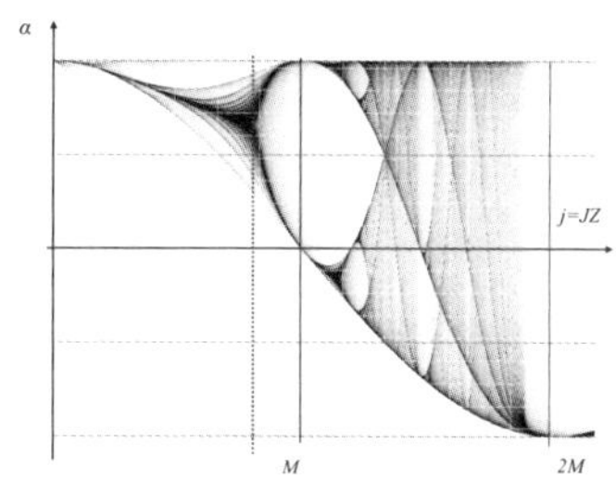

(b) MAP bifurcation diagram.[3,4] Smearing effects due to phase fluctuations.

Fig. 2.

2.6. *Powerball example*

Based on this analysis it was recently proposed[5] that the MAP type I conditions are present in the "Powerball" system.[6,7] The function of this device is based on a mechanical gear-type coupling between precession and spin frequency, see eq. (11). The rolling axis diameter at the end of the rotor shaft (#28 in [6]) is usually about 2 mm, the diameter of the groove (#24) or bearing ring element (#30) about 60 mm providing for $N \sim 30 =$ 60mm/2mm that counts how often the spin axis has to roll in the circular bearing to turn its axis by 360° with precession angle $\theta = \pi/2$, where $\omega_\Delta = \omega_\psi$. A wobbling by the hand introduces (an)holonomy and a shift by $\omega_{\bar{\theta}}$, see eqs. (6)-(10). The MAP angle is given by $\theta = \text{arcos}[(N-j)/N]$, see Fig. 2a, a Powerball with $N \sim 30$, $M \sim 12$ and $j = 1$ has $\theta(j=1) \sim 15°$, bifurcation starts near $j = JZ \sim 10$ or 75°.

3. Polynomal Quantum-Gravitational Connection (Type II)

Type I provides according to eq. (13) transcendental solutions, but curvature (by relativity or external accelerations) can support periodicity with rational or polynomial solutions which look like closed loop rolling cone paths. Under the Lorentz rotation group $SO(3)$ rolling cone paths can describe anholonomy and precession (see e.g. [1,2,5,11]) while providing for special periodicity conditions. In this case we get curves of constant precession (Darboux) in the Frenet description from the periodicity condition in the $SO(3)$ rotation matrix representation in [5], characterizing the cone rotation by $\cos(2\pi N\cos\theta) = \cos[2\pi(N-j)] = 1$ related to the cone geometry with $N \in [1, 2, ...] > j \in [1/2, 1, 3/2, \ldots]$ in a single-valued monochromatic loop.

3.1. *Tschebyscheff polynomials of the first kind*

With integral N, M, $2j$ we get periodicity and monocromatic precession, where eq. (11) provides for the closed loop conditions current condition

$$\cos\theta = \cos(n\theta),\ n = MN/j. \tag{15}$$

Eq.(15) provides for a finite order Tschebyscheff polynomial solution $T_n(x)$:

$$T_n(x) = \cos\left[n\arccos(x)\right], \quad \frac{d^2T_n}{d\theta^2} + n^2T_n = 0, \tag{16}$$

representing the MAP type II precession and geometric phase condition relating the first polynomial $T_1(x) = x$ and linear tilt/precession term as an amplitude/phase condition to a higher polynomial

$$T_n(x) = T_1(x) = x, \text{ where } T_{n+1}(x) = 2xT_n(x) - T_{n-1}(x) \qquad (17)$$

The polynomial can be written as a sum or a product,

$$T_n(x) = \sum_{i=0}^{Floor(n/2)} \binom{n}{2i} x^{n-2i} \left(x^2-1\right)^i = 2^{n-1} \prod_{k=1}^{n} \left\{ x - \cos\left[\frac{(2k-1)\pi}{2n}\right] \right\} \qquad (18)$$

where the zeros are with eqs. (11) and (18) at $k = (N-j)+\frac{1}{2}$. Mapping the x-current to a 2-dimensional y, z-precession in circular coordinates (voltage projection in Chua's circuit) we get a closed loop nutation pattern (see Fig. 4a):

$$T_n(z = \cos\theta, y = \sin\theta) = \sum_{i=0}^{Floor(n/2)} \binom{n}{2i} z^{n-2i}(-y^2)^i. \qquad (19)$$

With the phase advance proportional to frequency at a given time period, the Chebyshev relations above provide for a frequency filter connecting the 1-d current and the 2-d closed loop space-time holonomy and dynamics.

(a) Closed modulated surfaces[4] by 2d Tschebyscheff polynomials in (19).

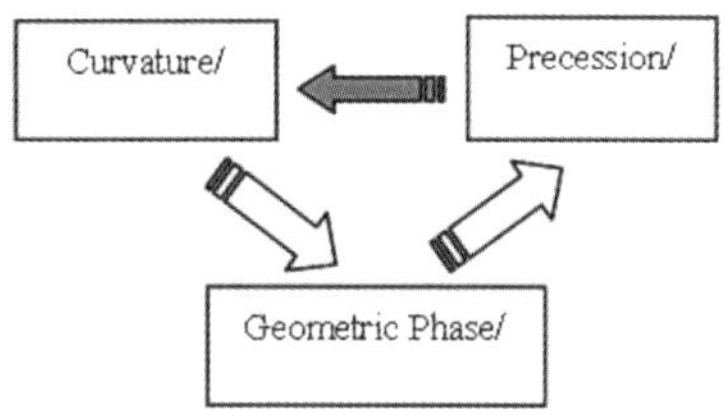

(b) MAP recursive chaotic relation in space-time adjusting (an)holonomy.

Fig. 3.

3.2. *MAP and relativity*

In General Relativity (an)holonomy can be directly related to curvature, see i.e. [1,10,11,14], where the precession induced by curvature and/or torsion angle is directly given by the coupling strength. W. de Sitter was the first who predicted geodetic precession in curved space-time,[14] where the gyroscope will be carried by parallel transport along the path of free fall.

Regarding MAP one could argue on a number theoretical level (15)-(19) that it is more economic to avoid transcendental and prefer rational or polynomial values based on (an)holonomy adjusted by recursion to periodicity, see Fig. 4b. This would mean that there could be a natural preference for curved space-time and anholonomy supporting closed loop conditions on scales comparable to the self-coherence length, which is large in superconductors. MAP is expected to occur on all scales, see the table of candidates in [12]. If we apply MAP as a coupling condition on the cosmic scale, the scale below the co-expanding unit scale appears to be contracting relative to the expanding unit scale, which is the inversion scale.[15]

4. MAP Spin-Network Connection (Type III)

Precession dynamics including coupling according to eq. (14) can lead to synchronization in a network of coupled spins J_i . A "charged" discrete-time recurrent neural network can be found with a number Z of MAP elements connected by matrix coefficients ω_{ij} , if the resulting precession angle is the sum of all weighted components of the previous stage

$$\theta_{j,t+1} = \pi J_j \cos\left(\sum_{i=0,i\neq j}^{Z-1} \omega_{ij}\theta_{i,t}\right), \tag{20}$$

where the sin/cosine transfer function is obviously exotic according to [16]. Defining a mean MAP by the time-discrete weighted sum of MAP elements

$$\begin{aligned} \theta_{is,t+1} = \sum_{j=0}^{Z-1} \omega_{ij}\theta_{j,t+1} &= \sum_{j=0,j\neq i}^{Z-1} \omega_{ij}\theta_{j,t+1} + \omega_{ii}\theta_{i,t} \\ &= \pi J_i \sum_{j=0}^{Z-1} \omega_{ij}\cos(\theta_{is,t} - \omega_{ii}\theta_{i,t}), \end{aligned} \tag{21}$$

analytical solutions requiring synchronization can be obtained with a simple homogeneous interaction model. With coupling coefficients $\omega_{ij} = M^{-1}$ supporting eq. (10) as the dynamic coupling strength, the global mean system $\theta_{is,t+1}$ behaves like a single MAP element in eq.(13) with charge Z

$$\theta_{is,t+1} = \frac{\pi J_i}{M}\sum_{j=0}^{Z-1} \cos\left(\theta_{js,t} - \omega_{jj}\theta_{j,t}\right) = \pi\frac{J_i Z}{M}\cos\left(\theta_{is,t}\right). \tag{22}$$

This requires some symmetry conditions in precession dynamics, which can be achieved by a pair wise cancellation of cosine terms in the sum of eq.

(22) with precession pairs $\cos\theta_1 + \cos\theta_2 = 0$ or by a symmetric distribution of local MAP angles around the mean $\theta_{is,t}$ with $\omega_{jj}\theta_{j,t} = \pi j/(Z-1)$.

Since $SU(2)$ is the spin double cover of the group $SO(3)$, MAP could be relevant to relativistic quantum systems like the spin-orbit Eigenstates of a relativistic electron and also to "charged" spin network systems in the solid or nuclear range.[3] Free fall precession in curved space (with Berry-Hannay connection) can generate at the center a magnetic monopole if the blue point in Fig. 1 is a charge.[9] Magnetic monopoles are field configurations that arise naturally in gauge field theories, even at the classical level. A monopole requires the non-integrability of the phase of the wave function leading to a singularity. In MAP these are the singularities, where the nonlinear cosine equals the linear extra-precession term (see Figs. 2b, 3). With $j = JZ$, charge Z, spin J, and MJ the Dirac monopole charge quantum number, MAP type III could be related to the nature of the electromagnetic field and the vector potential. $M = 137$, $J = 0.5$, $Z = 2$ (Cooper pairs) could be relevant to superconductivity with Berry's phase part of the fine structure in the Josephson effect. The first bifurcation for a closed loop spin-network system (heavy nuclear particle) would occur near $Z = 115.05275\ldots$, see Figs. 2b, 3b. Strong monopoles could be generated by MAP superconductor discs.[3,4]

5. Conclusions

It was found that a time-discrete chaotic system given by the cosine map can model a special recursive spin-spin coupling situation, where the precession angle of a rotating rotor is linearly related to the frequency shift induced by precession. The frequency or energy shift (precession forces) couple back to the internal or external acceleration (equivalent to a curvature) affecting holonomy and exciting j extra rotor loops with monopole charge M. If we allow for this chaotic feedback, the systems can auto-tune to an external precession starting bifurcations at $j/M > 0.84$ in experiments and simulation, see the Powerball device[6] and Chua's circuit approach.[13] The MAP dynamics carries the symmetry of the Lorentz group $SO(3)$, where degrees of freedom get frozen to $SO(2)$ or $U(1)$. The geometric phase shows a 2-d to 1-d mapping, where quantum-relativistic type II precession angles with space-time bending effects could generate closed loop periodicity conditions are supported by the Tschebyscheff polynomials. With synergetic effects given by a neural-type recursive network based on coupled MAP elements, a strong long range spin-network synchronization dynamics should be possible.

References

1. J. E. Marsden and T. Ratiu, *Introduction to Mechanics and Symmetry* (Springer-Verlag, 1994).
2. H. K. Chen, *J. Sound Vib.* **255**(4), 719 (2002).
3. B. Binder, Geometric phase locked in fine structure, in *Iterative Coupling and Balancing Currents*, (e-Book, 2003)
4. B. Binder, Magic angle precession, in *AIP Conf. Proc.*, 969, 1103(2008).
5. B. Binder, Magic angle precession at staif 2008, in *Quanics Presentation notes*, (www.quanics.com/MAPTalk20080214.pdf, 2008)
6. L. Mishler, *Gyroscopic Device* (United States Patent, 1973).
7. P. Heyda, *Am. J. Phys.* **70**, 1049 (2002).
8. W. F. Osgood, *Transactions of the American Mathematical Society* **23**(3), 240 (1922).
9. W. F. Osgood, *Mechanics* (MacMillan, New York, 1937).
10. F. Wilczek and A. Zee, *Phys. Rev. Lett* **52**, p. 2111 (1984).
11. A. Ishlinskii, *Mechanics of Special Gyroscopic Systems* (Izd. AN USSR, Kiev, 1952).
12. B. Binder, www.quanics.com/map_tableprocesses.pdf.
13. T. Matsumoto, L. Chua and M. Kumoro, *IEEE Transaction on Circuits and Systems* **CAS-32**(8), 798 (1985).
14. C. Misner, K.Thorne and J. Wheeler, *Gravitation* (Freeman, 1973).
15. B. Binder, Friedmann propulsion in an flat holographic universe, in *AIP Conf. Proc.*, 969, 1146(2008).
16. W. Duch and N. Jankowski, Transfer functions: Hidden possibilities for better neural networks, in *9th European Symposium on Artificial Neural Networks*, (2001).

Chaotic Phenomena at Czech Capital Market

D. Blatna* and J. Tresl†

University of Economics
Prague, Czech Republic
*Emails: *blatna@vse.cz, †tresl@vse.cz*

Several financial applications of methods developed originally in physics are demonstrated. First, the time development of cumulative strain prior to material failures and preceeding log-periodic oscillations have been observed in Telecom stock prices. Second, the search for chaotic patterns in PX index time series and its volatility was performed using correlation dimensions and BDS test. Finally, Hurst exponent computed for PX index was slightly greater than 0.5, which supported persistent behaviour with tendency to cycles creation.

Keywords: Log-periodic oscillations; stock prices; correlation dimension; Hurst exponent.

1. Log-Periodic Oscillations

First, let us mention very important notion of scale invariance.[1] Under a change of scale, say $x \to \lambda x$, scale invariance of a system means that a function f reproduces itself, i.e.

$$f(x) = \mu f(\lambda x) \Longrightarrow f(x) = Cx^{\alpha}$$

Solving this equation with respect to α, we arrive at

$$\mu\lambda^{\alpha} = 1 = \exp(2\pi n i) = \cos(2\pi n), \qquad \alpha_n = -\frac{\ln\mu}{\ln\lambda} + i\frac{2\pi n}{\ln\lambda} \tag{1}$$

On the other hand, it was found from many studies of material failures (ruptures, occurence of earthquakes etc.), the cumulative strain $\epsilon(t)$ depends on the stress acting as:[1,2]

$$\epsilon(t) = A + B(t_c - t)^{\alpha}\left[1 + \sum_{n=1}^{\infty} c_n \cos\left(\frac{2\pi n}{\ln\lambda}\ln(t_c - t)\right)\right]$$

$$A > 0, B < 0, 0 < \alpha < 1 \tag{2}$$

where t_c is a critical time, when a failure occurs. It means, cumulative strain is accompanied by log-oscillations, which can serve as failure predictors.[1] transferred these ideas also to financial crashes. Keeping only the leading term in log-oscillations, we modify the first equation to

$$\ln P(t) = A + B(t_c - t)^{\alpha}\Big\{1 + C\cos\big(\omega\ln(t_c - t) - \phi\big)\Big\}$$

where $P(t)$ denotes stock prices and ϕ is phase shift.

We have applied this tool to the time series of Telecom stock prices during the period of the growth from 21/10/1999 to 24/03/2000 (107 daily values). After final date, price decrease has been observed for nearly 400 days. Using regression analysis, we estimated parameters as follows:

$$\begin{array}{lll} A = 6.88757 & B = -0.02458 & \alpha = 0.73746 \\ C = 0.13346 & \omega = 4.55392 & \phi = -0.42953 \end{array} \tag{3}$$

The results obtained signalize the improvement of the model using log-oscillations: model MSE decreased from 0.212 (with $C = 0$) to 0.060 (with $C = 0.13346$).

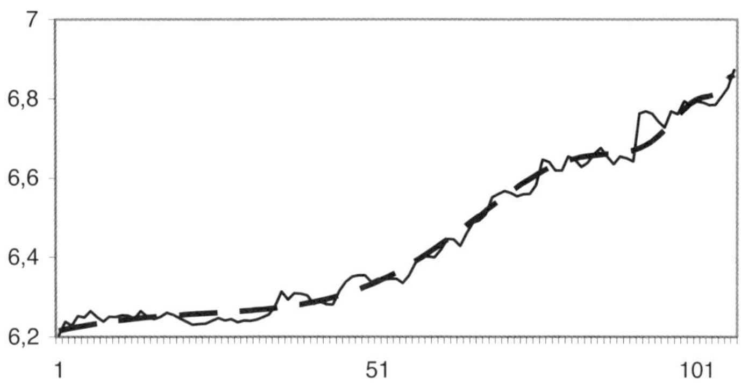

Fig. 1. Log-periodic model applied to TELECOM stock prices

2. Chaotic Behaviour

The reasons for the construction of quantitative characteristics of chaotic behavior are:

- quantificators can help to distinguish deterministic chaos from "noisy" behavior, produced by the action of external random influences
- quantificators can help to determine minimum number of variables needed for the construction of a dynamical model of the system
- quantificators can help to classify systems according to universally valid regularities
- changes of quantificators may signalize changes in qualitative behavior of a system

On principle, we can use two kinds of description. The first type stresses the dynamics of chaotic behavior and corresponding quantificators describe the system evolution and mutual position of neighbouring trajectories. The second type is based on the geometry of trajectories in state space. Here we shall confine ourselves to the first kind of the description. One of characteristic features of chaotic behavior is the divergence of neighbouring trajectories. Let us consider two close points $x_0, x_0 + \epsilon$.

Applying certain iterative function n-times, the distance between these points will be:

$$d_n = \left| f^{(n)}(x_0 + \epsilon) - f^{(n)}(x_0) \right|$$

In the case of chaotic behavior, we expect the exponential growth of this distance with the number of iterations:

$$\frac{d_n}{\epsilon} = \exp(\lambda n) \qquad \lambda = \frac{1}{n} \ln \left(\frac{d_n}{\epsilon} \right)$$

In limit approaching $\epsilon \to 0$ we obtain after small algebra:

$$\lambda = \frac{1}{n} \{ \ln |f'(x_0)| + \ln |f'(x_1)| + \cdots + \ln |f'(x_n)| \}$$

This *Lyapunov exponent* characterizes the speed of the divergence of neighbouring trajectories.[3] It is given as averaged natural logarithm of absolute value of the derivatives of iterative function at individual points of a trajectory. One-dimensional iterative function has chaotic trajectories, if average Lyapunov exponent is positive (the condition of divergence).

Having a time series of equidistant values $x_0, x_1, \ldots, x_n$, we can determine Lyapunov exponent using the following approach. Let us take two trajectories with starting points x_i, x_j and let us create the sequence of differences:

$$d_0 = |x_j - x_i| \qquad d_1 = |x_{j+1} - x_{i+1}| \qquad \cdots \qquad d_n = |x_{j+n} - x_{i+n}|$$

We shall assume time evolution in the form:

$$d_n = d_0 \exp(\lambda n) \quad \Longrightarrow \quad \lambda = \frac{1}{n} \ln \frac{d_n}{d_0}$$

For $\lambda > 0$, the behavior of trajectories is chaotic. However, generally, λ depends on the choice of x_i. Therefore, the more reliable is to compute *average Lyapunov coefficient* from large number N initial values regularly distributed over whole attractor.

[4] introduced the characteristic called *correlation dimension*, based on the behavior of so-called correlation sum. For the computation of correlation dimension, we need data about the evolution of a trajectory (in sum n values). For each i-th point of the trajectory, we seek relative frequency $p_i(r)$ of trajectory points, lying at distance less than r from the point i (except i-th point):

$$p_i(r) = \frac{n_i}{n-1}$$

Correlation sum is then computed as average relative frequency

$$C_1(r) = \frac{1}{n} \sum_{i=1}^{n} p_i(r)$$

Obviously $C_1(r) = 0$ if r is less than minimal distance among the points of a trajectory. On contrary $C_1(r) = 1$ means, the distances among individual points do not exceed r. Minimal possible non-zero value $C_1(r) = \frac{2}{n(n-1)}$ occurs in the case, only one distance is less than r. Relative frequency can be formally expressed using Heaviside step function:

$$p_i(r) = \frac{1}{n-1} \sum_{j=1, j \neq i}^{n} H\big(r - |x_i - x_j|\big)$$

Similarly, correlation sum can be written as:

$$C_1(r) = \frac{1}{n(n-1)} \sum_{i=1}^{n} \sum_{j=1, j \neq i}^{n} H\big(r - |x_i - x_j|\big)$$

In limit case $n \to \infty$, correlation sum is melted into *correlation integral.* Correlation dimension is then given by formula:

$$D_1 = \lim_{r \to 0} \frac{\log C_1(r)}{\log r}$$

A time series of single variable can be often sufficient for the determination of important characteristics of a multidimensional dynamical system. The groups of values

$$x_{i+1}, x_{i+2}, \ldots, x_{i+d} \qquad t = 0, 1, 2, \ldots, n-d$$

give the coordinates of a point in d-dimensional space. Then the sequence of these groups describes the time evolution of a system in d-dimensional *embedding space*. In this case, correlation sum can be written as

$$C_d(r) = \frac{1}{n(n-1)} \sum_{i=1}^{n} \sum_{j=1, j \neq i}^{n} H\big(r - |\mathbf{x}_i - \mathbf{x}_j|\big)$$

because it depends on embedded dimension d. Vector $\mathbf{x_i}$ of dimension d has components

$$\mathbf{x_i} = \big(x_i, x_{i+L}, x_{i+2L}, \ldots, x_{i+(d-1)L}\big)$$

where L is time lag between neighbouring values. The length of the difference of two vectors is mostly calculated as Euklid distance

$$\mathbf{x_i} - \mathbf{x_j} = \sqrt{\sum_{k=0}^{d-1} \big(x_{i+kL} - x_{j+kL}\big)^2}$$

Then it holds for correlation dimension:

$$D_d = \lim_{r \to 0} \frac{\log C_d(r)}{\log r}$$

[4] have studied the behavior of these characteristic. In the case of i.i.d. process, $D_d = d$. On the contrary, in the case of non-linear deterministic process the behavior of correlation sum is quite different. This tool was used also for our time series of PX returns (see Table 1). Clearly, up to $d = 5$, $D_d \approx d$ and the behaviour of PX returns seems to be similar to i.i.d. process. On the other hand, for $d > 5$, $D_d < d$. Thus, if this time series is chaotic, it should be generated by some high-dimensional process.

Brock, Dechert and Scheinkman (see [5]) have shown, for a finite r and i.i.d. process the following relation is valid:

$$C_d(r) = \big[C_1(r)\big]^d$$

and suggested test statistic:[5]

$$T_d(r,n) = \frac{C_d(r,n) - \big[C_1(r)\big]^d}{s_d(r,n)}$$

where $C_d(r,n)$, $C_q(r,n)$ are sample correlation sums and $s_d(r,n)$ is the estimate of the standard deviation of the difference in the previous expression. This statistic has asymptotically standard normal distribution $N(0,1)$ providing the validity of null hypothesis (i.i.d. process).

All test statistics given in the Table 2 exceed the critical value 1.96 and null hypothesis (i.i.d. process) is rejected in all cases.

Table 1. PX index logarithmic returns: Correlation dimensions computed for different embedded dimensions (d) and different lengths of time series.

length	d=1	d=2	d=3	d=4	d=5	d=6
225	1.032	1.915	2.176	4.232	5.132	—
450	1.036	2.099	2.822	4.148	—	5.382
675	1.052	2.091	2.995	3.999	4.720	—
900	1.016	2.105	3.069	4.010	4.790	—
1125	1.021	2.101	3.060	3.962	4.637	—
1350	1.014	2.122	3.057	3.886	—	5.013
1575	1.013	2.080	3.037	3.992	—	5.078
1800	1.025	2.082	3.003	3.917	4.573	4.913
2025	1.033	2.072	3.048	3.916	4.532	4.865
2250	1.034	2.078	3.022	3.892	4.549	4.918

Table 2. PX logarithmic returns: BDS test statistic computed for different embedded dimensions (d) and different lengths of time series.

length	d=2	d=3	d=4	d=5	d=6
225	—	3.02	3.89	4.36	5.21
450	3.00	3.80	4.52	5.15	6.40
675	4.56	5.67	6.37	6.94	8.23
900	3.69	4.53	5.45	6.07	7.08
1125	4.65	6.35	7.52	8.38	9.53
1350	5.49	7.43	8.66	9.63	10.86
1575	6.03	8.04	9.42	10.48	11.64
1800	7.08	9.46	11.14	12.48	14.03
2025	8.12	10.82	12.59	13.99	15.62
2250	8.68	11.40	13.36	14.90	16.67

3. Rescaled Range Analysis

The general procedure is the following:[6]

- given a time series containing values $x_1, x_2, \ldots, x_N$ registered in time points $t = 1, 2, \ldots, N$; thus, N time intervals with unit length arise
- subdivide the whole time series into m neighbouring intervals with length n, $N = mn$
- compute average value for each interval

$$\bar{x}_j = \frac{1}{n} \sum_{i=1}^{n} x_{ij} \qquad j = 1, 2, \ldots, m$$

- create the time series of cumulative deviations from average value

$$z_{kj} = \sum_{i=1}^{k} (x_{ij} - \bar{x}_j) \qquad k = 1, 2, \ldots, n$$

- compute the range of cumulated deviations for each interval

$$R_j = \max(z_{kj}) - \min(z_{kj}) \geq 0$$

- compute standard deviation for each interval

$$s_j = \sqrt{\frac{1}{n} \sum_{i=1}^{n} (x_{ij} - \bar{x}_j)^2}$$

- compute rescaled range R_j/s_j for each interval
- average value R/S for an interval with length n is then

$$(R/S)_n = \frac{1}{m} \sum_{j=1}^{m} (R_j/S_j)$$

Hurst assumed general type of the dependence of R/S on time in the form[6]

$$(R/S)_n = Cn^H$$

where C is a constant and H the *Hurst exponent.* In practice, Hurst exponent can be estimated using linear regression

$$\log\big[(R/S)_n\big] = \log C + H \log n$$

by virtue of the values $(R/S)_n$ computed for different n. Interpretation of Hurst exponent is the following: If $H = 0.5$, then is time series generated by i.i.d. process. In the range $0.5 < H < 1$ are corresponding processes called *persistent* and they are characterized by long memory. In the range $0 < H < 0.5$ so-called *antipersistent* processes are underlying, which alter signs more frequently than white noise.

As for PX logarithmic returns, Hurst exponent was computed for different lengths; the results are given in the Table 3:

All values of Hurst exponent are greater than 0.5 and weak tendency to persistent behaviour is proved in all segments of time series investigated.

4. Conclusions

With the use of real financial data, we demonstrated the power of certain tools originally developed in physics. First, log-periodic oscillations were detected prior to Telecom price decrease in March 2000. Corresponding

Table 3. PX index logarithmic returns: Hurst exponent computed for different lengths of time series.

length	225	450	675	900	1125
H	0.598	0.592	0.565	0.561	0.546
length	**1350**	**1575**	**1800**	**2025**	**2250**
H	0.552	0.558	0.556	0.552	0.548

regression model exhibited clear improvement after including this phenomena. Further, we investigated the possibility of chaotic behaviour in financial time series. If this is true, then only high-order chaos probably arises, because correlation dimensions computed for PX index are of the same order as embedded dimensions up to 5-th order. On the other hand, BDS test results clearly signalize departure from purely random behaviour. Moreover, the Hurst exponent computed for this time series is slightly greater than 0.5; thus, some kind of persistent behaviour is present and we expect weak tendency to cycles creation.

References

1. D. Sornette, *Critical phenomena in natural sciences* (Springer, Berlin, 2000).
2. J. Voit, *The statistical mechanics of financial markets* (Springer, Berlin, 2001).
3. K. Alligood, T. D. Sauer and J. A. Yorke, *Chaos: An introduction to dynamical systems* (Springer, Berlin, 2001).
4. P. Grassberger and I. Procaccia, *Phys. Rev. Letters* **50**, 346 (1983).
5. J. Campbell, A. W. Lo, A. C. MacKinlay and A. Y. Lo, *The econometrics of financial markets* (Princeton Univ. Press, Princeton, 1997).
6. E. E. Peters, *Fractal market analysis* (Wiley, New York, 1994).

Stability Study for Quark Systems using a Semi-Classical Billiard Model

C. C. Bordeianu*, C. Beşliu, Al. Jipa, I. V. Grossu and T. Eşanu

University of Bucharest, Faculty of Physics
Bucharest-Magurele, PO Box MG 11, 77125, Romania
*Email: *cristian.bordeianu@brahms.fizica.unibuc.ro*

D. Felea

Institute of Space Sciences, Laboratory of Space Research
Bucharest-Magurele, PO Box MG 23, 77125, Romania
Email: felea@venus.nipne.ro

We consider a semi-classical version of the motion of the particles and of the collective coordinates. Several interacting quarks are moving in a 2D potential well and hitting the vibrating surface. We use a Cornell type potential consistent with the indications of lattice QCD calculations. The numerical simulation is based on the solutions of the Hamilton equations which were solved using an algorithm of Runge-Kutta type (order 4-5) having an optimized step size, taking into account that the absolute error for each variable is less than 10^{-6}. Total energy is conserved with high accuracy, i.e. approx. 10^{-6} in absolute value. Herewith we analyze the chaotic behavior of the nonlinear dynamics system using phase-space maps, autocorrelation functions, power spectra and Lyapunov exponents. There is a transition from a chaotic motion corresponding to hadronic plasma to a more ordered behaviour of the system which could be revealed from the dependence of largest Lyapunov exponent (LLE) versus temperature.

Keywords: Lyapunov exponent; quark-gluon plasma.

1. Introduction

Over the last two decades an increasing number of papers have treated the study of the deterministic chaotic behaviour of Fermi nuclear systems. The relation between the order to chaos transition in the dynamics of independent classical particles in a container was first studied using computer simulations by Blocki et al.[1] They analyzed the behaviour of a gas of classical non-interacting particles enclosed on a multipole-deformed container which undergoes periodic shape oscillations and showed that higher multipolarities leads to chaotic motion. Another step in this direction was done by Bauer et al.[2] They have performed self-consistent calculations in semi classical approximation utilizing a multipole-multipole interaction of the Bohr-Mottelson type for quadrupole and octupole deformations. In both cases the dynamical evolution showed a regular undamped collective motion which coexists with a weak chaotic single-particle dynamics. Then Burgio et al.[3] considered a number of nucleons without spin and charge and with no internal structure which are moving in a 2D deep Woods-Saxon potential well and hitting the oscillating surface with a certain frequency.

In this paper we use a very similar model in order to study the critical behaviour of quarks undergoing phase transition to hadrons. This issue was previously studied in the framework of the Ising model.[4] Other studies[5] analyzed the largest Lyapunov exponents of SU(1) and SU(2) gauge field configurations on the lattice which are initialized by quantum Monte Carlo simulations. The results demonstrate that chaos is present when particles are confined, but it also persists into the Coulomb and quark-gluon plasma phase.

One of the most important challenges of the last years in physics, the Quark Gluon Plasma, brings us today the unique opportunity of reproducing, in laboratory, complex processes, similar with the evolution of early Universe at ages under one microsecond. The nucleus-nucleus collisions at relativistic and ultra-relativistic energies are the only experimental way to investigate this unusual state of matter, which exists for a very short time interval $(5-10fm/c)$, and whose creation supposes high densities and temperatures $(\rho = 2GeV/fm^3, T = 150 - 250MeV)$. For a better phenomenological understanding of the complexity involved, we analysed a many-body system, and we tried to find some connections with the existing knowledge on QGP.

2. Toy Model

We chose the system under study as an ensemble with N quarks interacting with each other only via a 2D Cornell potential well and hitting periodically the oscillating surface with a certain frequency. The Hamiltonian of this kind of interaction in polar coordinates is:

$$H(r_i, \theta_i, \alpha) = \sum_{i=1}^{N} \left(\frac{p_{r_i}^2}{2m} + \frac{p_{\theta_i}^2}{2mr_i^2} + V_{\text{int}}\left(r_i, R(\theta_i)\right) \right) + \frac{p_\alpha^2}{2M} + \frac{1}{2} M\Omega^2\alpha^2 \quad (1)$$

where $\{p_{r_i}, p_{\theta_i}, p_\alpha\}$ are the conjugate momenta of the i-th particle and of the collective coordinates $\{r_i, \theta_i, \alpha\}$, $m = 300MeV$, Ω is the oscillating frequency of the collective variable α and $M = mNR_0^2$ is the Inglis mass.

The colored and flavored quarks of u and d types are treated as semi classical particles interacting via a Cornell potential with color matrices C_{ij} where V_{int} is given by the function:

$$V_{\text{int}} = \frac{1}{2}\sum_{i,j} C_{ij} V_C\big(|r_i - r_j|\big) \tag{2}$$

V_C is the Cornell potential given by the function:

$$V_C\big(r_i, R(\theta_i, \alpha)\big) = \sigma\big(r_i - R(\theta_i, \alpha)\big) - \frac{\alpha_0}{r_i - R(\theta_i, \alpha)} \tag{3}$$

where $\sigma = 900MeV/fm$ is the string tension and $\alpha_0 = 1.25$ is the QCD fine structure constant.[6]

This potential is phenomenologically motivated to mimic the soft gluon part of a quark-gluon plasma accordingly with the qMD (quark Molecular Dynamics) model.[7] In Table 1 we present the elements of the color matrices for different elementary color combinations of the quarks and antiquarks.

Table 1. Color matrix elements of different elementary color combinations of the quarks and antiquarks.[7]

C_{ij}	R	G	B	$\overline{B}$	$\overline{G}$	$\overline{R}$
R	-1	$+\frac{1}{2}$	$+\frac{1}{2}$	$-\frac{1}{2}$	$-\frac{1}{2}$	$+1$
G	$+\frac{1}{2}$	-1	$+\frac{1}{2}$	$-\frac{1}{2}$	$+1$	$-\frac{1}{2}$
B	$+\frac{1}{2}$	$+\frac{1}{2}$	-1	$+1$	$-\frac{1}{2}$	$-\frac{1}{2}$
$\overline{B}$	$-\frac{1}{2}$	$-\frac{1}{2}$	$+1$	-1	$+\frac{1}{2}$	$+\frac{1}{2}$
$\overline{G}$	$-\frac{1}{2}$	$+1$	$-\frac{1}{2}$	$+\frac{1}{2}$	-1	$+\frac{1}{2}$
$\overline{R}$	$+1$	$-\frac{1}{2}$	$-\frac{1}{2}$	$+\frac{1}{2}$	$+\frac{1}{2}$	-1

The quarks were initially randomly distributed in the billiard with the radius R_0 and the corresponding momenta were generated according to Maxwell distribution. The vibrating surface of the potential well can be written as usually[8] depending on the collective variable and Legendre polynomials $P_L(cos(\theta_i))$:

$$R(\theta_i, \alpha) = R_0\big[1 + \alpha \cdot P_L(cos(\theta_i))\big] \tag{4}$$

where $R_0 = 6.0\,fm$ and $L = 0$ (monopole), $L = 1$ (dipole), $L = 2$ (quadrupole), and $L = 3$ (octupole).

The numerical simulations are based on the solution of a system containing $4N + 2$ nonlinear Hamilton equations coupled by the collective variable α :

$$\dot{r}_i = \frac{p_{r_i}}{m} \tag{5}$$

$$\dot{p}_{r_i} = \frac{p_{\theta_i}^2}{mr_i^3} - \frac{\partial V_{int}}{\partial r_i} \tag{6}$$

$$\dot{\theta}_i = \frac{p_{\theta_i}}{mr_i^2} \tag{7}$$

$$\dot{p}_{\theta_i} = -\frac{\partial V_{int}}{\partial R} \cdot \frac{\partial R}{\partial \theta_i} \tag{8}$$

$$\dot{\alpha} = \frac{p_\alpha}{M} \tag{9}$$

$$\dot{p}_\alpha = -M\Omega^2\alpha - \frac{\partial V_{\rm int}}{\partial R} \cdot \frac{\partial R}{\partial \alpha} \tag{10}$$

We solved this system of ordinary differential equations using an algorithm of Runge-Kutta type (order 4-5) having an optimized step size[9,10] and taking into account that the absolute error for each variable is less than 10^{-6}.

We developed our codes using Scilab 3.1.1. Scilab,[11] developed by researchers from INRIA and ENPC, is a scientific software package for numerical computations providing a powerful open computing environment for engineering and scientific applications. We chose Scilab, because it provides various functions for ordinary differential equation solving, Fast Fourier Transform, autocorrelation and excellent 2D and 3D graphical capabilities.

3. Numerical Study and Preliminary Conclusions

The chaos analysis we performed includes phase portraits, power spectra, autocorrelation functions and Lyapunov exponents. We studied systems having the number of quarks between 2 and 10.

One of the various ways for studying the dynamics of the system is to calculate the largest Lyapunov exponent (LLE) which is a measure of the sensitivity of the system to initial conditions.

The LLE was previously used to study the solid to liquid phase transitions in Lennard-Jones drops by Dorso and Bonasera.[12] They showed the presence of a maximum in LLE which indicates the transition from a chaotic regime to a more regular one. In the chaotic regime, the system is composed mainly of a liquid drop while the regular one corresponds to almost freely flowing particles and small clusters.

We calculate the multi-dimensional Lyapunov exponents using the formula:

$$\lambda(t) = \lim_{d_0 \to 0} \ln \frac{\sqrt{\sum_{i=1}^{N} \left[(r-r_0)^2 + (\theta-\theta_0)^2 + (p_r - p_{r_0})^2 + (p_\theta - p_{\theta_0})^2\right]_i}}{d_0} \tag{11}$$

where the sums are taken over all N quarks of the system and the quantities from the parentheses refer to two trajectories in the quark-quark phase-space which initially differ only by an infinitesimal quantity $d_0 = 10^{-6}$. We take integration times about to $500 fm/c$ sufficiently large to avoid errors on the calculation of the Lyapunov exponents. In Figure 1 we present the

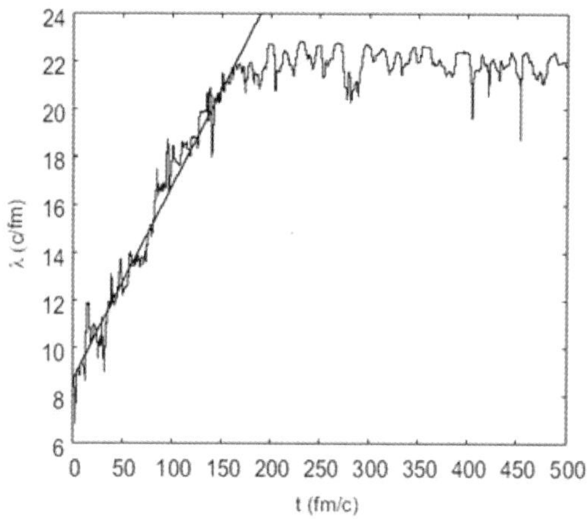

Fig. 1. The multi-dimensional Lyapunov exponent as a function of time for two quarks system. The straight line is a fit whose slope equals the largest Lyapunov exponent (LLE).

dependence of multi-dimensional Lyapunov exponent versus time. The saturation behavior and the appearance of a plateau can be seen. The straight line represents a fit whose slope is the largest Lyapunov exponent(LLE).

The preliminary results for two quarks systems are presented in Figure 2. All the calculations are done for L = 2 (dipole). It can be seen that LLE's have positive values, which is an evidence of a chaotic behaviour of the system. Moreover, we believe that the dependence of the Lyapunov exponents versus temperature can give us qualitative and quantitative information

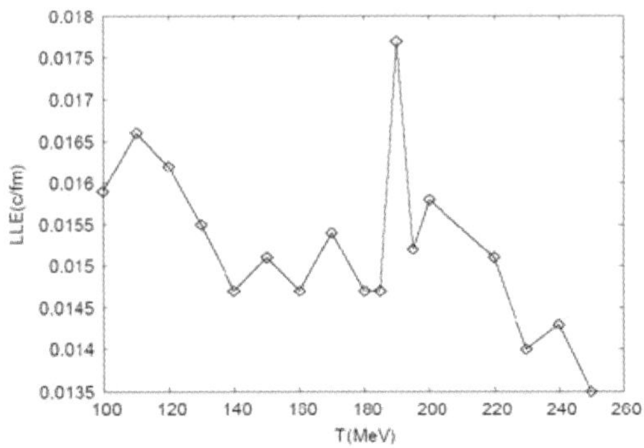

Fig. 2. The largest Lyapunov exponent (LLE) as a function of temperature for two quarks system. (The lines which connect points are only for eye orientation).

about a phase transition from the hadronic plasma to quark-gluon plasma. Thus the presence of the maximum in the LLE signals a transition from a chaotic motion corresponding to hadronic plasma to a more ordered behaviour of the system, which could be associated with the above mentioned phase transition.

A similar correlation was made by using quantum Monte Carlo numerical simulations for nucleons systems by Zhang, Wu and Lee.[13] Accordingly, the temperature corresponding to the maximal LLE is named "critical temperature" which signals a transition from a chaotic to a more ordered motion.

The critical temperature we found, $T = 195MeV$, corresponding to the maximum of the LLE, is in a good agreement with the expected value for QGP $(150 - 250MeV)$. Another numerical simulation we did,[14] using this time a relativistic many-body system and a two-particles potential based on the quark bag model, gives a similar result for the critical temperature. We believe that future works, considering an increasing number of quarks, will offer valuable information about the phase transition to quark-gluon plasma. Further analysis along those lines is currently in progress.

References

1. J. Blocki, J. Shi and W. Swiatecki, *Nucl. Phys. A* **554**, p. 387 (1993).
2. W. Bauer, D. McGrew, V. Zelevinsky and P. Schuck, *Phys. Rev. Lett.* **72**, p. 377 (1994).
3. G. F. Burgio, F. M. Baldo, A. Rapisarda and P. Schuck, *Phys.Rev. C* **52**, p. 2475 (1995).

4. R. C. Hwa, *hep-ph/9611372v2* (1996).
5. T. S. Biro, N. Hormann, H. Markum and R. Pullirsch, *hep-ph/9909309* (1999).
6. Y. Akimura, T. Maruyama, N. Yoshinaga and S. Chiba, *Eur. Phys. J. A* **25**, 405 (2005).
7. S. Scherer, Modeling ultra-relativistic heavy ion collisions with the quark molecular dynamics qmd, PhD thesis, University Johann Wolfgang Goethe Frankfurt(2005).
8. A. deShalit and H. Feshbach, *Theoretical Nuclear Physics* (Wiley, New York, 1974).
9. R. Landau, M. Paez and C. Bordeianu, *Computational Physics. Problem Solving with Computers*, 2nd edn. (Wiley VCH, 2007).
10. R. Landau, M. Paez and C. Bordeianu, *A Survey of Computational Physics. Introductory Computational Science* (Princeton University Press, 2008).
11. C. C. Bordeianu, C. Besliu, A. Jipa, D. Felea and I. V. Grossu, *Computer Physics Communications* **178**, p. 788 (2008).
12. C. O. Dorso and A. Bonasera, *Eur. Phys. J. A* **11**, p. 421 (2001).
13. Y. Zhang, X. Wu and Z. Li, *Phys. Rev. C* **69**, p. 044609 (2004).
14. I. V. Grossu, C. Besliu, A. Jipa, D. Felea and C. C. Bordeianu, *hep-ph/0702245v2* (2007).

Cross-Frontal Chaotic Transport in Oceanic Jet Currents

M. V. Budyansky, M. Yu. Uleysky and S. V. Prants*

V.I.Il'ichev Pacific Oceanological Institute
of the Russian Academy of Sciences
43 Baltiyskaya ulitsa, 690041 Vladivostok, Russia
*Email: *prants@poi.dvo.ru*
www.dynalab.poi.dvo.ru

Transport of water masses across oceanic jet currents is considered in kinematic and dynamic models of chaotic advection of passive scalars. From the point of view of dynamical systems theory it is a problem of breakdown of the last transport barrier with changing control parameters. We find numerically two scenarios for the breakdown of the transport barrier in a kinematic model of a western boundary meandering jet current like the Gulf Stream in the Atlantic ocean and the Kuroshio in the Pacific ocean and discuss causes for such a breakdown. The cross-frontal chaotic transport is shown to occur as well in a dynamic model of chaotic advection in the Antarctic circumpolar current with two Rossby waves. In both the models the breakdown of the transport barrier may occur at comparatively small values of the amplitudes of perturbations.

Keywords: Chaotic advection; oceanic jet currents; cross-frontal transport barrier.

1. The Dynamical Systems Approach to Large-Scale Mixing and Transport in the Ocean

Mixing of water masses and their properties in the ocean is a complicated process with the scales ranging from molecular to the planetary ones. We consider in this paper mesoscale mixing and transport and restrict ourselves by advection only. The action of small-scale turbulence is not studied. The turbulent diffusion may affect a fine structure of deterministic advective processes developing for a sufficiently long time and should be included into consideration in order to estimate correctly water fluxes. Temperature, salinity, pollutants, phytoplankton and even fish eggs may be treated as advected passive tracers, the equations of motion for which are very simple in two-dimensional incompressible flows

$$\frac{dx}{dt} = u(x,y,t) = -\frac{\partial \Psi}{\partial y}, \quad \frac{dy}{dt} = v(x,y,t) = \frac{\partial \Psi}{\partial x}, \tag{1}$$

where u and v are zonal and meridional velocities, respectively, the stream function Ψ plays the role of a Hamiltonian, and the coordinates (x, y) of a tracer are canonically conjugated variables. It is a Hamiltonian dynamical system with the phase space being the physical space of a two-dimensional flow. The motion is integrable in steady flows with time-independent stream functions and may be chaotic if Ψ depends on time (quasi)periodically. The phenomenon, known as chaotic advection,[1] is a hypostasis of Hamiltonian chaos in the fluid motion.[2] Chaotic advection has been observed in laboratory[3] where it is possible to create strictly periodic flows.

Geophysical flows in the ocean and the atmosphere are not periodic or quasiperiodic, they are rather stochastic. In some cases one can model a geophysical flow as a (quasi)periodic one with a weak stochastic component and study the contribution of chaotic advection, produced by the (quasi)periodic perturbation, to a large-scale mixing of water masses. The ideas and methods of the dynamical systems approach to large-scale mixing and transport in the ocean and the atmosphere have been elaborated for the last 15 years (for recent reviews on this topic see [4,5]).

The mesoscale coherent features in the ocean are major jet currents, like the Gulf Stream, the Kuroshio, the Antarctic circumpolar current and many others, and eddies. Mesoscale and submesoscale moving eddies widely occur in the ocean. Quasistationary submesoscale topographic eddies, arising in the oceanic flows over seamounts, hollows and ridges, are much more rare features. The daily tides perturb the background flow periodically and may produce chaotic mixing by the topographic eddies (for reviews on chaotic advection in kinematic and dynamic models of topographic eddies see [4,6]). Since the diameter of the topographic eddies is of the order of a few tens of kilometers the chaotic mixing and transport there are of a local significance.

The topic of interest in this paper is a cross-frontal chaotic transport in major jet oceanic currents which is much more important from the practical point of view because it means breaking of a transport barrier between the waters with different characteristics. Whether or not the jet axis provides an effective barrier is a crucial question in physical oceanography (and in the physics of the atmosphere). Observations of float and drifter trajectories for many years indicate that transport does occur across the Gulf Stream and Kuroshio, and deeper floats have been found more frequently to cross the streams than shallow ones.[7] There are, at least, two approaches to solving that problem. In the approach based on full-physics numerical models, one can generate the velocity field as an outcome of a basin or subbasin model and solve the advection equations (1) for a large number of initial conditions. It is possible, in principle, to know how strong is a jet barrier and estimate a flux across the jet (if any). There are, of course, many technical problems in solving a great number of partial differential equations on a grid with sufficiently fine resolution, in acquiring input meteorological data, in interpolating the outcome velocity field between the grid points and many others. Moreover, even if the fact of the cross-frontal transport would be established, it would be difficult to specify physical causes for that among the possible mechanisms: ring detachment, variations in the meanders amplitude, change in the flow topology at the depth, Rossby waves, etc.

In the theoretical approach, one postulates a kinematic model of the meandering jet current or solve dynamic equations to construct a dynamic model of the jet flow. In both the cases, one specifies a stream function and solve the corresponding advection equations (1). Simulation with different values of control parameters and a large number of tracers will show whether the cross-jet transport exists or not and what is a mechanism for that. Moreover, one can trace the process of breakdown of the transport barrier and study different scenarios for that. However, it is hardly possible to estimate correctly the cross-transport fluxes with idealized models.

2. Cross-Frontal Chaotic Transport in a Kinematic Model of a Meandering Jet

In kinematic studies one does not solve any dynamical equations but simply postulates the stream function of a jet current based on known features of the current. As to the western boundary currents in the ocean, a kinematic model has been proposed by Bower[8] and Samelson[9] to explain water exchange in the meandering Gulf Stream. Chaotic mixing and transport in this model have been studied in detail in a number of papers.[9–12] Numerical results indicate that exchange of water masses across the jet core (meridional exchange) is more difficult to achieve than mixing along each side of the jet. Unrealistically large values of the meander perturbation amplitudes have been found in previous studies to break a meridional transport barrier.[9,13] In this section we are interested mainly in two questions: can the cross-frontal chaotic transport occur at sufficiently small and realistic values of the meander perturbation amplitudes and what is a scenario for that.

As a kinematic model of the meandering current in the ocean, we take the Bickley jet with a running wave (a meander) imposed with the following normalized stream function in the frame moving with the phase velocity of the meander:[8,10]

$$\Psi = -\tanh\left(\frac{y - A\cos x}{L\sqrt{1 + A^2\sin^2 x}}\right) + Cy, \tag{2}$$

where the jet's width L, meander's amplitude A and its phase velocity C are the control parameters. The phase portrait of the flow (2) shown in Figure 1(a) consists of three different regions: the central eastward jet J, chains of the northern and southern circulations C and the peripheral westward currents P. In Figure 1(b) we plot a frequency map $f(x_0, y_0)$ that shows by color the value of the frequency f of a particles with initial positions (x_0, y_0) in the unperturbed system. The maximal value of the frequency, $f_{\max} = 1.278$, have the particles moving in the central jet.

Under even a small modulation of the meander's amplitude, $A = A_0 + \varepsilon\cos\omega t$, the separatrices, connecting saddle points, are destroyed and transformed into stochastic layers. The strength of chaos depends strongly on both the perturbation parameters, the perturbation amplitude ε and frequency ω. It is easily to understand what happens when ε grows while ω is fixed. At small ε, there are two thin stochastic layers around the northern and southern separatrices of the unperturbed system. Those layers are separated by the central jet, and particles cannot cross the jet. As ε grows, the width of the layers grow as well until at some critical value they merge. In terminology of dynamical systems theory it is an onset of global chaos. In physical oceanography the phenomena may be called "a cross-frontal chaotic transport" or "a meridional chaotic transport". The crucial point is that global chaos may occur in Hamiltonian systems at unusually small values of the perturbation amplitude ε.[14,15] The reason for that is an overlapping of resonances of the same order and their overlapping in energy with stochastic layers of the unperturbed separatrices.

The following numerical procedure has been chosen to establish the fact of the cross-frontal transport in the kinematic model of the meandering jet current. The advection equations for given values of the perturbation amplitude ε and frequency ω and with twenty particles placed nearby the northern saddle point have been integrated up to a time instant when one of the particles was detected to cross the straight line $y = y_s^{(s)}$ passing through the southern saddle. The ε–ω diagram in Figure 2(a) shows the parameters range for which the cross-frontal chaotic transport exists (white color

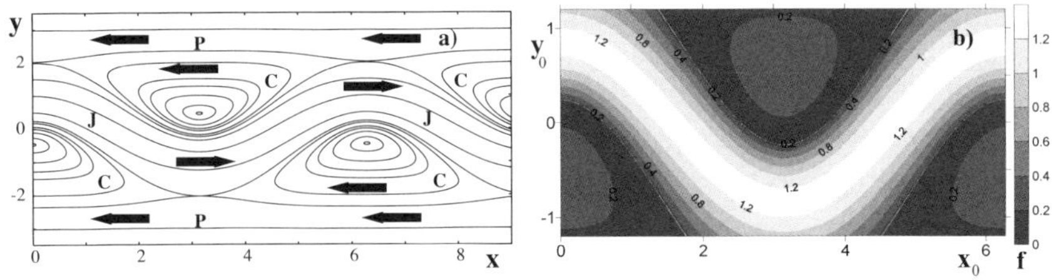

(a) The phase portrait of the unperturbed flow (2) in the frame moving with the meander's phase velocity. Streamlines in the circulation (C), jet (J) and peripheral currents (P) regions are shown.

(b) The frequency map represents by color the values of the frequency f of the particles with initial positions (x_0, y_0) advected by the unperturbed flow.

Fig. 1.

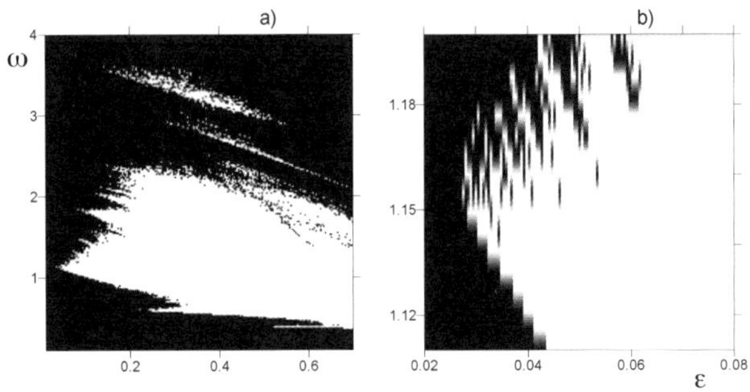

Fig. 2. The perturbation amplitude-frequency diagram ε–ω showing the parameters range in the kinematic model for which the cross-frontal chaotic transport exists (white color zones). (a) The whole diagram and (b) its fragment near the amplitude minimum.

zones). There are a number of the frequency values for which the transport occurs at surprisingly small values of the perturbation amplitude ε. The diagram has a number of cusp-like local minima. The absolute minimal value of the perturbation amplitude, at which the cross-frontal chaotic transport occurs, $\varepsilon_{\min} = 0.0268 \approx A_0/29$, corresponds to the frequency $\omega = 1.15$ which is close to the natural frequencies of the particles moving in the central jet. The fragment of the diagram near $\omega = 1.15$ (see Fig. 2(b)) demonstrates a self-similar cusp-like structure indicating to an intermittent character of the breakdown of the transport barrier. At comparatively large values $\varepsilon > 0.16$, a pitchfork bifurcation occurs when each saddle orbit bifurcates into an elliptic orbit and a pair of two new saddle orbits.

To study topology of the breakdown of the transport barrier in the kine-

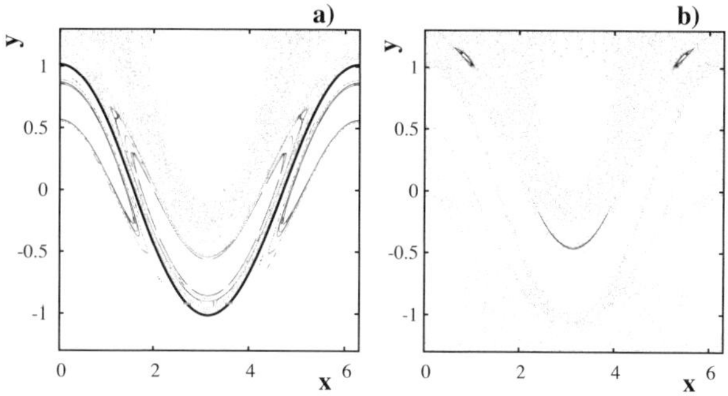

Fig. 3. The first scenario for the breakdown of the transport barrier in the kinematic model at $\omega = 1.15$. Poincaré sections at (a) $\varepsilon_1 = 0.025$ just before the breakdown and at (b) $\varepsilon_2 = 0.03$ just after the breakdown of the barrier.

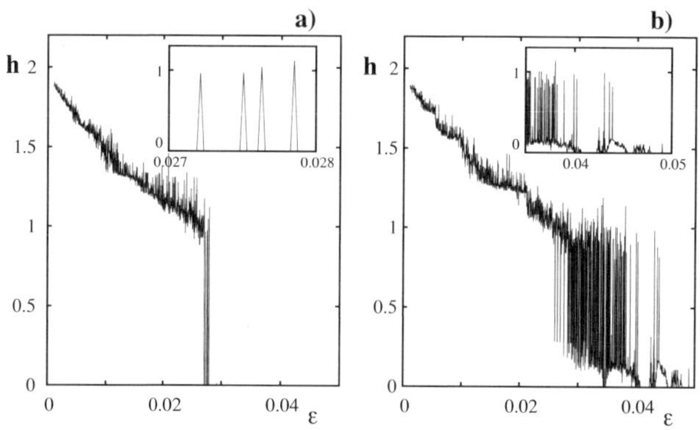

Fig. 4. Dependences of the width of the central transport barrier h on the perturbation amplitude ε. (a) The sudden breakdown of the barrier at $\omega = 1.15$ in the first scenario without change in topology of the last barrier curve. (b) The breakdown of the barrier at $\omega = 1.2$ in the second scenario with change in topology of the last barrier curve. Insets show an intermittent character of the breakdown with increasing the perturbation amplitude.

matic model, we compute Poincaré sections at a given value of the perturbation frequency but varying the perturbation amplitude ε. Two topologically different scenarios for such a breakdown have been found. In Figure 3 we plot with a number of initial conditions the Poincaré sections at $\omega = 1.15$ with the value $\varepsilon_1 = 0.025$ just before the breakdown (the panel (a)) and at $\varepsilon_2 = 0.03$ just after the breakdown of the barrier (the panel (b)). The black thick curve on the panel (a) is an invariant curve which we call "the last barrier curve". In fact, it is the last KAM invariant torus in the central jet

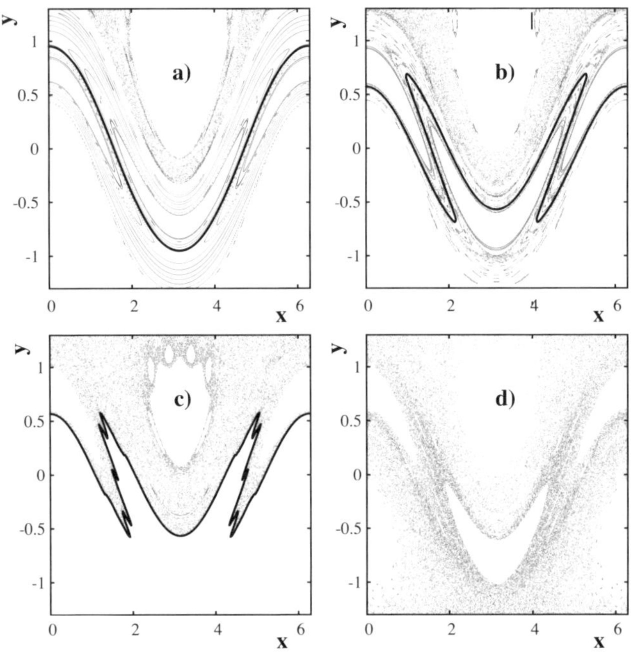

Fig. 5. The second scenario for the breakdown of the transport barrier in the kinematic model at $\omega = 1.2$. Poincaré sections at (a) $\varepsilon_1 = 0.01$, (b) $\varepsilon_2 = 0.02$ and (c) $\varepsilon_3 = 0.04$ before the breakdown and at (d) $\varepsilon_4 = 0.041$ after the breakdown of the barrier.

which breaks down under a perturbation. The other invariant curves in the figure represent regular ballistic trajectories in the jet. The clouds of points to the north off the last barrier curve are tracks of a chaotic trajectory with an initial condition chosen near the northern saddle orbit. In the first scenario the breakdown of the last barrier curve occurs at a critical value of the perturbation amplitude without change of its topology (see Fig. 3(b)). Under the perturbation at the frequency $\omega = 1.15$, there appears a ballistic resonance 1:1 and resonances of higher orders in the region of the jet. In Figure 3(a) the resonance 1:1 is represented by large resonant islands separated by a central transport barrier consisting of nested invariant curves with the last barrier curve inside. The width of the central transport barrier decreases with increasing the perturbation amplitude because of increase of the widths of the stochastic layers around the resonance 1:1 and overlapping and destroying the higher-order resonances.

Dependences of the width of the central transport barrier h on the perturbation amplitude ε with different scenarios of its breakdown are plotted in Figure 4. In the first scenario at $\omega = 1.15$ we see a sudden breakdown of the barrier (the left panel) whereas in the second scenario at $\omega = 1.2$ (the right panel) the breakdown is a much more complicated process.

The second scenario is illustrated in Figure 5 where we plot the Poincaré sections at $\omega = 1.2$ before (the panels (a), (b) and (c)) and just after the breakdown of the barrier (the panel (d)). In difference from the first scenario, we see that before the breakdown the last barrier curve changes its topology becoming more and more meandering curve with subsequent increasing the perturbation amplitude ε.

3. Cross-Frontal Chaotic Transport by Rossby Waves in a Dynamic Model of a Jet Current

In the preceding section we have studied the cross-frontal chaotic transport in the kinematic model of a meandering jet current where the velocity field was imposed as a given function of space and time. Here we aim to demonstrate this phenomenon in a dynamical model of a jet current where the flow is a self-consistent solution of dynamical equations. In the β-plane approximation on the rotating Earth, the shallow-water wave motion is governed by the conservation of potential vorticity.[16] Seeking solutions as a sum of a basic zonal flow and a small perturbation, it is possible to find a stream function of the linearized version of this equation[17] as a superposition of the zonal flow and two neutral modes which are known as Rossby waves, the westward propagating planetary waves intensifying jet currents.

To be specific we model in this section chaotic transport across the Antarctic circumpolar current (ACC), the most intensive current in the World ocean that flows from west to east around Antarctica (see Fig. 6). In the reference frame moving at the phase velocity of the Rossby wave with the amplitude A_1, the normalized stream function is

$$\Psi(x, y, \tau) = Cy - \tanh y + A_1 \operatorname{sech}^2 y \cos x + A_2 \operatorname{sech}^2 y \cos(kx + \sigma\tau), \quad (3)$$

where $C \equiv c_1/u_0$ is the normalized phase velocity of the first Rossby wave, $k \equiv k_2/k_1$, $\sigma \equiv k(c_1 - c_2)/u_0$, and u_0 is a maximal zonal velocity in the jet. The stream function (3) is, of course, an oversimplified model of the ACC which, in fact, is not a single jet but consists of a number of fronts. But it may be considered as a satisfactory model if one aims to study generic features of the cross-frontal chaotic transport in dynamic models.

Let us first consider the steady flow with the stream function

$$\Psi_0(x, y) = Cy - \tanh y + A \operatorname{sech}^2 y \cos x \quad (4)$$

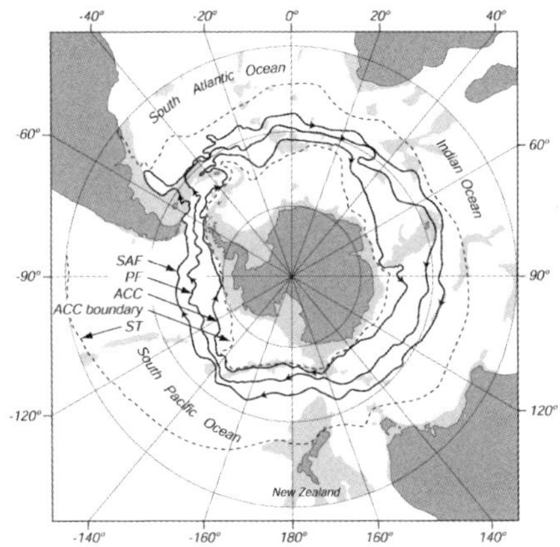

Fig. 6. The schematic view of the Antarctic circumpolar current and its fronts.

and try to find stationary points and their bifurcations. After equaling to zero the right hands of the corresponding advection equations, one gets the following algebraic equation:

$$2Az^3 + z^2 - 2Az + C - 1 = f(z) = 0. \tag{5}$$

The function $f(z)$ has two extrema at

$$z_{\max} = -\frac{\sqrt{1+12A^2}+1}{6A} < 0, \quad z_{\min} = \frac{\sqrt{1+12A^2}-1}{6A} > 0. \tag{6}$$

Let us consider two cases.

(1) $C > 1$. In this case $f(0) > 0$ and Eq. (5) has one negative root and two or zero positive ones. The negative root has no physical sense because $f(-1) = C > 0$. If $f(z_{\min}) < 0$, then two positive roots exist, and they are always smaller then one because $f(1) = C > 0$. The condition $f(z_{\min}) < 0$ corresponds to the following condition:

$$C < C_{\text{crit}} = \frac{(1+12A^2)^{\frac{3}{2}} + 36A^2 - 1}{54A^2}. \tag{7}$$

The right part of the inequality (7) is always greater than one.

(2) $C < 1$. Because of $f(\pm 1) = C > 0$ and $f(0) = C - 1 < 0$, there exist one positive and two negative roots. One of the negative roots is smaller than one and, therefore, has no physical sense.

Thus, if the condition (7) is satisfied, then the advection equations with the stream function (4) have four stationary points and no stationary points otherwise. Stability analysis has shown that at $x = 0$ ($x = \pi$) the upper point is a center(a saddle) and the lower one is a saddle (a center).

At $C = C_{\text{crit}}$, a saddle-center bifurcation occurs. The bifurcation known as "a separatrix reconnection" is possible in our system. Reconnection means merging of two separatrices and may be defined as a topological rearrangement of streamlines in which stationary points do not change the type of their stability. As a result of the separatrix reconnection, the phase portrait changes from a homoclinic type to a heteroclinic one. We did not find any analytical condition for the separatrix reconnection, but this may be down numerically using the condition $\Psi_{s0}(A,C) = \Psi_{s\pi}(A,C) = 0$, where $\Psi_{s0}(A,C)$ and $\Psi_{s\pi}(A,C)$ are the values of the stream function at the saddles $x = 0$ and $x = \pi$, respectively. The bifurcation diagram on the plane of the control parameters, the amplitude A and the phase velocity C of the Rossby wave, is shown in Figure 7(a).

The unsteady flow with two Rossby waves (3) is known to produce chaotic advection of passive particles in some ranges of the control parameters.[17] Being motivated by the study of the cross-frontal transport in the ACC, we choose a circular geometry of the jet with the heteroclinic connection (zone I) and the realistic values of its maximal velocity $u_0 = 0.442165$ m/s and width $L = 100206$ m. The values of the phase velocities and wave numbers, corresponding to the ACC parameters, can be obtained from dispersion relations to be $C = 0.593912$ and $k = 0.35$.

In Figure 7(b) we plot the diagram of the cross-frontal transport on the amplitude plane A_1–A_2 computed with the same algorithm as the amplitude-frequency diagram for the kinematic model (see Fig. 2). The detailed analysis and discussion of the breakdown of the transport barrier in the ACC will be published in a specialized journal. Here we demonstrate only two Poincaré sections of the flow before and after the breakdown (see Fig. 8).

4. Conclusion

The geophysical jet currents in the ocean and atmosphere are robust structures whose characteristics change in space and time in a very complicated way. In this paper we have studied the cross-frontal transport of water masses with two simple models of jet currents. The kinematic model of a periodically meandering western boundary current, like the Gulf Stream or the Kuroshio, has been shown to demonstrate, at least, two scenarios of the breakdown of the transport barrier. In the first one the last barrier curve breaks down suddenly and without change of its topology, and in the

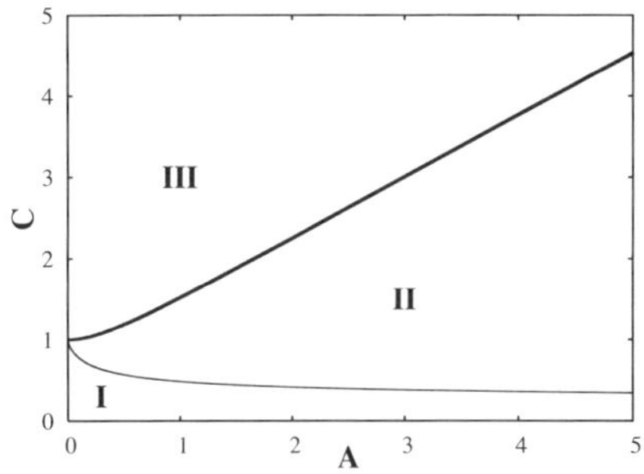

(a) Bifurcation diagram on the parameter plane A–C. Zone I — four stationary points with heteroclinic separatrices (eastward jet). Zone II — four stationary points with homoclinic separatrices (westward jet). Zone III — no stationary points (westward current).

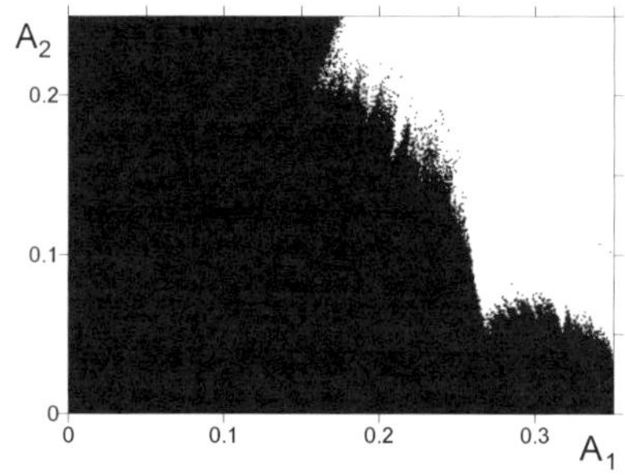

(b) The diagram for the dynamic model of the ACC showing the amplitude range of two Rossby waves for which the cross-frontal chaotic transport occurs (white color zones).

Fig. 7.

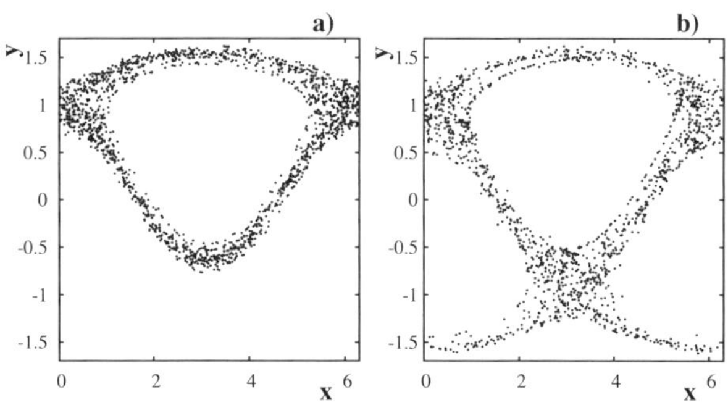

Fig. 8. The breakdown of the transport barrier in the dynamic model of the ACC at $C = 0.593912$, $k = 0.35$ and $\sigma = 0.39608$. Poincaré sections at (a) $A_1 = 0.32$, $A_2 = 0.055$ before the breakdown and at (b) $A_1 = 0.325$, $A_2 = 0.055$ after the breakdown.

second one it changes considerably the form before the breakdown. In the dynamical model of the Antarctic circumpolar current the chaotic cross-frontal transport is caused by two Rossby waves. To quantify the range of the control parameters, in which the transport may occur, we computed the corresponding diagrams (see Figs. 2 and 7(b)). The common feature of both the models is that the chaotic cross-frontal transport may occur at surprisingly small values of the perturbation amplitudes. The result obtained may explain in part observations of float and drifter trajectories crossing the Gulf Stream[7] and other oceanic jet currents. The method developed in this paper can be applied to more realistic models of jet currents in order to quantify cross-frontal fluxes.

Acknowledgments

The work was supported by the Program "Fundamental Problems in Nonlinear Dynamics" of the Russian Academy of Sciences, the Program for Basic Research of the Far Eastern Division of the Russian Academy of Sciences, and the Russian Foundation for Basic Research (grant no 09-06-98520).

References

1. H. Aref, *J. Fluid Mech.* **143**, 1 (1984).
2. G. Zaslavsky, *Hamiltonian chaos and fractional dynamics* (Oxford University Press, Oxford, 2005).
3. J. Ottino, *The kinematics of mixing: stretching, chaos, and transport* (Cambridge University Press, Cambridge, 1989).
4. K. Koshel and S. Prants, *Uspekhi Fisicheskikh Nauk* **176**, 1178 (2006).
5. K. Koshel and S. Prants, *Chaotic advection in the ocean* (Institute for Computer Science, Moscow-Izhevsk, 2008).
6. Yu. Izrailsky, V. Kozlov, and K. Koshel, *Phys. Fluids* **16**, 3173 (2004).
7. A. Bower and H. Rossby, *J. Phys. Oceanogr.* **19**, 1177 (1989).
8. A. Bower, *J. Phys. Oceanogr.* **21**, 173 (1991).
9. R. Samelson, *J. Phys. Oceanogr.* **22**, 431 (1992).
10. S. Prants, M. Budyansky, M. Uleysky, and G. Zaslavsky, *Chaos* **16** 033117, (2006).
11. M. Uleysky, M. Budyansky, and S. Prants, *Chaos* **17**, 024703 (2007).
12. M. Uleysky, M. Budyansky, and S. Prants, *J. Phys. A: Math. Theor.* **41**, 215102 (2008).
13. F. Raynal and S. Wiggins, *Physica D* **223**, 7 (2006).
14. S. Soskin, O. Yevtushenko, and R. Manella, *Phys. Rev. Lett.* **90**, 174101 (2003).
15. S. Soskin, R. Manella, and O. Yevtushenko, *Phys. Rev. E* **77**, 036221 (2008).
16. J. Pedlosky, *Geophysical fluid dynamics* (Spinger Verlag, N.Y., 1987).
17. D. del-Castillo-Negrete and J. Pedlosky, *Phys. Fluids A* **5**, 948 (1993).

A Novel Monolithic Integrated Optical Chaos Emitter: Experiment, Data Analysis and Numerical Predictions

K. E. Chlouverakis*, A. Argyris*, A. Bogris*, M. Hamacher† and D. Syvridis*

*Department of Informatics and Telecommunications, University of Athens
†Fraunhofer Institute for Telecommunications, Heinrich-Hertz-Institute
10587 Berlin, Germany

A novel photonic integrated device has been designed and evaluated as a transmitter for chaos applications in optical communications. The device consists of a semiconductor laser integrated with a semiconductor optical amplifier, a phase section and a waveguide, forming thus an integrated external cavity laser with adjustable optical feedback and phase. Under diverse operating parameters the device behaves in different modes, from stable operation to limit cycles and hyperchaotic broadband behavior. Chaos data analysis from experimental data is performed in order to quantify the complexity of the experimental reconstructed attractors under the influence of noise. Calculation of the correlation dimension and Kolmogorov entropy are compared with numerical predictions from well-known theoretical models.

Keywords: Optical chaos; semiconductor lasers; Lyapunov exponents; correlation dimension; Kolmogorov entropy; noise filtering; integrated photonics.

1. Introduction

The need nowadays for increased security in data exchange using the attractive technology of optical communication networks has directed a considerable part of research to physical layer data encryption techniques.[1] Synchronization in non-linear and chaotic systems[2,3] has established a new approach in securing data communications over fiber transmission, by encoding messages over chaotic optical carriers. Based on optical feedback,[4] optical injection[5] and optoelectronic feedback,[6] various configurations of transmitters have been proposed and implemented, providing high-dimensional chaotic carriers capable of message encryption.[1] The latter take advantage of the off-the-shelf fiber-optic technology resulting in rather cumbersome devices, impractical for commercial use.

The miniaturization of such configurations through photonic integration appears very attractive, considering the efficiency of specifically designed photonic integrated circuits to generate intense non-linear dynamics.[7,8] Albeit the aforementioned device presents attractive nonlinear behavior, it is not suitable for generating high-dimensional broadband chaotic light for message encryption applications due to the ultra-short size of the external cavity. The ultimate objective of our work is to exploit the complex dynamics of such devices in a way that will fulfill the criteria for efficient data encryption on the physical layer.

In this paper we present a novel photonic integrated device, based on an external-cavity DFB (Distributed Feedback Laser) semiconductor laser, for chaos applications in fiber communication systems. The proposed device resembles the fundamental scheme of all-optical feedback with stand-alone components and is mainly dedicated to offer a compact, in terms of operation, solution for chaos-based encryption in future optical communication networks. The configuration of the device has been designed appropriately in order to fully control all the crucial parameters that lead the device to operate in a chaotic manner as it will be shown next.

2. The Photonic Device

2.1. *Design*

The active region of the integrated device consists of three successive sections: a DFB InGaAsP semiconductor laser (Fig. 1a) operating at 1561nm followed by a semiconductor optical amplifier (SOA - Fig. 1b) and a phase section (PHS - Fig. 1c). The output facet of the laser section is anti-reflective coated. The external cavity laser is formed between the active region, and specifically the DFB laser, and the edge of a 1-cm long straight optical waveguide (WG - Fig. 1d) which is coated with a highly reflective mirror ($R \approx 96\%$ - Fig. 1e), and thus the observed dynamics correspond to the short-cavity regime.[9]

Criterion of the selection of the cavity length is the ability of the device to provide complex dynamics. The above length corresponds to a delay of approximately 240ps round-trip time of the electric field in the external cavity; this value is confirmed by numerical simulations[4] to raise complex dynamics and to be efficient for the oscillator to enter the coherence collapse regime. The necessity of integrating SOA results from the requirement to control the optical feedback strength from the external cavity. For more details see [8].

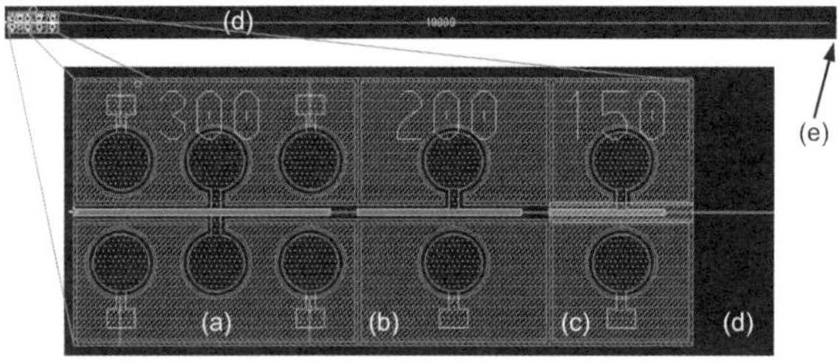

Fig. 1. Mask design of the integrated device: (a) InGaAsP laser, (b) SOA (or VOA), (c) phase section, (d) optical waveguide. The end of the waveguide is coated with a highly reflective mirror, (e) The lengths of the different sections are in m.

2.2. *Experimental results and time-series analysis*

In order to investigate the dynamics of the experimentally resolved time-series, chaos data analysis is performed since the prevalent idea is to identify the complexity of the strange attractors and substantiate their suitability for encryption applications. Laser series are always noise-contaminated and may lead to false claims regarding chaos identification or even to erroneous estimates of the correlation dimension and Kolmogorov entropy calculation.[10] In order to visualize the substantial dynamics for every recorded time-series, it is prudent to enroll noise reduction methods. For complex chaotic data, state-space averaging is considered as one of the most apt and intricate methods (in contrast to linear filtering) in order to remove non-deterministic components from a series, ameliorate visually the geometrical structure of the reconstructed attractor and illuminate the underlying mechanisms. Following this method, one has to postdict or hindcast with a posteriori knowledge of all N points, by averaging the segments of the data in which a window of $\pm m$ points on each side of X_n (the matrix containing each experimental data) is close to it in state-space:[11]

$$\tilde{X}_n = \frac{\sum_{k=m}^{N-m} X_n e^{-\frac{1}{\sigma_N^2}\sum_{i=-m}^{m}(X_{k-i}-X_{n-i})^2}}{\sum_{k=m}^{N-m} e^{-\frac{1}{\sigma_N^2}\sum_{i=-m}^{m}(X_{k-i}-X_{n-i})^2}} \tag{1}$$

The values of m and σ_N used herein were[11] $m = 9$ and $\sigma_N = 0.2$. A value for m greater that 7 will provide reliable results, taking into account numerical predictions already performed for the short-cavity all-optical feedback scenario.[4] The noise-filtered attractors are superimposed to the initially recorded ones with color in Fig. 2. The delay τ is defined as the time value for which the autocorrelation function decays to the inverse Euler constant[10] e^{-1}. The applied noise-reduction method herein follows the rationale that the geometrical topology and dynamics maintain their unmitigated complexity, the variance of noise is smaller than the signal and that the latter lies on a smooth manifold.

The correlation dimension D_2 and Kolmogorov entropy K_2 were calculated with the Grassberger-Procaccia method that is considered a very trustworthy tool since the results herein were found to comply with numerical predictions.[4] In Fig. 3 (black line) the depicted values of D_2 and K_2 are for the noise-filtered attractors. The calculation of complexity is patently arduous and misleading for the original attractors especially when periodic states are considered. Due to the noise presence resulting in the augment of loss of correlation between neighbor points, the periodic states are incorrectly evaluated as chaotic ($D_2 > 2$ and $K_2 > 0$). Therefore, the latter are omitted and D_2 is plotted (Fig. 3 – red line) only for the chaotic states.

The different dynamic states, that include limit cycles and chaotic states, are depicted in Fig. 2 and emerge for different phase values. In some narrow phase current regions two intense peaks are observed at 3.3GHz and 6.6GHz, while the rest spectral components lay almost just above the noise floor (Fig. 2a,e).

In other phase current regions, various limit cycle conditions arise: the first peak may be suppressed entirely or additional peaks of moderate power may come up (Fig. 2b). All the above cases that correspond to limit cycle dynamics are characterized by a low correlation dimension D_2 – approximately close to unity, taking into account the stochastic noise sources – and zero entropy K_2, as depicted in Fig. 3. There is also a phase current values region (from 5.4mA to 5.9mA) in which the device operates in the coherence collapse regime providing a chaotic output, as revealed from the phase plots of Figs. 2c and 2d. In this operating region the radio-frequency spectrum of the output signal becomes broadband and extends up to the cutoff frequency of the optical receiver (around 8 GHz). The frequency peaks are now much less intense however distinguishable from the rest strengthened spectral components. The correlation dimension and the Kolmogorov entropy now are considerably increased up to $D_2 = 6$ (Fig. 3 – red line) and K_2=8.8ns^{-1} (Fig. 3 – dash-dotted line) respectively.

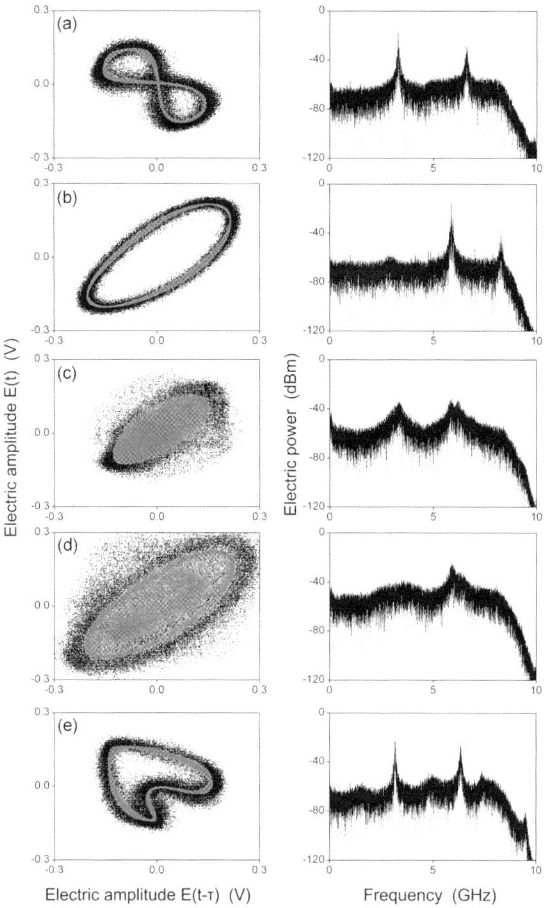

Fig. 2. Feedback phase dependence: Experimental phase plots (left) and corresponding microwave spectra (right) of the device output electric field, for $I = 3I_{\mathrm{TH}}$, I_{SOA}=0mA and from (a) to (e) I_{PHS}=3.3, 4.8, 5.5, 5.9 and 6.9mA. The colored trace denotes the filtered attractor.

3. Numerical Model and Predictions

3.1. *Lyapunov spectra and complexity calculations*

A semiconductor laser subjected to all-optical feedback from an external cavity is modeled with the well-known Lang-Kobayashi rate equations for the complex slowly varying amplitude of the electrical field $E(t)$ and the

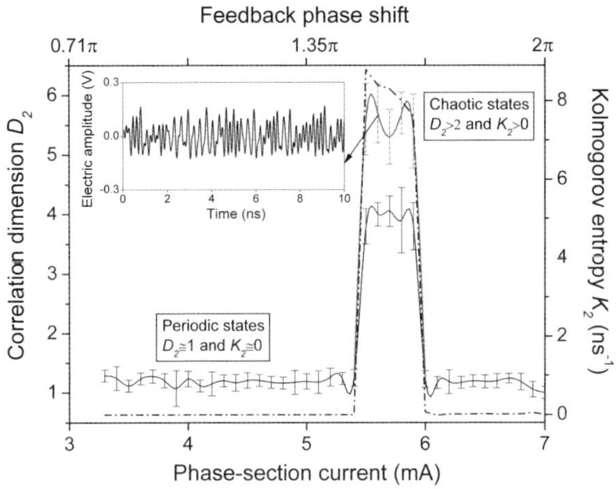

Fig. 3. Correlation dimension D_2 and Kolmogorov entropy K_2 as the feedback phase varies. Red color denotes the value of D_2 for the non-filtered original series, solid line D_2 for the filtered series and dash-dotted line K_2 for the filtered series. $I = 3I_{TH}$ and $I_{SOA} = 0$mA.

carrier number $N(t)$ inside the cavity.

$$\begin{aligned}\dot{E}(t) &= \frac{(1+i\alpha)}{2}(G(t) - \frac{1}{t_{ph}})E(t) \ + \kappa E(t-\tau)e^{-i\Omega\tau} \\ \dot{N}(t) &= \frac{I}{e} - \frac{1}{t_n}N(t) - G\left|E(t)\right|^2 \\ G(t) &= \frac{g(N(t)-N_0)}{1+s\left|E(t)\right|^2}\end{aligned} \tag{2}$$

where α is the linewidth enhancement factor, g is the gain parameter, s is the gain saturation coefficient, t_{ph} is the photon lifetime, t_n is the carrier lifetime, N_0 is the carrier number at transparency, e is the electron charge, Ω is the frequency of the free running laser, κ is the feedback coefficient and τ is the external cavity round-trip. The two latter parameters determine the external cavity features.

With integrating Eq.(2) and applying specific algorithms[12] it is straightforward to calculate the Lyapunov exponents' spectrum. Since the configuration is described with delay differential equations, it is prudent to derive the latter to finitely many discrete variables that advance in small discrete time steps taking also in mind the short-time delay.

Comparing Figs. 3 and 4, it is clearly noted that the comparison between the substansial experimental complexity and the numerical one is

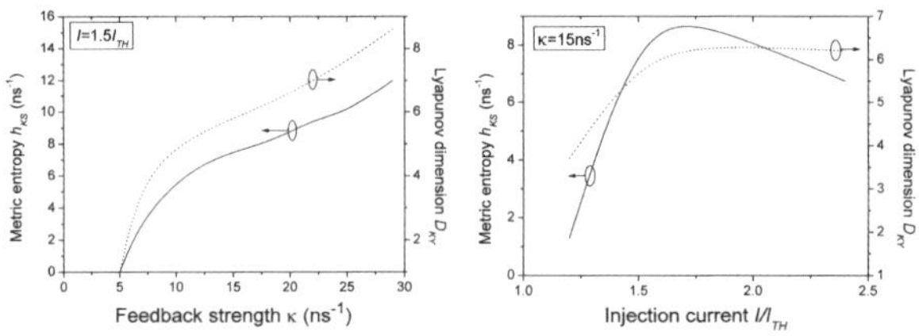

Fig. 4. Metric entropy and Lyapunov dimension for varying the feedback strength and the injection current.

adequate and satisfying, considering the fact that the rate-equations (2) are only approximative. Furthermore it should be reminded that D_2 is always less than the Lyapunov dimension D_{KY} together with $K_2 < h_{\mathrm{KS}}$ (metric entropy) as well. Nevertheless, such a device is more suitable than electronic circuits as a chaotic transmitter for encryption applications in the physical layer, due to the higher bandwidth and complexity. The feedback strength in the numerical simulations is in units of inverse seconds as shown in Fig. 4, and direct comparison with the current of the optical attenuator I_{SOA} is not applicable, albeit the primary scope herein is the quantitative predictions for the complexity and chaoticity of the specific dynamical system.

3.2. *Phase dependence*

Semiconductor lasers subjected to optical feedback from a short-external cavity are known to exhibit extreme dependence on the feedback phase C_p of the complex electric field E relative to the cavity length.[4] Hence, changing the phase during an experiment is of great importance as depicted in Figs. 2 and 3 and can easily be achieved with a phase shifter (Fig. 1). In the next diagrams we present the effect of phase, numerically, as we increase the injection current I and the feedback strength.

In Fig. 5 we plot the extrema of the complex field amplitude. It can be easily seen that in order to avoid the windows of periodicity, high injection currents are necessary, together with high values of the feedback strength κ. Hence, it is concluded that in order to obviate the phase dependence that may annihilate the desired chaotic carriers, it is a necessity to carefully tune the feedback phase during an experiment, or more preferably to operate the laser in high injection currents $I > 2I_{\mathrm{TH}}$ and high feedback strengths $\kappa > 15\mathrm{ns}^{-1}$ as also verified from our experiments.

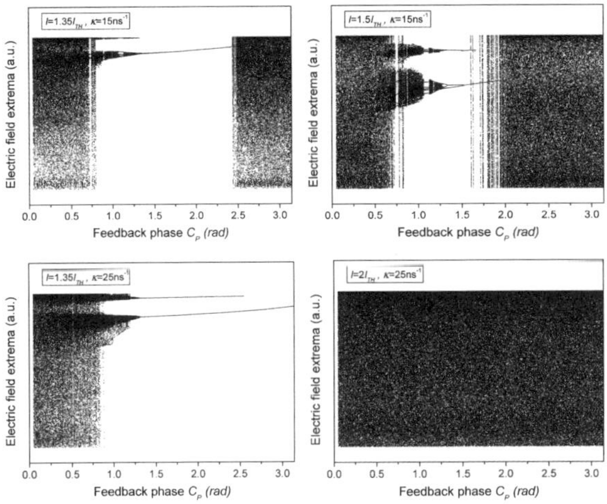

Fig. 5. Bifurcation diagrams of the complex electric field amplitude of Eq.(2) for different values of feedback strength and injection current, revealing the phase dependence of the short-cavity laser subjected to optical feedback.

4. Conclusions

A novel photonic integrated circuit that is capable to produce chaotic dynamics was designed, fabricated and investigated in terms of complexity and chaoticity. Dynamics may be easily controlled experimentally via the phase current, the feedback strength and the injection current, establishing therefore this device as a compact chaos emitter. Complexity was calculated proclaiming adequate numerical values for the attractor dimension and Kolmogorov entropy from experimental data analysis resulting in excellent aggrement with numerical predictions using simple theoretical models. Since the dynamics are well-identified, the advantages of the proposed photonic integrated device can be fully exploited to our benefit as a broadband chaos transmitter with applications to secure chaos-encoded optical communications.

Acknowledgments

This work has been supported under EC Project PICASSO IST-2006-34551.

References

1. H.-F. Chen and J.-M. Liu, *Phys. Rev. E* **71**, p. 046216 (2005).

2. K. Chlouverakis and J. C. Sprott, *Physica D* **200**, p. 156 (2005).
3. K. M. Cuomo, A. V. Oppenheim and S. H. Strogatz, *IEEE Trans. Circuits Syst. II* **40**, p. 626 (1993).
4. A. Argyris, D. Syvridis, L. Larger, V. Annovazzi-Lodi, P. Colet, I. Fischer, J. Garcia-Ojalvo, C. R. Mirasso, L. Pesquera and K. A. Shore, *Nature* **438**, p. 343 (2005).
5. A. Argyris, M. Hamacher, K. Chlouverakis, A. Bogris and D. Syvridis, *Phys. Rev. Lett.* **100**, p. 194101 (2008).
6. K. Chlouverakis, A. Argyris, A. Bogris and D. Syvridis, *Physica D* **237**, p. 568 (2008).
7. M. Yousefi, Y. Barbarin, S. Beri, E. A. J. M. Bente, M. K. Smit, R. Nötzel and D. Lenstra, *Phys. Rev. Lett.* **98**, p. 044101 (2007).
8. R. Vicente, J. Dauden, P. Colet and R. Toral, *IEEE. J. Quantum Electron.* **41**, p. 541 (2005).
9. T. Heil, I. Fischer, W. Elsässer, B. Krauskopf, K. Green and A. Gavrielides, *Phys. Rev. E* **67**, p. 066214 (2003).
10. L. M. Pecora and T. L. Carroll, *Phys. Rev. Lett.* **68**, p. 821 (1990).
11. M. Peil, L. Larger and I. Fischer, *Phys. Rev. E* **76**, p. 045201 (2007).
12. T. Schreiber, *Phys. Rev. E* **48**, p. R13 (1993).

Self-Similarity/Memory-Length Parameter Estimation for Non-Gaussian Hermite Processes via Chaos Expansion

A. Chronopoulou

Department of Statistics, Purdue University
West Lafayette, IN 47907-2067, USA
Email: achronop@purdue.edu

We study the behavior of Hermite processes of order q with self-similarity index $H \in (\frac{1}{2}, 1)$. Using Wiener-Itô multiple stochastic integrals and Malliavin calculus we develop a class of consistent estimators for the self-similarity index based on the asymptotic behavior of the filtered process.

Keywords: Hurst index estimation; Hermite process; Wiener chaos expansion; Malliavin calculus.

1. Introduction

A large number of phenomena in different scientific fields, such as economics, biology, medicine, traffic network, hydrology exhibit long-range dependence, i.e. long memory. A class of stochastic processes, known as self-similar stochastic processes, can be used to effectively model such phenomena. In such a process any part of the trajectory is invariant under time scaling. Self-similar processes are parametrized by the self-similarity (or Hurst) index, which is in fact the parameter that describes the memory and consequently the behavior of the phenomena they describe.

Hermite processes are a family of self-similar processes that encompass the well-known fractional Brownian motion, and also include many non-Gaussian processes. In our work we study the behavior of the quadratic variation for such processes and we apply the corresponding results in the estimation of the self-similarity index.

Hermite processes were initially introduced by Taqqu[1] and Dobrushin and Major.[2] We denote by $\left(Z_t^{(q,H)}\right)_{t\in[0,1]}$ the Hermite process with self-similarity index H of order q. There are two ways to define it. One way is defining it as a multiple integral with respect to fractional Brownian motion with an appropriate Hurst index. An alternative definition is to define it as a multiple integral with respect to a standard Wiener process $(W_t)_{t\in[0,1]}$. In our analysis we adopt the second definition.

The Hurst or self-similarity or long-memory parameter H describes the memory and characterizes all the important properties of a Hermite process. Therefore, it is of great interest to estimate it efficiently. So far many estimators have been proposed using other techniques such as maximum likelihood (exact and approximate), variograms, wavelets. A complete reference of those methods can be found in the book by Beran.[3] The approach we consider is based on the discrete variations of the filtered process.

Assume that we observe the process $(Z_t^{(q,H)})_{t\in[0,1]}$ at discrete times $\{0, \frac{1}{N}, \frac{2}{N}, \ldots, 1\}$ and let α be a filter of length $(\ell+1)$ and order $p \geq 1$; that is α is an $(\ell+1)$-dimensional vector such that $\sum_{k=0}^{\ell} \alpha_k k^r = 0$ for $0 \leq r \leq p-1$ and $\sum_{k=0}^{\ell} \alpha_k k^p \neq 0$. For instance, a filter could be constructed by differencing the process; the order of differencing is the same as the order and the length of the filter. Another filter could be a wavelet-type filter, such as Daubechies or Coiflets. We obtain the filtered process by the following scheme

$$|V_\alpha\,(i/N)\,| = \sum_{k=0}^{\ell} \alpha_k Z_{\frac{i-k}{N}} \tag{1}$$

The quadratic variation statistic associated with the filter α is defined as

$$V_N(2,\alpha) = \frac{1}{N-\ell} \sum_{i=\ell}^{N-1} \left\{ \frac{|V_\alpha\,(i/N)\,|^2}{E|V_\alpha\,(i/N)\,|^2} - 1 \right\}. \tag{2}$$

Our analysis is an extension of results for the fractional Brownian motion and Rosenblatt process for a filter of order 1 by Tudor and Viens.[4] We apply Wiener-Itô chaos expansions as well as recent results on the convergence of multiple integrals in order to study the asymptotic behavior of the proposed estimators.

The article is structured as follows: In Section 2 we present a short introduction on fractional analysis. In Section 3 the main results for the Hermite processes are presented, while Section 4 contains the results for the Rosenblatt process with the use of higher order filter. Finally, in Section 5 we apply our results to the estimation of the Hurst parameter.

2. Preliminaries on Fractional Calculus

Let $(W_t)_{t\in[0,1]}$ be a classical Wiener process on a standard Wiener space $(\Omega, \mathcal{F}, \mathbf{P})$. If $f \in L^2([0,1]^n)$ with $n \geq 1$ integer, we introduce the multiple

Wiener-Itô integral of f with respect to W. For more details on the construction and properties of this integral we refer to [5].
Define $\mathcal{S}$ to be the set of elementary functions, which means that if $f \in \mathcal{S}$ then it can be expressed as

$$f = \sum_{i_1,\dots,i_m} c_{i_1,\dots i_m} 1_{A_{i_i} \times \dots \times A_{i_m}}$$

where the coefficients satisfy $c_{i_1,\dots,i_m} = 0$ if $i_k - i_l$ and the sets $A_i \in \mathcal{B}([0,1])$ are disjoint. For such a function f we define

$$I_n(f) = \sum_{i_1,\dots,i_m} c_{i_1,\dots i_m} W(A_{i_1}) \cdots W(A_{i_m}),$$

where $W([a,b]) = W_b - W_a$. The mapping I_n from $\mathcal{S}$ to $L^2(\Omega)$ is an isometry on $\mathcal{S}$, which means that

$$\mathbf{E}\left[I_n(f) I_m(g)\right] = n! \langle f, g \rangle_{L^2([0,1]^n)} \text{ if } m = n \tag{3}$$

and zero otherwise.

Since the set of step functions, $\mathcal{S}$, is dense in $L^2([0,1]^n)$ the mapping I_n can be extended to an isometry from $L^2([0,1]^n)$ to $L^2(\Omega)$ and the above properties hold true for this extension. We say that the mapping I_n belongs to the Wiener chaos of order n. For more details about this notion we refer to [5]. Moreover, observe that I_n can be viewed as an iterated stochastic integral

$$I_n(f) = n! \int_0^1 \int_0^{t_n} \cdots \int_0^{t_2} f(t_{t_1}, \dots, t_n) dW_{t_1} \dots dW_{t_n};$$

here the integrals are of Itô type. It is easy to see that if $f \in L^2([0,1]^n)$ and $g \in L^2([0,1])^m$ then their product is expressed as a sum of integrals belonging in chaoses of orders up to $m \wedge n$.

$$I_n(f) I_m(g) = \sum_{l=0}^{m \wedge n} l! C_l^m C_l^n I_{m+n-2l}(f \otimes_l g). \tag{4}$$

The contraction $f \otimes_l g$ belongs to $L^2([0,1]^{m+n-2l})$ for $l = 0, 1, \dots, m \wedge n$ and it is given by

$$\begin{aligned} &(f \otimes_l g)(s_1, \dots, s_{n-l}, t_1, \dots, t_{m-l}) \\ &= \int_{[0,1]^l} f(s_1, \dots, s_{n-l}, u_1, \dots, u_l) g(t_1, \dots, t_{m-l}, u_1, \dots, u_l) du_1 \dots du_l \end{aligned} \tag{5}$$

Having defined the multiple Wiener-Itô integral, we can define rigorously the Hermite proccess of order q.

Definition 2.1. The Hermite process $(Z_t^{(q,H)})_{t\in[0,1]}$ of order $q \geq 1$ and with self-similarity index $H \in (\frac{1}{2}, 1)$, $t \in [0,1]$ is given by

$$Z_t^{(q,H)} =$$

$$d(H)\int_{[0,t]^q} dW_{y_1}\dots dW_{y_q}\left(\int_{y_1\vee\dots\vee y_q}^{t} \partial_1 K^{H'}(u,y_1)\dots\partial_1 K^{H'}(u,y_q)du\right),$$

where $K^H(t,s) = c_H s^{\frac{1}{2}-H}\int_s^t (u-s)^{H-\frac{3}{2}} u^{H-\frac{1}{2}} du$ when $t > s$ is the kernel of the fBm, $c_H = \left(\frac{H(2H-1)}{\beta(2-2H, H-\frac{1}{2})}\right)^{\frac{1}{2}}$, $H' = 1 + \frac{H-1}{q}$ and $\beta(\cdot,\cdot)$ is the Beta function.

Some basic properties of this family are the following:

- the Hermite process $Z^{(q,H)}$ is H-selfsimilar and has stationary increments.
- the covariance function of $Z^{(q,H)}$ is identical to that of fBm, namely
$$E\left[Z_s^{(q,H)} Z_t^{(q,H)}\right] = \frac{1}{2}(t^{2H} + s^{2H} - |t-s|^{2H}). \tag{6}$$
- the mean square of the increment is given by $\mathbf{E}\left[\left|Z_t^{(q,H)} - Z_s^{(q,H)}\right|^2\right] = |t-s|^{2H}$. This implies that $Z^{(q,H)}$ has Hölder-continuous paths of any order $\delta < H$.
- it exhibits long-range dependence in the sense that $\sum_{n\geq 1}\mathbf{E}[Z_1^{(q,H)}(Z_{n+1}^{(q,H)} - Z_n^{(q,H)})] = \infty$. In fact, the summand in this series is of order n^{2H-2}. This property is identical to that of fBm since the processes share the same covariance structure.
- if $q = 1$ then $Z^{(1,H)}$ is a fractional Brownian motion with Hurst parameter H while for $q \geq 2$ this stochastic process is not Gaussian. In the case $q = 2$ it is known as *Rosenblatt process*.

3. Variations of the Hermite Process

In this section we consider the case of a filter of order 1, i.e. $\alpha = \{-1, +1\}$. Therefore, the filtered process is just the increment process

$$V_\alpha\left(\frac{i}{N}\right) = \left(Z_{\frac{i+1}{N}}^{(q)} - Z_{\frac{i}{N}}^{(q)}\right). \tag{7}$$

The fundamental idea in our approach starts from the fact that we can find an explicit expansion of V_N (defined in Eq. (2)) in the Wiener chaos. Indeed, we can show that

$$Z_{\frac{i+1}{N}}^{(q,H)} - Z_{\frac{i}{N}}^{(q,H)} = I_q(f_{i,N}),$$

where $f_{i,N}(y_1,\ldots,y_q) = L_{\frac{i+1}{N}} - L_{\frac{i}{N}}$ Having this as a starting point the steps we follow are:

(1) We write the Wiener Chaos Expansion of V_N:

$$V_N = N^{2H-1}\sum_{k=0}^{q-1} I_{2q-2k}\left(\sum_{i=0}^{N-1} f_{i,N}\otimes_k f_{i,N}\right) = \sum_{n=1}^{q} c_{2n}T_{2n}, \quad (8)$$

where I_{2q-2k} is a multiple Wiener-Itô integral with respect to W and $f_{i,N}(y_1,\ldots,y_q)$ is defined as above.

(2) We estimate the L^2 norm of the terms appearing in the chaos expansion of V_N. The term $T_2 := N^{2H-1}I_2\left(\sum_{i=0}^{N-1} f_{i,N}\otimes_{q-1} f_{i,N}\right)$ is the dominant term. Therefore, the behavior of V_N is determined by the behavior of T_2. We prove that

$$\lim_{N\to\infty} \mathbf{E}\left[c_{1,H}^{-1}N^{(2-2H')2}c_2^{-2}V_N^2\right] = 1.$$

(3) We determine the asymptotic distribution of V_N. First of all we start by checking if we have convergence to a Normal limit, by using the characterization of $\mathcal{N}(0,1)$ by Nualart and Ortiz-Latorre ([6], Theorem 4). If this is not satisfied, then we need to identify the limiting distribution, using another approach taking advantage of the isometry property of the Wiener-Itô integral.

We summarize our results in the following theorem:

Theorem 3.1. *Let $H\in(1/2,1)$ and $q\in\mathbf{N}\setminus\{0\}$. Let $Z^{(q,H)}$ be a Hermite process of order q and self-similarity parameter H. Define*

$$c = (q!)^{-1}\frac{(4H'-3)^{1/2}(4H'-2)^{1/2}}{2d(H)^2(H'(2H'-1))^{q-1}}.$$

Then $cN^{(2-2H)/q}V_N$ converges in $L^2(\Omega)$ to a standard Rosenblatt random variable with parameter $H'' := \frac{2(H-1)}{q}+1$, that is, to the value at time 1 of a Hermite process of order 2 and self-similarity parameter H''.

In previous results about the fractional Brownian motion (see [7] and [4]), the limiting distribution was Normal for certain values of H. Here instead, we observe that this is not true. The limiting distribution is proved to be Rosenblatt always.

By studying the behavior of the remaining terms in the decomposition of V_N (the non-dominant terms) we discovered what we call the **reproduction property**: *From a Hermite process of order $q>2$ we can construct*

Hermite processes of lower orders. The results are summarized in the following theorem:

Theorem 3.2. *Let $Z^{(q,H)}$ be a Hermite process, as in the previous theorem. Let T_{2n} be the term of order $2n$ in the Wiener chaos expansion of V_N: this is a multiple Wiener integral of order $2n$, and we write $V_N = \sum_{n=1}^{q} c_{2n}T_{2n}$ where $c_{2q-2k} = k!(C_k^q)^2$.*

- *For every $H \in (1/2, 1)$ and for every $k = 1, \ldots, q-1$ we have convergence of the expression $(z_{k,H})^{-1} N^{(2-2H')k} T_{2k}$ in $L^2(\Omega)$ to $Z^{r,K}$, a Hermite random variable of order $r = 2k$ with self-similarity parameter $K = 2k(H'-1)+1$.*
- *For every $H \in (1/2, 3/4)$, with $k = q$, we have convergence of $x_{1,H}^{-1/2}\sqrt{N}T_{2q}$ to a standard normal distribution.*
- *For every $H \in (3/4, 1)$, with $k = q$, we have convergence of $x_{2,H}^{-1/2}N^{2-2H}T_{2q}$ in $L^2(\Omega)$ to $Z^{2q,2H-1}$, a Hermite r.v. of order $2q$ with parameter $2H-1$.*
- *For $H = 3/4$, with $k = q$, we have convergence of $\sqrt{N/\log N}x_{3,H}^{-1/2}T_{2q}$ to a standard normal distribution.*

The terms $z_{k,H}$, $x_{1,H}$, $x_{2,H}$, $x_{3,H}$ are constants that depend only on H.

More details can be found in Ref. 8.

4. Variations of the Rosenblatt Process

The Rosenblatt process is a Hermite process of order $q = 2$ belonging in the second Wiener chaos. Therefore, it admits the following representation

$$Z(t) = \int_0^t \int_0^t L_t(y_1, y_2) dW_{y_1} dW_{y_2}, \tag{9}$$

where $L_t(y_1, y_2)$ is the kernel

$$L_t(y_1, y_2) = d(H) \int_{y_1 \vee y_2}^{t} \frac{\partial K^{H'}}{\partial u}(u, y_1) \frac{\partial K^{H'}}{\partial u}(u, y_2) du \tag{10}$$

with $H' = \frac{H+1}{2}$ and $d(H) = \frac{1}{H+1}\left(\frac{H}{2(2H-1)}\right)^{-1/2}$. In this situation we are interested in studying the behavior of the statistic V_N (defined in Eq. (2)) for a longer filter of arbitrary order $p \geq 2$.

Using the definition of the filter, we can compute the covariance of the filtered process $V^\alpha\left(\frac{i}{N}\right)$

$$\pi_H^\alpha(j) = -\frac{N^{-2H}}{2}\sum_{k,r=0}^{\ell}\alpha_k\alpha_r|j+k-r|^{2H}. \tag{11}$$

and rewrite the variation statistic as

$$V_N = \frac{2N^{2H}}{c(H)(N-\ell)}\sum_{i=\ell}^{N-1}\left[\left|V^\alpha\left(\frac{i}{N}\right)\right|^2 - \pi_H^\alpha(0)\right], \tag{12}$$

where $c(H) = -\sum_{k,r=0}^{\ell}\alpha_r\alpha_k|k-r|^{2H}$.

Using the Wiener chaos expansion of V_N and following the same steps as before we get the following theorem

Theorem 4.1. *For $H \in (1/2, 1)$*

$$c_H^{-1/2}\, N^{1-H}\, V_N \to^{\mathcal{D}} \mathcal{Z}_1 \tag{13}$$

where c_H is a constant depending only on H and the filter coefficients.

Therefore, unlike the fractional Brownian motion case, we cannot obtain convergence to a normal limit when we use a longer filter on our data.

5. Estimators for the Self-Similarity Index *H*

In this section we construct an estimator for the self-similarity index based on the discrete observations of the process. Suppose that the observed process Z is a Hermite process of order q. Consider now the quadratic variation statistic for a filter of order p:

$$S_N(2,\alpha) = \frac{1}{N}\sum_{i=\ell}^{N}\left(\sum_{k=0}^{\ell}\alpha_k Z\left(\frac{i-k}{N}\right)\right)^2. \tag{14}$$

Recall that $E\left[S_N(2,\alpha)\right] = -\frac{N^{-2H}}{2}\sum_{k,r}^{\ell}\alpha_k\alpha_r|k-r|^{2H}$. By estimating $E\left[S_N(2,\alpha)\right]$ with $S_N(2,\alpha)$ we can construct an estimator of H. We have an analytic expression for $\hat{H}_N$ only when the filter is of order 1 and this is $\hat{H}_N = -\left(\log S_N\right)/\left(2\log N\right)$. For a higher order filter we can compute the estimator of $\hat{H}_N$ numerically by solving the following non-linear equation:

$$-\frac{N^{-2\hat{H}_N}}{2}\sum_{q,r}^{\ell}\alpha_q\alpha_r|q-r|^{2\hat{H}_N} - S_N(2,\alpha) = 0$$

Such a solution always exists for $H > 1/2$.

Using the results about the asymptotic distribution of the variations of the Hermite process we can obtain its asymptotic distribution.

Theorem 5.1.

(i) *The estimator* $\hat{H}_N$ *is strongly consistent, i.e.* $\lim_{N\to\infty} \hat{H}_N = H$ *almost surely.*

(ii) *Let* $Z^{(q,H)}$ *be a Hermite process of order* q *with self-similarity index* H *and assume that we have a filter of order 1. There exists a standard Rosenblatt random variable* R *with self-similarity parameter* $1+2(H-1)/q$ *such that*

$$\lim_{N\to\infty} \mathbf{E}\left[\left|2N^{2(1-\hat{H}_N)/q}\left(H-\hat{H}_N\right)\log N - q!c_{H,q}^{1/2}R\right|\right] = 0,$$

where $c_{H,q}$ *is a constant that depends on* q *and* H.

(iii) *Let* $Z^{(2,H)}$ *be a Rosenblatt process with self-similarity index* H *and assume that we have a filter of order* $p \geq 2$. *Then, there exists a standard normal random variable* Y *such that*

$$\lim_{N\to\infty} \mathbf{E}\left[\left|2\, c_H^{-1/2} N^{1-\hat{H}_N}\log N\left(\hat{H}_N - H\right) - Y\right|\right] = 0,$$

where c_H *is a constant depending only on* H *and the filter coefficients.*

References

1. M. Taqqu, *Z. Wahrscheinlichkeitstheorie verw. Gebiete* **31**, 287 (1975).
2. R. Dobrushin and P. Major, *Z. Wahrscheinlichkeitstheorie verw. Gebiete* **50**, 27 (1979).
3. J. Beran, *Statistics for Long-Memory Processes* (Chapman and Hall, 1994).
4. C. Tudor and F. Viens, Variations and estimators for the selfsimilarity order through malliavin calculus, (2007).
5. D. Nualart, *Malliavin Calculus and Related Topics* (Springer, 2006).
6. D. Nualart and S. Ortiz-Latorre, Central limit theorems for multiple stochastic integrals and malliavin calculus, To appear in Stochastic Processes and their Applications, (2006).
7. J. Coeurjolly, *Statistical Inference for Stochastic Processes* **4**, 199 (2001).
8. A. Chronopoulou, C. Tudor and F. Viens, Self-similarity parameter estimation and reproduction property for non-gaussian hermite processes, (2008).

The Pasinetti-Solow Growth Model with Optimal Saving Behaviour: A Local Bifurcation Analysis

P. Commendatore* and C. Palmisani†

*Dipartimento di Teoria Economica e Applicazioni
Università di Napoli "Federico II"
via Rodinò 22, I-80138 Naples, Italy
Email: commenda@unina.it

†Dipartimento di Matematica e Statistica
Università di Napoli "Federico II"
via Cinthia 26, I-80126, Naples, Italy
Email: cpalmisani@unina.it

We present a discrete time version of the Pasinetti-Solow economic growth model. Workers and capitalists are assumed to save on the basis of rational choices. Workers face a finite time horizon and base their consumption choices on a life-cycle motive, whereas capitalists behave like an infinitely-lived dynasty. The accumulation of both capitalists' and workers' wealth through time is reduced to a two-dimensional map whose local asymptotic stability properties are studied. Various types of bifurcation emerge (flip, Neimark-Sacker, saddle-node and transcritical): a precondition for chaotic dynamics.

Keywords: Economic growth; optimal choice; wealth distribution; dynamical systems; bifurcation analysis.

1. Introduction

In this paper we put forward a two-dimensional economic growth model close to the one developed by Samuelson and Modigliani [1]. Following Pasinetti's [2] suggestion, these authors extended the well-known Solow [3] model by distinguishing in the economy two types of agents or "classes", capitalists and workers, characterised by different saving behaviour. Our model departs from the standard Pasinetti-Solow model for two crucial assumptions. First, capitalists and workers saving decisions are based on optimal behaviour. To model such behaviour we follow Michl ([4] and [5]) assuming that an infinitely lived "dynasty" of capitalists saves on the basis of an altruistic motive, whereas workers are driven by a life-cycle motive. Second, we assume discrete time. Accordingly, various types of bifurcations occur. In particular period-doubling or Neimark-Saker bifurcations, a precondition for the emergence of chaotic dynamics, can occur by reducing the elasticity of substitution between the factors of production and by varying suitably workers and capitalists' time discount factors. We show how these bifurcations can occur via a diagrammatical tool known as "stability triangle" (Azariadis [6]).

2. The Model

We frame the analysis in discrete time denoting by t the time unit. Moreover, we express all the relevant variables in terms of labour units.

We consider a single good economy. Production involves only two factors, capital and labour, and a production function with constant elasticity of substitution (CES) between capital and labour:

$$f(k_t) = [\alpha + (1-\alpha)k_t^{\rho}]^{\frac{1}{\rho}} = [\alpha k_t^{-\rho} + (1-\alpha)]^{\frac{1}{\rho}} k_t \tag{1}$$

where k_t is the capital/labour ratio, $0 < \alpha < 1$ the distribution coefficient, $-\infty < \rho < 1$ ($\rho \neq 0$) the substitution coefficient and $(1-\rho)^{-1}$ the elasticity of substitution. The only sources of income in the economy are wages and profits and capital the only asset. For each short-run equilibrium, perfectly competitive labour and capital markets ensure equality between the wage rate w_t and the marginal product of labour $f(k_t) - f'(k_t)k_t$ and between the profit rate r_t and the marginal product of capital $f'(k_t)$. That is,

$$w_t = f(k_t) - f'(k_t)k_t \tag{2}$$

$$r_t = f'(k_t) \tag{3}$$

There are two distinct groups of agents, capitalists and workers. Both save and accumulate capital, $k_{c,t}$ and $k_{w,t}$ representing respectively capitalists and workers' capital, where $0 \leq k_{c,t} \leq k_t$, $0 \leq k_{w,t} \leq k_t$ and $k_t = k_{c,t} + k_{w,t}$.

We represent capitalists as a single dynasty with an infinite time horizon. Each generation cares about its offspring and saves for a bequest motive behaving like an infinitely-lived household (Barro [7]). Given initial wealth $k_{c,0}$ and time discounting factor $0 < \beta_c < 1$, at beginning of period 0 capitalists choose consumption quantities $(c_{c,0}, c_{c,1}, c_{c,2}, \ldots, c_{c,t}, \ldots) = (c_{c,t})_0^{\infty}$ to solve the following constrained utility maximisation problem:

$$max(1-\beta_c)\sum_{t=0}^{t=\infty} \beta_c^t \ln c_{c,t}$$

subject to $c_{c,t} + k_{c,t+1} \leq (1 + r_t - \delta)k_{c,t}$; where $k_{c,t}$ represents capitalists' wealth and δ the rate of capital depreciation.

Solutions satisfy the condition $c_{c,t+1} = (1 - \beta_c)(1 + r_t - \delta)k_{c,t}$. The difference between profits, capitalists only source of income, and consumption gives capitalists' saving:

$$s_{c,t} = \left[\beta_c(1 + r_t - \delta) - (1 - \delta)\right]k_{c,t} \tag{4}$$

The population of workers has an overlapping generation structure. Each generation of workers faces a finite time horizon composed of two periods. Each individual is active, working and earning a wage, only when "young", i.e., in the first period of her life; while she is in retirement when old, i.e., in the second period of life. Each young worker inelastically supplies one unit of labour at the wage w_t. In contrast to capitalists, workers save out of wages only to be able to consume during retirement according to a life-cycle motive. There is no intergenerational transfer of wealth (Diamond [8]). Given the time discount factor $0 < \beta_w < 1$, at time t a young worker chooses the quantities of current consumption $c_{w,t}$ and future consumption $c_{w,t+1}$ in order to solve the following constrained utility maximisation problem:

$$\max(1 - \beta_w)\ln c_{w,t} + \beta_w \ln c_{w,t+1}$$

subject to $c_{w,t} + \frac{c_{w,t+1}}{1+r_t-\delta} \leq w_t$.

The solution is $c_{w,t} = (1 - \beta_w)w_t$, from which the individual worker saving can be easily deduced:

$$s_{w,t} = w_t - c_{w,t} = \beta_w w_t \tag{5}$$

Assuming that the labour force grows at the constant rate n, capitalists and workers accumulate capital according to the rules

$$k_{c,t+1} = \frac{1}{1+n}(1 - \delta)k_{c,t} + i_{c,t} \tag{6}$$

$$k_{w,t+1} = \frac{1}{1+n}i_{w,t} \tag{7}$$

where $i_{c,t}$ and $i_{w,t}$ represent, respectively, capitalists and workers' gross investment.

In a short-run equilibrium, gross investment equals saving. Using (2), (3), (4), (5) and discarding the time subscripts, (6) and (7) become

$$G_c(k_c, k_w) = \frac{1}{1+n}\beta_c\big(1 - \delta + f'(k)\big)k_c \tag{8}$$

$$G_w(k_c, k_w) = \frac{1}{1+n}\beta_w\big(f(k) - f'(k)k\big) \tag{9}$$

Recalling that $k = k_c + k_w$, the two-dimensional dynamical system (8) and (9) fully describes the time evolution of workers and capitalists' wealth.

3. Existence and Properties of Long Run Equilibria

The long run (or steady growth) equilibria correspond to the fixed points of the two-dimensional map (8) and (9) obtained by imposing $G_c(k_c, k_w) = k_c$ and $G_w(k_c, k_w) = k_w$. There exist three different types of equilibria: a Pasinetti equilibrium involves capitalists owing a positive share of capital; a dual equilibrium, instead, allows only workers to own capital; finally, in a trivial equilibrium, the overall capital is zero. Denoting by $e_f(k) = f'(k)k/f(k) = (1-\alpha)(\alpha k^{-\rho} + 1 - \alpha)^{-1}$ the capital-output elasticity, where $0 \leq e_f(k) < 1$, the three types of equilibria, can be expressed as:

Pasinetti equilibrium $(\bar{k}_c, \bar{k}_w)$ where

$$\bar{k}_c = \bar{k} - \bar{k}_w, \ \frac{\bar{k}_w}{\bar{k}} = \frac{\beta_w \bar{r}}{1+n}\left(\frac{1-e_f(\bar{k})}{e_f(\bar{k})}\right) \text{ and } f'(\bar{k}) = \bar{r} \equiv \frac{1+n}{\beta_c} - (1-\delta).$$

A Pasinetti equilibrium exists iff:

$$0 < \frac{\beta_w \bar{r}}{1+n+\beta_w \bar{r}} < e_f(\bar{k}) < 1 \tag{10}$$

Dual equilibrium $(\tilde{k}_c, \tilde{k}_w)$ where

$$\tilde{k}_c = 0, \ \tilde{k}_w = \tilde{k} \text{ and } \frac{f(\tilde{k})}{\tilde{k}} = \frac{1+n}{\beta_w(1-e_f(\tilde{k}))}.$$

Considering (1), a dual equilibrium solves $\left(\frac{1-\alpha}{e_f(\tilde{k})}\right)^{\frac{1}{\rho}} = \frac{1+n}{\beta_w\left(1-e_f(\tilde{k})\right)}$. The latter equation can be transformed into

$$h(x) = (1-\alpha)\left(\frac{1+n}{\beta_w}\right)^{-\rho} \tag{11}$$

where $h(x) = \frac{x}{(1-x)^\rho}$ and $x = e_f(\tilde{k})$. The shape of $h(x)$ depends on the sign of ρ. When $0 < \rho < 1$, $h(x)$ is strictly increasing and convex for all $0 \leq x < 1$. By the intermediate value theorem, there exists a unique x which solves (11). Instead, when $\rho < 0$, $h(x)$ is strictly increasing for $0 \leq x < x^* \equiv \frac{1}{1-\rho}$; and strictly decreasing for $x^* < x < 1$. By the intermediate value theorem, iff $(1-\alpha)(\frac{1+n}{\beta_w})^{-\rho} \leq h(x^*) = (-\rho)^\rho(1-\rho)^{\rho-1}$ there exist one (equality) or two dual equilibria (strict inequality).

Trivial equilibrium $(k_c^0, k_w^0) = (0, 0)$

The trivial equilibrium exists and is unique iff $f(0) = 0$.

4. Bifurcation Analysis

To investigate the bifurcation process involved in the system (8) and (9), we use the geometrical method based on the "stability triangle" (see Azariadis [6]). Briefly, a fixed point (y, z) is locally stable if the trace and the determinant of the Jacobian matrix $J(y, z)$ evaluated at (y, z), that is, $T(y, z)$ and $D(y, z)$, lie within the boundaries of the isosceles triangle ABC shown in Figure 1. That is, if the conditions (i) $1 + T(y, z) + D(y, z) > 0$, (ii) $D(y, z) < 1$ and (iii) $1 - T(y, z) + D(y, z) > 0$ hold. The system loses stability if, varying a parameter, trace or determinant move outside the triangle. According to which side of the triangle is crossed, a particular type of local bifurcation occurs. If the interior of the segment AB is crossed, the system undergoes a flip or period doubling bifurcation. Along this segment, the smallest eigenvalue of $J(y, z)$ is smaller than -1. If the segment BC is crossed, the system undergoes a Neimark-Sacker bifurcation. Along this segment, the two eigenvalues are complex conjugates with modulus equal to 1. Finally, if the segment AC is crossed, the system may undergo a saddle-node or, alternatively, a transcritical bifurcation. Along this segment, the largest eigenvalue is equal to 1. Applying the linearization theorem, the Jacobian evaluated at the Pasinetti equilibrium $(\bar{k}_c, \bar{k}_w)$ is

$$J(\bar{k}_c, \bar{k}_w) = \begin{pmatrix} \frac{\bar{r}}{\bar{r}+1-\delta} e_{f'}(\bar{k}) \frac{\bar{k}_c}{\bar{k}} & 1 + \frac{\bar{r}}{\bar{r}+1-\delta} e_{f'}(\bar{k}) \frac{\bar{k}_c}{\bar{k}} \\ -\frac{\beta_w \bar{r}}{1+n} e_{f'}(\bar{k}) & -\frac{\beta_w \bar{r}}{1+n} e_{f'}(\bar{k}) \end{pmatrix}$$

The corresponding trace and the determinant are:

$$T(\bar{k}_c, \bar{k}_w) = 1 + D(\bar{k}_c, \bar{k}_w) + \frac{\bar{r}}{\bar{r} + 1 - \delta} e_{f'}(\bar{k}) \frac{\bar{k}_c}{\bar{k}} \tag{12}$$

$$D(\bar{k}_c, \bar{k}_w) = -\frac{\beta_w \bar{r}}{1 + n} e_{f'}(\bar{k}) \tag{13}$$

where $e_{f'}(k) = f''(k)k/f'(k) = -\alpha(1-\rho)[\alpha + (1-\alpha)k^{\rho}]^{-1}$ is the marginal productivity elasticity with respect to capital.

The quantity $e_{f'}(k) < 0$ measures the curvature of the production function: the closer is this elasticity to 0 $(-\infty)$, the larger (smaller) is the degree of substitutability between capital and labour. This variable has a crucial importance in determining the local dynamics properties of the system. Indeed, using (12) and (13), conditions (i) to (iii) can be rewritten as

(i) $e_{f'}(\bar{k}) > \Omega \equiv \frac{2}{\frac{2\beta_w \bar{r}}{1+n} - \frac{\bar{r}}{\bar{r}+1-\delta} \frac{\bar{k}_c}{\bar{k}}}$ if $\Omega < 0$. It is always satisfied otherwise.

Note that for $0 < \bar{k}_c < \bar{k}$, we have $\frac{1+n}{\beta_w \bar{r}} < \Omega < \infty$ or $-\infty < \Omega < -\frac{1}{\bar{r}} \frac{1+n}{\beta_c - 2\beta_w} < -1$, it follows that $e_{f'}(k) < -1$ is a necessary condition for a flip bifurcation to occur.

(ii) $e_{f'}(k) > -\frac{1+n}{\beta_w \bar{r}}$.
Note that if the discount rate of capitalists is larger or not too smaller than the one of workers, i.e. $\beta_w/\beta_c < 1 + \frac{1-\delta}{1+n}$, from (ii) it follows that $e_{f'}(k) < -1$ represents a necessary condition for a Neimark-Saker bifurcation to occur;

(iii) $e_f(\bar{k}) > \frac{\beta_w \bar{r}}{1+n+\beta_w \bar{r}}$.

Condition (iii) is included in condition (10). The violation of condition (iii) involves a transcritical bifurcation characterised by two coexisting equilibra exchanging stability: that is, the Pasinetti equilibrium loses stability in favour of a dual equilibrium.

For the bifurcation analysis we choose the workers' discount rate β_w as control parameter. It is easy to verify that trace and determinant in (12) and (13), redefined as $T(\beta_w)$ and $D(\beta_w)$ respectively, depend linearly on β_w. More precisely, varying β_w within the interval $[0, 1]$ allows to plot in the Trace-Determinant plane a segment S with length λ and slope μ , where $0 < \mu \equiv \frac{[\bar{r}+1-\delta]e_f(\bar{k})}{\bar{r}+(1-\delta)e_f(\bar{k})} < 1$ and $\lambda \equiv \sqrt{(T(1) - T(0))^2 + (D(1) - D(0))^2}$ and where $(T(0), D(0))$ and $(T(1), D(1))$ represent the start and end points of S.

The following proposition spells out the local stability properties of the Pasinetti equilibrium and clarifies the conditions that should hold for the emerging of a specific type of bifurcation:

Proposition 1. *Given $(T(0), D(0))$ and $(T(1), D(1))$ in the Trace-Determinant plane, the segment S satisfies the following properties:*

Case 1 *when $-1 < T(0) < 1$*

- *it never intersects the AB-side;*
- *it intersects the BC-side if $\mu^* \leq \mu < 1$ and $\lambda \geq \sqrt{10}$;*
- *it intersects the AC-side if $0 < \mu \leq \mu^* \equiv \frac{1}{2-T(0)} < 1$ and $\lambda \geq \sqrt{10}$.*

Case 2 *when $-3 \leq T(0) \leq -1$*

- *it intersects the AB-side if $\mu > 0$ and $\lambda \geq 2$;*
- *it intersects the BC-side if $\mu^* \leq \mu < 1$ and $\lambda \geq \sqrt{17}$;*
- *it intersects the AC-side if $0 < \mu \leq \mu^*$ and $\lambda \geq \sqrt{17}$.*

Case 3 *when $T(0) < -3$*

- *it intersects the AB-side if* $0 < \mu \leq \mu^{**} \equiv -\frac{1}{2+T(0)}$ *and* $\lambda \geq \lambda^{*} \equiv -(T(0)+1)$ *;*
- *it intersects the BC-side if* $\mu^{*} \leq \mu \leq \mu^{**} < 1$ *and* $\lambda \geq \lambda^{**} \equiv \sqrt{1+(T(0)-2)^2}$*;*
- *it intersects the AC-side if* $0 < \mu \leq \mu^{*}$ *and* $\lambda \geq \lambda^{**}$*.*

Figure 1 illustrates Proposition 1. Since $T(0) = 1 + e_{f'}(k)\frac{\bar{r}}{\bar{r}+1-\delta} < 1$ and $D(0) = 0$, S starts on the horizontal axis, on the left of 1. S_1' and S_2' exemplify case 1. Both start at $-1 < T'(0) < 1$ and have sufficient length to exit the stability triangle ABC. S_1' satisfies the inequality $\mu < \mu^{*}$ while the inequality $\mu > \mu^{*}$ holds for S_2'. As the parameter β_w is increased from 0 to 1, moving along S_1' the Pasinetti equilibrium loses stability via a transcritical bifurcation; instead, along S_2' stability loss occurs via a Neimark-Saker bifurcation.

S_1'' and S_2'' exemplify case 2. Both start at $-3 \leq T''(0) \leq -1$ and have sufficient length to cut through ABC. $\mu < \mu^{*}$ holds for S_1'' and $\mu > \mu^{*}$ for S_2''. As the parameter β_w is increased from 0 to 1, S_1'' or S_2'' enter the triangle ABC and the Pasinetti equilibrium gains stability via a (reversed) flip bifurcation; as β_w is further increased stability loss may occur via a transcritical bifurcation, along S_1''; or via a Neimark-Saker bifurcation, along S_2''.

Finally, exemplifying case 3, S_1''' and S_2''' start at $T'''(0) < -3$. Their behaviour and properties are similar to those of S_1'' and S_2''.

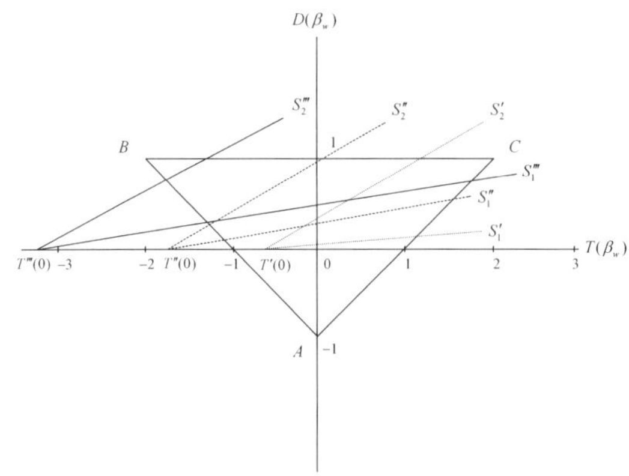

Fig. 1. Stability Triangle.

We conclude this section discussing briefly the stability conditions for dual and trivial equilibria. The Jacobian evaluated at a dual equilibrium $(\tilde{k}_c, \tilde{k}_w) = (0, \tilde{k}_w)$ is

$$J(\tilde{k}_c, \tilde{k}_w) = \begin{pmatrix} \frac{\tilde{r}+1-\delta}{\bar{r}+1-\delta} & 0 \\ -\frac{\beta_w \bar{r}}{1+n} e_{f'}(\tilde{k}) & -\frac{\beta_w \bar{r}}{1+n} e_{f'}(\tilde{k}) \end{pmatrix}$$

where $\tilde{r} = f'(\tilde{k})$. Defining by Λ_1 and Λ_2, the eigenvalues of the lower triangular matrix $J(\tilde{k}_c, \tilde{k}_w)$, trace and determinant can be expressed as $T(\tilde{k}_c, \tilde{k}_w) = \Lambda_1 + \Lambda_2$ and $D(\tilde{k}_c, \tilde{k}_w) = \Lambda_1 \Lambda_2$, where $\Lambda_1 \equiv \frac{\tilde{r}+1-\delta}{\bar{r}+1-\delta}$ and $\Lambda_2 \equiv -\frac{\beta_w \bar{r}}{1+n} e_{f'}(\tilde{k})$ are both real and positive. A dual equilibrium is stable as long as $\Lambda_1 < 1$ and $\Lambda_2 < 1$. The violation of the first inequality leads to a transcritical bifurcation involving an exchange of stability in favour of the Pasinetti equilibrium. Indeed, it is possible to show that this inequality corresponds to the reversed of the central inequality in (10). Moreover, the violation of the second inequality leads to a saddle-node bifurcation involving the collision and disappearance of two dual equilibria, one stable and the other unstable.

Finally, it is straightforward to show that the trivial equilibrium is always unstable as long as $\bar{k} > 0$.

5. Conclusions

In this paper we presented a two-dimensional growth model in which two groups of economic agents, capitalists and workers, follow different saving rules. This model represents a discrete time version of the one developed in various stages by Solow [3], Pasinetti [2] and Samuelson and Modigliani [1]. A crucial difference is the assumption of optimal saving behaviour. We identified three different types of long-run equilibria and verified their local stability properties. We verified also the emergence of various types of local bifurcations. In particular, flip and Neimark-Saker bifurcations can occur by reducing the degree of substitutability between productive factors and choosing suitably capitalists and workers' time discount factors. Simulations, not presented here, show that the time evolution of capitalists and workers' wealth accumulation may undertake several types of complex behaviour.

References

1. P. A. Samuelson and F. Modigliani, *Quarterly Journal of Economics* **33**, 269 (1966).

2. L. Pasinetti, *The Review of Economic Studies* **34**, 267 (1962).
3. R. Solow, *Quarterly Journal of Economics* **70**, 65 (1956).
4. T. R. Michl, *Review of Political Economy* **18**, 449 (2006).
5. T. R. Michl, *Metroeconomica* **58**, 244 (2007).
6. C. Azariadis, *Intertemporal Macroeconomics* (Blackwell Publisher, Cambridge Massachusetts, 1993).
7. R. J. Barro, *The Journal of Political Economy* **82**, 1095 (1974).
8. P. Diamond, *American Economic Review* **55**, 1126 (1965).

A Note on Quantum Chaology and Gamma Approximations to Eigenvalue Spacings for Infinite Random Matrices

C. T. J. Dodson
School of Mathematics
University of Manchester
Manchester M13 9PL, UK

Quantum counterparts of certain classical systems exhibit chaotic spectral statistics of their energy levels; eigenvalues of infinite random matrices model irregular spectra. Eigenvalue spacings for the Gaussian orthogonal ensemble (GOE) of infinite random real symmetric matrices admit a gamma distribution approximation, as do the hermitian unitary (GUE) and quaternionic symplectic (GSE) cases. Then chaotic and non-chaotic cases fit in the information geometric framework of the manifold of gamma distributions, which has been the subject of recent work on neighbourhoods of randomness for general stochastic systems.

Keywords: Random matrices; GOE; GUE; GSE; quantum chaotic; eigenvalue spacing; statistics; gamma distribution; randomness; information geometry.

1. Introduction

Berry introduced the term quantum chaology in his 1987 Bakerian Lecture[1] as the study of semiclassical but non-classical behaviour of systems whose classical motion exhibits chaos. He illustrated this with statistics of energy levels, following his work with Tabor[2] and developments from the study of various systems. A regular spectrum for a bound system with $n \geq 2$ degrees of freedom and n constants of motion, has the energy levels labelled by n quantum numbers; these quantum numbers of nearby energy levels may be very different. For an irregular spectrum, such as an ergodic system with only energy conserved, quantum number labelling fails. So, energy level spacing distributions were used for comparisons among different spectra.[2] Porter[3] had shown that the spacings between energy levels of complex nuclei and atoms with n large are modelled by the eigenvalue spacings of random matrices and that the Wigner distribution[4] gives a very good fit. Indeed, the spacing distributions for generic regular systems are negative exponential, that is random; but for irregular systems the distributions are skew and unimodal, at the scale of the mean spacing. Forrester's online book[5] gives analytic details on the mathematics and physics for the three ensembles of particular interest: Gaussian orthogonal (GOE), unitary (GUE) and symplectic (GSE), being the real, complex and quaternionic cases, respectively. Deift[6] illustrates how random matrix theory has significant links to a wide range of mathematical problems in the theory of functions as well as to mathematical physics. The matrices in these ensembles are respectively invariant under the appropriate orthogonal, unitary and symmetric transformation groups, and moreover in each case the joint density function of all independent elements is controlled by the trace of the matrices and is of form[5]

$$p(X) = A_n \, e^{-\frac{1}{2}\,\mathrm{Tr}\, X^2} \tag{1}$$

where A_n is a normalizing factor.

Gamma distributions provide approximations to eigenvalue spacing distributions for the GOE distribution comparable to the Wigner distribution at the scale of the mean and for the GUE and GSE distributions, except near the origin. This may be useful in the study of irregular spectra of more general real systems and their perturbations because the gamma distribution has a well-understood and tractable information geometry (Arwini and Dodson[7–9]) as well as the following important uniqueness property:

Theorem 1.1 (Hwang and Hu[10]). *For independent positive random variables with a common probability density function f, having independence of the sample mean and the sample coefficient of variation is equivalent to f being the gamma distribution.*

The non-chaotic case has an exponential distribution of spacings between energy levels. Also, the sum of n independent identical exponential random variables follows a gamma distribution and the sum of n independent identical gamma random variables follows a gamma distribution; moreover, the product of gamma distributions is well-approximated by a gamma distribution. Information geometry provides a unique maximum likelihood metric on the space of distributions and so allows comparison of trajectories through differing states during perturbations of parameters.

Monte Carlo methods of Caër et al.[11] established the best fit of GOE, GUE and GSE unit mean distributions, for spacing $s > 0$, using the generalized gamma probability density (pdf) which is:

$$g(s; \beta, \omega) = a(\beta, \omega)\, s^{\beta}\, e^{-b(\beta,\omega) s^{\omega}} \quad \text{for } \beta, \omega > 0 \tag{2}$$

where

$$a(\beta,\omega) = \frac{\omega\Big[\Gamma\big((2+\beta)/\omega\big)\Big]^{\beta+1}}{\Big[\Gamma\big((1+\beta)/\omega\big)\Big]^{\beta+2}} \quad \text{and} \quad b(\beta,\omega) = \left[\frac{\Gamma\big((2+\beta)/\omega\big)}{\Gamma\big((1+\beta)/\omega\big)}\right]^{\omega}.$$

Table 1 gives the best fits of the parameter values in (2) accurate to within $\sim 0.1\%$ of the true distributions from Forrester.[5] The exponential distribution is recovered by the choice $g(s;0,1) = e^{-s}$. Figure 1 shows these along with corresponding fits of the gamma distribution. More details of the study reported here and of the differential geometry of manifolds of probability density functions in application to near-random and other scenarios will be found in the book by Arwini and Dodson.[7]

Table 1. The best fit from Caër et al.[11] of GOE, GUE and GSE unit mean distributions, for eigenvalue spacing $s > 0$, using the generalized gamma density.

Ensemble	β	ω	Variance
GOE	1	1.886	0.2856
GUE	2	1.973	0.1868
GSE	4	2.007	0.1100

2. Eigenvalues of Random Matrices

The mean spacing between eigenvalues of infinite symmetric real random matrices, the real Gaussian Orthogonal Ensemble (GOE), is bounded and therefore it is convenient to normalize the distribution to have unit mean; the same is true for the complex unitary (GUE) and quaternionic symplectic (GSE) cases. Wigner[4,12,13] had conjectured that the cumulative probability distribution function at the scale of the mean spacing should be of the form:

$$W(s) = 1 - e^{-\frac{\pi s^2}{4}} \tag{3}$$

which has unit mean and variance $\frac{4-\pi}{\pi} \approx 0.273$ with probability density function (pdf)

$$w(s) = \frac{\pi}{2} s\, e^{-\frac{\pi s^2}{4}}. \tag{4}$$

Wigner's surmise gave a good fit with numerical computation of the true GOE distribution ([14], Appendix A.15) and with a variety of observed data from atomic and nuclear systems.[1,2,4,14]

From [14], p. 171, we have bounds on the cumulative probability distribution function P for the spacings between eigenvalues of infinite symmetric real random matrices:

$$L(s) = 1 - e^{-\frac{1}{16}\pi^2 s^2} \leq \; P(s) \; \leq U(s) = 1 - e^{-\frac{1}{16}\pi^2 s^2} \left(1 - \frac{\pi^2 s^2}{48}\right) \tag{5}$$

Here the lower bound L has mean $\frac{2}{\sqrt{\pi}} \approx 1.13$ and variance $\frac{4(4-\pi)}{\pi^2} \approx 0.348$, and the upper bound U has mean $\frac{5}{3\sqrt{5}} \approx 0.940$ and variance $\frac{96-25\pi}{9\pi^2} \approx 0.197$. The probability densities for the bounds are, respectively,

$$l(s) = \frac{\pi s}{2} e^{-\frac{\pi s^2}{4}}, \quad u(s) = \frac{\pi^2 s(64 - \pi^2 s^2)}{384} e^{-\frac{1}{16}\pi^2 s^2}. \tag{6}$$

The probability density function for a gamma distribution with dispersion parameter $\kappa > 0$ and mean $\kappa/\nu > 0$ for positive random variable s is given by

$$f(s; \nu, \kappa) = \nu^\kappa \frac{s^{\kappa-1}}{\Gamma(\kappa)} e^{-s\nu}, \quad \text{for } \nu, \kappa > 0 \tag{7}$$

with variance $\frac{\kappa}{\nu^2}$. Then the subset having unit mean is given by

$$f(s; \kappa, \kappa) = \kappa^\kappa \frac{s^{\kappa-1}}{\Gamma(\kappa)} e^{-s\kappa}, \quad \text{for } \kappa > 0 \tag{8}$$

with variance $\frac{1}{\kappa}$. These parameters ν, κ are called natural parameters because they admit a presentation of the family (7) as an exponential family (Amari and Nagaoka[15]) and thereby provide an associated natural affine immersion in $\mathbb{R}^3$ (Dodson and Matsuzoe[16])

$$h : \mathbb{R}^+ \times \mathbb{R}^+ \to \mathbb{R}^3 : \begin{pmatrix} \nu \\ \kappa \end{pmatrix} \mapsto \begin{pmatrix} \nu \\ \kappa \\ \log \Gamma(\kappa) - \kappa \log \nu \end{pmatrix}. \tag{9}$$

Arwini and Dodson[7,8] used this affine immersion to present tubular neighbourhoods of the 1-dimensional subspace consisting of exponential distributions ($\kappa = 1$), thus giving neighbourhoods of random processes. The maximum entropy case has $\kappa = 1$ and corresponds to an underlying Poisson random event process and so it can model spacings in the spectra for non-chaotic systems; for $\kappa > 1$ the distributions are skew unimodular, like chaotic spectral spacings. The unit mean gamma distribution fit to the true GOE distribution from Mehta[14] has variance ≈ 0.379 and hence $\kappa \approx 2.42$.

In fact, κ is a geodesic coordinate in the Riemannian 2-manifold of gamma distributions with Fisher information metric; arc length along this coordinate from $\kappa = a$ to $\kappa = b$ is given by

$$\left| \int_a^b \sqrt{\frac{d^2 \log(\Gamma(\kappa))}{d\kappa^2} - \frac{1}{\kappa}}\, d\kappa \right| . \tag{10}$$

See Arwini and Dodson[7,8] for more details and related properties; also, it provides details of information geometry in neighbourhoods of uniform processes and neighbourhoods of independence for bivariate processes.

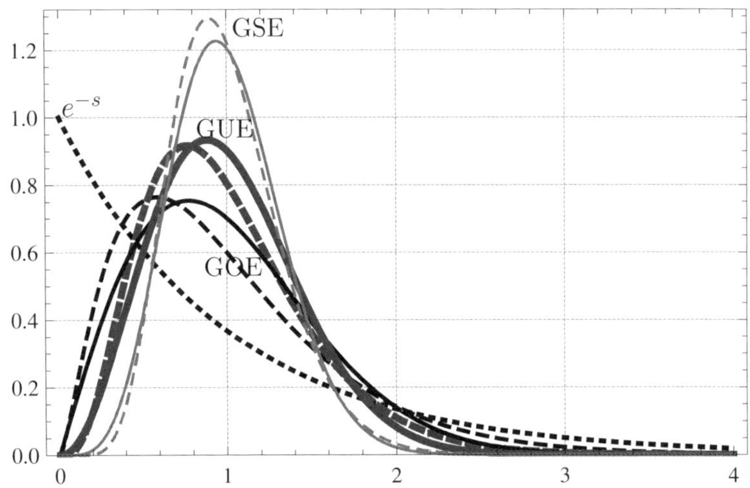

Fig. 1. Probability density functions for the unit mean gamma distributions (dashed) and generalized gamma distribution (solid) fits to the true variances for the GOE, GUE and GSE cases. The two types coincide in the random exponential case, e^{-s}, shown dotted.

3. Deviations

Near the origin, the GOE and Wigner distributions are linear; the unitary ensemble (GUE) $\sim s^2$ and the symplectic ensemble (GSE) $\sim s^4$. At unit mean the gamma density behaves like $s^{\kappa-1}$ near the origin, so linearity would require $\kappa = 2$ with variance $\frac{1}{\kappa} = \frac{1}{2}$; the GOE fitted gamma distribution has $\kappa \approx 2.42$ and hence variance ≈ 0.379. Figure 1 shows the probability density functions for the unit mean gamma distributions (dashed)

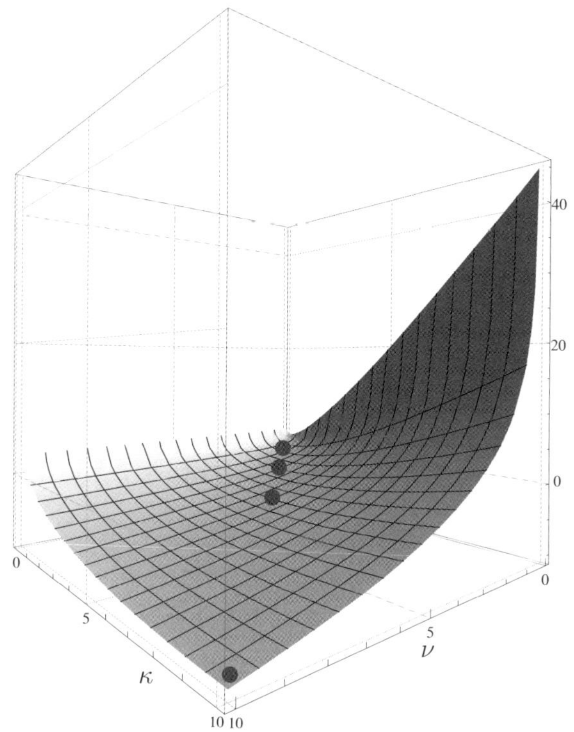

Fig. 2. The unit mean gamma distributions corresponding to the random (non-chaotic) case, $\kappa = \nu = 1$ and those with exponent $\kappa = \nu = 2.420,\ 4.247,\ 9.606$ for the best fits to the true variances of the spacing distributions for the GOE, GUE and GSE cases, as points on the affine immersion in $\mathbb{R}^3$ of the 2-manifold of gamma distributions.

and generalized gamma distribution (solid) fits to the true variances for the GOE, GUE and GSE cases; the two types coincide in the random case, which is exponential, e^{-s}, shown dotted. Figure 2 shows gamma best fits as points on the affine immersion in $\mathbb{R}^3$ of the 2-manifold of gamma pdfs.[7,16] The information metric provides information distances on the gamma manifold and so could compare real data on eigenvalue spacings fitted to gamma pdfs; this may allow identification of qualitative properties and represent trajectories during structural changes of systems.

Figure 3 shows with unit mean the probability distribution for spacings between the first 2,001,052 zeros of the Riemann zeta function (large points), that for the true GUE distribution (medium points) and the gamma fit to the true GUE (small points), which has $\kappa \approx 4.247$. The grand mean spacing between zeros from the data was ≈ 0.566, the coefficient of variation ≈ 0.422 and variance ≈ 0.178.

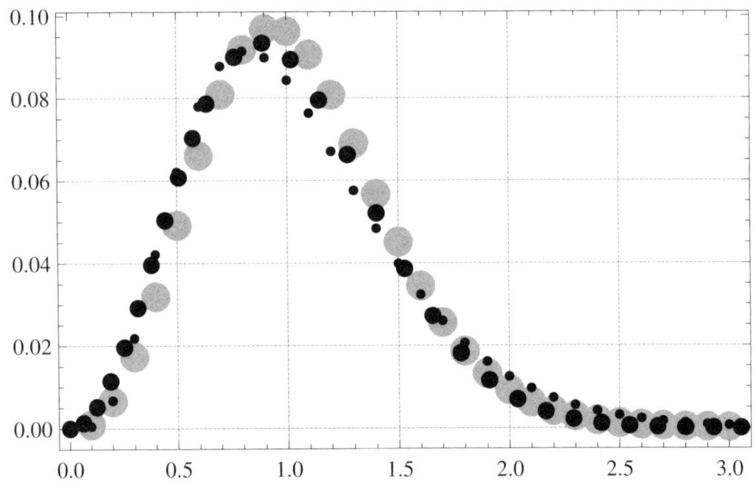

Fig. 3. Probability plot with unit mean for the spacings between the first 2,001,052 zeros of the Riemann zeta function (large points), that for the true GUE distribution (medium points) and the gamma fit to the true GUE (small points).

The effect of location on the statistical data for spacings in the first ten consecutive blocks of 200,000 zeros of the Riemann zeta function normalized with unit grand mean and the effect of sample size were investigated in Arwini and Dodson.[7] For gamma distributions we expect the coefficient of variation to be independent of sample size and location, by Theorem 1.1, and this was approximately the case.

4. Conclusion

The gamma distribution provides approximations to the true distributions for the spacings between eigenvalues of infinite random matrices for the GOE, GUE and the GSE cases. However, it is clear that gamma distributions do not precisely model the analytic systems discussed here, and do not give correct asymptotic behaviour at the origin, as is evident from the results of Caër et al.[11] who obtained excellent approximations for GOE, GUE and GSE distributions using the generalized gamma distribution (2). The differences may be seen in Figure 1 which shows the unit mean distributions for gamma (dashed) and generalized gamma (solid) fits to the true variances for the Poisson, GOE, GUE and GSE ensembles. The generalized gamma distributions do not have tractable information geometry and so gamma distribution approximations may be useful in studies of qualitative generic properties for data from real systems. It would be interesting to investigate the extent to which data from real atomic and nuclear systems has generally the qualitative property that the sample coefficient of variation is independent of the mean. That, by Theorem 1.1, is an information-theoretic distinguishing property of the gamma distribution.

References

1. M. V. Berry, *Proc. Roy. Soc. London A* **413**, 183 (1987).
2. M. V. Berry and M. Tabor, *Proc. Roy. Soc. London A* **356**, 373 (1977).
3. C. F. Porter, *Statistical Theory of Spectra: Fluctuations* (Academic Press, London, 1965).
4. E. P. Wigner, *SIAM Review* **9**(1), 1 (1967).
5. P. J. Forrester, *www.ms.unimelb.edu.au/ matpjf/matpjf.html* (2008).
6. P. Deift, *arXiv:0712.0849v1* (2007).
7. K. Arwini and C. T. J. Dodson, *Information Geometry: Near Randomness and Near Independence* (LNM 1953, Springer-Verlag, New York, Berlin, 2008).
8. K. Arwini and C. T. J. Dodson, *Sankhya: Indian Journal of Statistics* **66**(2), 211 (2004).
9. C. T. J. Dodson, *arxiv.org/abs/astro-ph/0608511* (2006).
10. T.-Y. Hwang and C.-Y. Hu, *Annals Inst. Statist. Math.* **51**(4), 749 (1999).
11. C. M. G. Le Caër and R. Delannay, *Physica A* **386,387**, 190 (2007, 2008).
12. E. P. Wigner, *Annals of Mathematics* **62**(3), 548 (1955).
13. E. P. Wigner, *Annals of Mathematics* **67**(2), 325 (1958).
14. M. L. Mehta, *Random Matrices* (Academic Press, London, 2004).
15. S.-I. Amari and H. Nagaoka, *Methods of Information Geometry* (American Mathematical Society, Oxford University Press, Oxford, 2000).
16. C. T. J. Dodson and H. Matsuzoe, *InterStat* **2**, 1 (2002).

One Dimensional Fractal Interpolation: Determination of the Vertical Scaling Factors using Convex Hulls

V. Drakopoulos* and P. Manousopoulos†

Department of Informatics and Telecommunications
University of Athens
Panepistimioupolis, 157 84, Athens, Greece
*Emails: *vasilios@di.uoa.gr, †polyman@di.uoa.gr*

Fractal interpolation functions are very useful in capturing data that exhibit an irregular (fractal) structure. A new method to calculate the vertical scaling factors of such a function using convex hulls is presented. Comparative results with existing methods are also presented establishing the proposed one as an attractive alternative to them. In particular, it clearly outperforms previous methods when the interpolation is sparse, i.e. when high compression ratios are involved.

Keywords: Fractal interpolation; iterated function systems; vertical scaling factors; Hausdorff distance.

1. Introduction

Fractal interpolation as defined in [1,2] is an alternative to traditional interpolation techniques, aiming mainly at data which present detail at different scales or some degree of self-similarity. It is based on the theory of *iterated function systems* (IFS's) and provides a constructive way to describe the data, rather than the descriptive one employed by most traditional methods. This means that we do not store any information about the data but only a procedural way to reconstruct them.

The accuracy of fit of a 1D fractal interpolation function is determined mainly by the vertical scaling factors, its only free parameters, and therefore their determination is crucial. However, there is no direct way to find the optimal values of these factors and the most popular approaches[3–6] employ algebraic or geometric methods to estimate them. In the first method, an analytical expression for the vertical scaling factors is obtained by minimizing (using derivatives) the sum of squared vertical distances between respective original and reconstructed points. In the second one, the factors are obtained by calculating ratios of vertical distances between the data points and the lines connecting the endpoints of the whole data or interpolation intervals. Our motivation is to create an alternative methodology for calculating the vertical scaling factors using convex hulls of appropriately chosen data points.

The paper is organised as follows. In Section 2 we briefly review fractal interpolation functions and in Section 3 we describe in detail our method for calculating the vertical scaling factors. Section 4 contains the results of the application of our method as well as comparisons to existing methods. Finally, Section 5 summarises our conclusions.

2. Fractal Interpolation Functions in 1D

Let us represent the given set of *data points* as $P = \{(u_m, v_m) \in \mathbb{R}^2\colon m = 0, 1, \ldots, M\}$ and as $Q = \{(x_i, y_i) \in \mathbb{R}^2\colon i = 0, 1, \ldots, N < M\}$ a subset of them, the *interpolation points*. Both sets are linearly ordered with respect to the abscissa, i.e. $u_0 < u_1 < \cdots < u_M$ and $u_0 = x_0 < x_1 < \cdots < x_N = u_M$. The interpolation points partition the set of data points into N *interpolation intervals* and may be chosen equidistantly or not. The data points within the n-th interpolation interval are represented as $P_n = \{(u_i, v_i) : i \in M_n\}$, $n = 1, 2, \ldots, N$, where M_n is an index set of P_n, such that $\bigcup_{n=1}^{N} M_n = \{0, 1, \ldots, M\}$ and $P = \bigcup_{n=1}^{N} P_n$.

Let $\{\mathbb{R}^2; w_n, n = 1, 2, \ldots, N\}$ be an *iterated function system* with affine transformations

$$w_n \begin{bmatrix} x \\ y \end{bmatrix} = \begin{bmatrix} a_n & 0 \\ c_n & s_n \end{bmatrix} \begin{bmatrix} x \\ y \end{bmatrix} + \begin{bmatrix} d_n \\ e_n \end{bmatrix}$$

constrained to satisfy

$$w_n \begin{bmatrix} x_0 \\ y_0 \end{bmatrix} = \begin{bmatrix} x_{n-1} \\ y_{n-1} \end{bmatrix} \quad \text{and} \quad w_n \begin{bmatrix} x_N \\ y_N \end{bmatrix} = \begin{bmatrix} x_n \\ y_n \end{bmatrix}$$

for every $n = 1, 2, \ldots, N$. Solving the above equations results in

$$\begin{aligned} a_n &= \frac{x_n - x_{n-1}}{x_N - x_0} \\ d_n &= \frac{x_N x_{n-1} - x_0 x_n}{x_N - x_0} \\ c_n &= \frac{y_n - y_{n-1}}{x_N - x_0} - s_n \frac{y_N - y_0}{x_N - x_0} \\ e_n &= \frac{x_N y_{n-1} - x_0 y_n}{x_N - x_0} - s_n \frac{x_N y_0 - x_0 y_N}{x_N - x_0} \end{aligned}$$

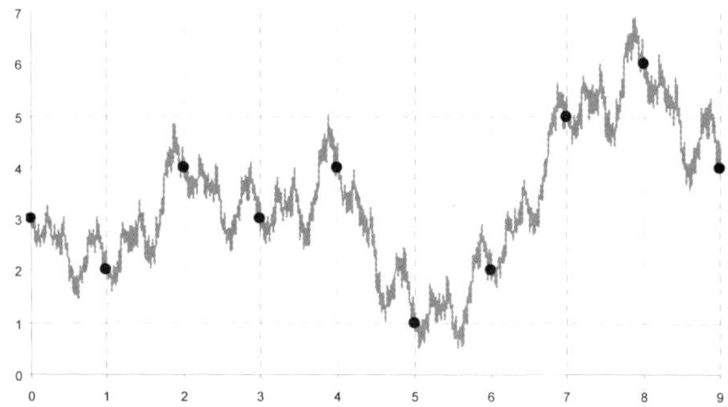

Fig. 1. A fractal interpolation function (gray) for a set of ten interpolation points (black).

i.e. the real numbers a_n, d_n, c_n, e_n are completely determined by the interpolation points, while the s_n are *free* parameters of the transformations satisfying $|s_n| < 1$, in order to guarantee that the IFS is hyperbolic with respect to an appropriate metric. The *attractor* of a hyperbolic IFS is the unique set $A_\infty = \lim_{k\to\infty} W^k(A_0)$ for every starting set A_0, where $W(A) = \bigcup_{n=1}^{N} w_n(A)$ for all $A \in \mathcal{H}(X)$ and $\mathcal{H}(X)$ is the metric space of all non-empty compact subsets of X with respect to some metric, e.g. the Hausdorff metric. The transformations w_n are *shear transformations*: line segments parallel to the y-axis are mapped to line segments parallel to the y-axis contracted by the factor $|s_n|$. For this reason, the s_n are called *vertical scaling* (or *contractivity*) *factors*.

It is well known (see e.g. [2]) that the attractor $G = \bigcup_{n=1}^{N} w_n(G)$ of the aforementioned IFS is the graph of a continuous function $f\colon [x_0, x_N] \to \mathbb{R}$ that passes through the interpolation points (x_i, y_i), $i = 0, 1, \ldots, N$. This function is called *fractal interpolation function (FIF)* corresponding to these points. It is a *self-affine* function since each affine transformation w_n maps the function to its *section* between the interpolation points (x_{n-1}, y_{n-1}) and (x_n, y_n) for all $n = 1, 2, \ldots, N$. An example is depicted in Figure 1, where a FIF is constructed for the set of interpolation points $Q = \{(0,3),$ $(1,2)$, $(2,4)$, $(3,3)$, $(4,4)$, $(5,1)$, $(6,2)$, $(7,5)$, $(8,6)$, $(9,4)\}$ using vertical scaling factors $s_n = 0.35$, for every $n = 1, 2, \ldots, 9$.

Although the FIF passes by definition through the interpolation points, this is not necessarily the case for the remaining data points $P \setminus Q$. The accuracy of fit can be measured as the squared error between the ordinates of the original and the reconstructed points, i.e. $\varepsilon = \sum_{m=0}^{M}(v_m - G(u_m))^2$,

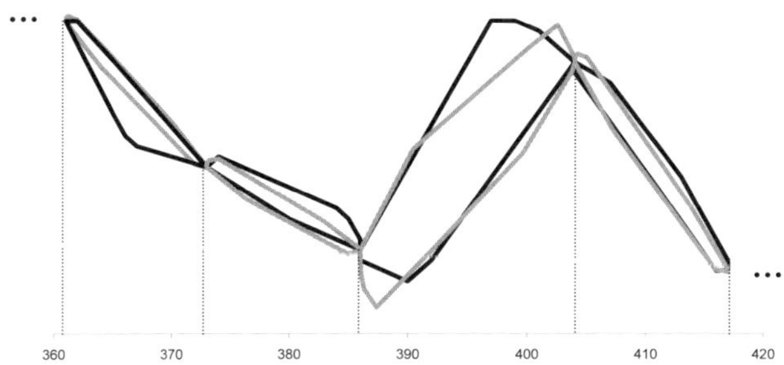

Fig. 2. The convex hulls of points within the interpolation intervals (black) and the convex hulls of the transformations of all points (gray).

or, alternatively, as the *Hausdorff distance* between the two sets, i.e. $\varepsilon = h(\{(u_m, v_m)\}, G)$. The first error measure is employed by existing methods in order to calculate the vertical scaling factors and evaluate their accuracy. The second one is adopted in this paper since it is considered more appropriate in the case of FIF's, as defined previously.

3. Determination of the Vertical Scaling Factors

Each affine transformation w_n of the interpolating IFS transforms the set of data points P to one of its subsets P_n. We propose to work with the bounding sets of P and P_n such that the transformed points $w_n(P)$ best approximate the data points within P_n. By selecting the value of s_n that results in the maximum overlap of the respective bounding sets, we are able to produce a better approximation of the data points. The advantage of this approach is that, for suitably chosen bounding sets, we are able to obtain the proper s_n using either analytical expressions or efficient algorithms, while this would be not possible if we used directly the data points.

The convex hull is the tightest reasonable bounding set, since non-convex bounding sets are likely to result in inefficient algorithms without necessarily improving the results. In Figure 2, the black polygons represent the convex hulls of points in four successive interpolation intervals, the gray polygons represent the transformation of the convex hull of all points under the respective affine transformations (using arbitrary values for s_n) and the dotted lines mark the interpolation intervals. As shown in this figure, the convex hulls can provide a satisfactory bounding set of the data points.

We want to minimize the area of the nonoverlapping parts of the convex hull of the points in the n-th interpolation interval and the transformation of the convex hull of all points under w_n. This is

$$Ar(s_n) = Ar\{CH(P_n)\} + Ar\{CH(w_n(P))\} - 2Ar\{CH(P_n \cap w_n(P))\} \quad (1)$$

where $Ar\{\cdot\}$ is the area of a polygon and $CH(\cdot)$ is the convex hull of a set of points. The calculation of the optimal $\hat{s}_n$ cannot be performed analytically, since it is not possible to derive an analytical formula for Ar as a function of s_n. As implied by Eq. 1, the calculation of $Ar(s_n)$ is algorithmic and involves the computation of convex hulls, polygon intersections and areas. An efficient approach is to use Brent's method,[7] which is a *bracketing method* with parabolic interpolation. In order for the method to converge to the optimal positive $\hat{s}_n$ (denoted by $\hat{s}_n^+$), we must initially bracket it, i.e. provide three constants s_a, s_b, s_c such that (a) $0 \le s_a < s_b < s_c \le 1$, (b) $Ar(s_a) > Ar(s_b) < Ar(s_c)$ and (c) $s_a < \hat{s}_n < s_c$. A good approximation is to estimate s_b such that $Ar\{CH(w_n(P))\} \approx Ar\{CH(P_n)\}$ and set s_a, s_c to be smaller and bigger respectively by a factor, i.e. $s_b = Ar\{CH(P_n)\}/(a_n Ar\{CH(P)\})$, $s_a = s_b/c$ and $s_c = \min\{c\, s_b, 1\}$ for some $c > 1$. The constant c can be determined a priori, e.g. $c = 2$, or more safely using an iterative procedure of checking successively larger values until a suitable bracketing triplet is found. It is possible to have also negative vertical scaling factors. The optimal negative $\hat{s}_n$ (denoted by $\hat{s}_n^-$) can be calculated similarly using the bracketing triplet $s'_b = -s_b$, $s'_a = \max\{c\, s'_b, -1\}$ and $s'_c = s'_b/c$. The algorithm for finding the optimal vertical scaling factors can be outlined as follows:

(1) Compute the convex hull of P.
(2) For each interpolation interval $n = 1, \ldots, N$:
   (a) Calculate the known parameters of w_n, i.e. a_n, c_n, d_n, e_n.
   (b) Compute the convex hull of P_n.
   (c) Find the optimal $\hat{s}_n^+$ that minimizes the area of nonoverlapping parts of $CH(P_n)$ and $CH(w_n(P))$ using the bracketing triplet $(s_b/c, s_b, \min\{cs_b, 1\})$, where

$$s_b = Ar\{CH(P_n)\}/(a_n Ar\{CH(P)\})$$

   and $c > 1$, yielding mimimum non-overlapping area Ar^+.
   (d) Find the optimal $\hat{s}_n^-$ that minimizes the area of nonoverlapping parts of $CH(P_n)$ and $CH(w_n(P))$ using the bracketing triplet $(\max\{cs'_b, -1\}, s'_b, s'_b/c)$, where $s'_b = -s_b$ and $c > 1$, yielding mimimum non-overlapping area Ar^-.
   (e) If $Ar^+ \le Ar^-$ then $\hat{s}_n = \hat{s}_n^+$, else $\hat{s}_n = \hat{s}_n^-$.

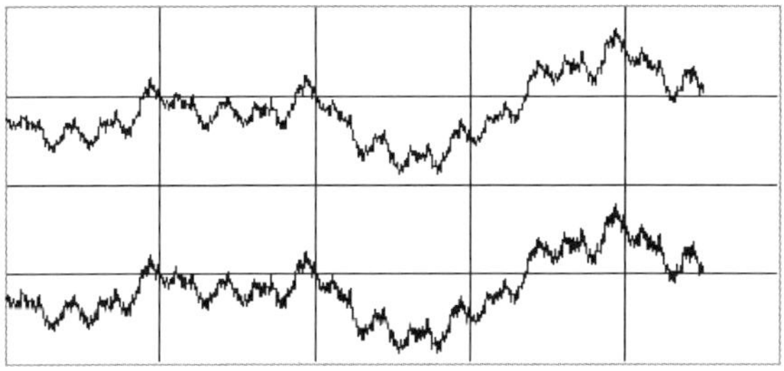

Fig. 3. The FIF constructed by the convex hull method (top) and the original data points (below).

The algorithm may seem computationally demanding, but actually it is possible to achieve linear complexity as well as efficient implementation. Specifically, the intersection of convex polygons and the calculation of a polygon's area require $O(n)$ time.[8] The calculation of the convex hull of a set of points on the plane generally requires $O(n \log n)$ time.[8] However, exploiting the fact that the data points are ordered and form a simple (i.e. non self-intersecting) polyline, it is possible to achieve $O(n)$ time using either Melkman's algorithm[9] or a generic convex hull algorithm such as Andrew's algorithm[10] without the initial sorting phase. Brent's method takes a constant number of steps, since it depends only on the required tolerance and not on the number of data points. Therefore, the overall algorithm complexity is linear. Moreover, the calculated convex hulls have relatively few vertices in practice, which implies that the polygon operations are inexpensive.

4. Results

In the lower part of Figure 3 a data set of 10000 points representing a synthesized function is presented. In the upper part, the FIF constructed by the proposed method using interpolation intervals of 50 points is shown. As can be concluded from the figure, the FIF interpolates the data points quite successfully despite the sparsity of the interpolation points. This is more clear in Figure 4 that depicts both functions in an interval whose length is about 1% of the length of the whole function. The FIF follows closely the original data exploiting the self-affinity when possible. Note that this figure has been expanded in the x-axis in order to be clearer.

Table 1 contains the Hausdorff distance between the original and the

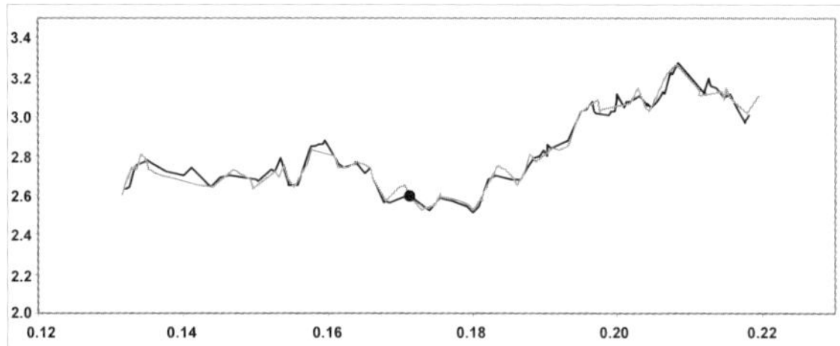

Fig. 4. The original data (gray) and the FIF (black) in an interval of length of about 1% of the whole function. The single interpolation point that falls within this interval is marked with a dot.

reconstructed data points[a]. We have compared three methods, the geometric and the algebraic of Mazel and Hayes,[6] and the method proposed in this paper. The first two methods are the best known, and moreover they generalize other approaches. Therefore we consider them to be representative enough for our test. The interpolation intervals have been chosen with fixed length L of 10 to 100, i.e. by taking every 10th to 100th point as interpolation point, in each case. The proposed method performs better than the algebraic method in all cases, and better than the geometric method in 70% of the cases. It is interesting that the proposed method outperforms both others for the larger interpolation intervals, i.e. for higher compression ratios. We also notice that in some cases the increase in the length L decreases the Hausdorff distance between the original and the reconstructed data for all methods. This is reasonable since a larger interpolation interval may sometimes be able to exploit better the self-affinity of the data.

5. Conclusions and Further Work

We have presented a novel method for determining the vertical scaling factors of 1D fractal interplation functions using an efficient algorithmic procedure. Our method aims at reducing the Hausdorff distance between the original and the reconstructed data points. The results show that it achieves its goal yielding better results than existing methods, particularly for sparse interpolation points. The proposed method can be efficiently combined with an algorithm for finding the optimal interpolation intervals, such as the iterative algorithm of Mazel and Hayes.[6] In this case, it will be able to achieve even better results by exploiting more efficiently the possible

[a]See e.g. [11] for an efficient algorithm for calculating the Hausdorff distance.

Table 1. The Hausdorff distance between the original and the reconstructed data using the three methods for various interpolation interval lengths.

L	Geometric method	Algebraic method	Convex hulls method
10	0.12363	0.17953	0.17948
20	0.15363	0.22113	0.16557
30	0.16292	0.29791	0.24828
40	0.24217	0.29801	0.21950
50	0.33169	0.31969	0.26135
60	0.31148	0.34923	0.27653
70	0.38921	0.37602	0.30572
80	0.33348	0.40622	0.29530
90	0.34332	0.45251	0.33512
100	0.43865	0.48039	0.28051

self-affinity of the data. Moreover, it can be directly extended to piecewise self-affine FIF's.[6] The only necessary modification is to use, for each affine transformation, the points in the current *address interval* instead of all data points when calculating the transformed convex hull.

Acknowledgements

This research project is co-financed by E.U.-European Social Fund (75%) and the Greek Ministry of Development-GSRT (25%), under the PENED programme 70/3/8405.

References

1. M. F. Barnsley, *Constr. Approx.* **2**, 303 (1986).
2. M. F. Barnsley, *Fractals everywhere*, 2nd edn. (Academic Press Professional, San Diego, 1993).
3. M. A. Marvasti and W. C. Strahle, *Signal Processing* **41**, 191 (1995).
4. D. S. Mazel, *IEEE Trans. Signal Processing* **42**, 3269 (1994).
5. D. S. Mazel and M. H. Hayes, Hidden-variable fractal interpolation of discrete sequences, in *Proc. Int. Conf. ASSP*, 1, 3393 (Toronto, Ont., Canada, 1991).
6. D. S. Mazel and M. H. Hayes, *IEEE Trans. Signal Processing* **40**, 1724 (1992).
7. W. H. Press, S. A. Teukolsky, W. T. Vetterling and B. P. Flannery, *Numerical Recipes in C, The Art of Scientific Computing*, 2nd edn. (Cambridge University Press, Cambridge, 1992).
8. J. O'Rourke, *Computational Geometry in C*, 2nd edn. (Cambridge University Press, Cambridge, UK, 1998).
9. A. A. Melkman, *Inform. Process. Lett.* **25**, 11 (1987).
10. A. M. Andrew, *Inform. Process. Lett.* **9**, 216 (1979).
11. V. Drakopoulos and N. Nikolaou, *Applied Mathematical Sciences* **1**, 145 (2007).

Effect of Parametric Dichotomic Markov Noise on the Properties of Chaotic Transitions in Dynamical Systems

J. M. Gac* and J. J. Żebrowski†

Warsaw University of Technology
Faculty of Physics, 00-662Warsaw, Poland
Emails: *gac@if.pw.edu.pl, †zebra@if.pw.edu.pl

A chaotic transition occurs when a continuous change of one of the parameters of the system causes a discontinuous change in the properties of the chaotic attractor of the system. Such phenomena are present in many dynamical systems, in which a chaotic behavior occurs. The best known of these transitions are: the period-doubling bifurcation cascade, intermittency and crises. The effect of dichotomous Markov noise (DMN) on the properties of systems with chaotic transitions is discussed. DMN is a very simple two-valued stochastic process, with constant transition rates between the two states. In spite of its simplicity, this kind of noise is a very powerful tool to describe various phenomena present in many physical, chemical or biological systems. Many interesting phenomena induced by DMN are known. However, there is no research on the effect of this kind of noise on intermittency or crises. We present the change of the mean laminar phase length and of laminar phase length distribution caused by DMN modulating the parameters of a system with intermittency and the modification of the mean life time on the pre-crisis attractor in the case of a boundary crisis. The results obtained analytically are compared with numerical simulations for several simple dynamical systems.

Keywords: Dichotomous Markov noise; chaotic transitions.

1. Introduction

Phenomena, in which a continuous change of one of the parameters of the system causes a discontinuous change in the properties of the chaotic attractor of the system are called "chaotic transitions".[1] These phenomena are present in many dynamical systems in which chaotic behavior occurs. The best known of these transitions are: intermittency, crisis and the period-doubling bifurcation cascade which is a typical route to chaos for many dynamical systems. During intermittency, irregularly long periods of the time occur during which the statistical properties of the trajectory differ strongly from those in other periods. One example is type I intermittency[2,3]

— one of three kinds of intermittency originally introduced by Pomeau and Mannevile.[2] The last of the above mentioned chaotic transitions, the crises,[1] can be observed in such systems in which the chaotic attractors — depending on the type of the crisis — suddenly grow, vanish or merge with another chaotic attractor when the control parameter of the system is changed continuously. The first of these chaotic transitions is known as the interior crisis, the second — the boundary crisis and the last one — the attractor merging crisis. Crisis-induced intermittency is associated with the interior crisis.

In both cases i.e. in Pomeau-Mannevile intermittency and in crises, to describe the behavior of the system in which they are present various statistical measures are used. The mean laminar phase length or the laminar phase length distribution are used in the case of a Pomeau-Mannevile intermittency.[3] The mean life time of the trajectory on the pre-crisis attractor and distribution of this time are used in the case of the boundary crisis. These statistics are characteristic for the specific type of chaotic transition and allow us to identify such a transition in real physical systems, when the form of equations describing the system is not known.

However, in real systems, noise usually is present. Due to noise the above mentioned statistics measured in real systems differ from these found in simple models. In the case of additive white noise, the effect on the laminar phase length distribution and mean laminar phase length was investigated and described by Hirsch.[3] The main conclusions were that in the presence of noise there are more shorter laminar phases than in the absence of noise and, consequently, the mean laminar phase length is also shorter. The effect of colored Gaussian noise on some chaotic transitions, e.g. on the period-doubling bifurcation cascade,[4] was also investigated.

In real physical systems, however, we should consider the effect of noise with a distribution different from a Gaussian. Some examples are: the Levy processes and dichotomic Markov noise (DMN).[5] In this paper, we want to concentrate on the latter, especially on the effect of this noise on the properties of chaotic transitions.

DMN is a very simple two-valued $A_\pm$ stochastic process, with constant transition rates $k_\pm$ between the two states.[5,6] One can easily check that the waiting time in the two states (i.e. time spent by system without change of value of the noise) is exponentially-distributed:

$$P\left(\tau_{+}\right) = k_{\pm} \exp\left(-k_{\pm}\tau_{\pm}\right) \tag{1}$$

and the autocorrelation function of DMN decays exponentially,[5] so this is colored noise. The stationary probabilities are equal:

$$P(\xi = A_+) = k_- \tau_c \tag{2}$$

$$P(\xi = A_-) = k_+ \tau_c \tag{3}$$

where the correlation time is expressed as follows:

$$\tau_c = \frac{1}{k_+ + k_-} \tag{4}$$

A very important case is the symmetric DMN, where we have:

$$A_+ = -A_- = A \tag{5}$$

$$k_+ = k_- = k \tag{6}$$

This kind of noise is very simple to simulate both in discrete and in continuous time problems. Also, many problems with DMN can be easily solved analytically. On the other hand, DMN mimics the effect of a finite correlation time of noise. These are the basic reasons, for which DMN becomes more and more popular as a stochastic forcing in many theoretical models of various phenomena.

In spite of its simplicity, such a kind of noise may lead to many interesting and non-trivial phenomena. E.g. Bena[5] considers a genetic model driven by DMN:

$$\frac{dx}{dt} = \left(\frac{1}{2} - x\right) + \lambda x (1 - x) \tag{7}$$

When the value of λ is constant, this equation has a stable fixed solution. E.g. for $\lambda = 0$ this fixed point is equal to 0.5. However, if this parameter is a DMN symmetric noise, the situation gets more complicated and the behavior of the system depends on the parameters of DMN: its amplitude and transition rate. In some cases, the histogram of the value of x has one, more or less thin, peak, while in other cases three or even four peaks appear. All of them are discussed in.[5]

Presently the effect of DMN on mechanical systems is well known but there are very few results for chaotic systems driven for this kind of noise. The effect of parametric DMN on the properties of the logistic was investigated in.[7] But the properties of chaotic transitions with parametric DMN were not discussed.

The paper is organized as follows. In Section 2 the effect of parametric DMN on the properties of the systems with type-I intermittency is presented. In Section 3 we analyse effect of such noise on systems with the boundary crisis and the discussion is in Section 4.

2. Effect of Dichotomous Markov Noise on Type I Intermittency

2.1. *General remarks*

Consider a system in which type-I intermittency appears at the critical value a_c of the control parameter of the system a. For $a > a_c$ the stable fixed point (or a periodic orbit) exists while for $a < a_c$ intermittent chaos is obtained.

If the parameter a is driven by a symmetric DMN, some noise-induced transitions, similar to those described in,[5] are obtained. Let the parameter a alternate between two values $a_1 = a_0 + \Delta a$ and $a_2 = a_0 - \Delta a$, both smaller than a_c. The changes in the dynamics of the system result as changes of the shape of laminar phase length distribution. In the noise-free case, this distribution has a well-known shape with two peaks: one close to zero and one at the maximal phase length. If a is a DMN, but a_1 and a_2 are nearly the same (i.e. the difference between the maximum laminar phase length obtained for both of these parameter values is of order of one) we observe a broadening of the right peak. A similar effect is observed, when the difference between a_1 and a_2 is not so small, but the transition rate k is relatively large. In this case, the laminar phase distribution is very similar to that obtained in the case of a Gaussian white noise driving the system — in good agreement with analytical results showing that in the limit of large values of amplitude and transition rate DMN reduces to a Gaussian white noise.[8]

A completely novel effect is observed when the difference between a_1 and a_2 is relatively large and the value of k is small (namely, ten times or even smaller than the inverse of the maximum laminar phase length obtained for the stationary case). In this case the histogram of the laminar phase length takes the form of a simple sum of two histograms, obtained in the stationary cases for a equal to a_1 and a_2, respectively. Then the right peak of thelaminar phase length distribution is split into two.

2.2. *An example: The logistic map*

As an example of a discrete time system consider the well-known logistic map:[1]

$$x_{n+1} = ax_n (1 - x_n) \tag{8}$$

It is well known that when the value of a is equal to $a_c = 1 + \sqrt{8} \cong 3.828427125\ldots$ the saddle-node bifurcation occurs and for larger values of

a a stable period-3 orbit appears. When a is smaller than a_c, the system exhibits type-I intermittency.

Let us consider the situation when the value of a is driven by DMN, i.e.:

$$a(n) = a_0 + \xi_{DMN} \tag{9}$$

Let us now assume that the k is much smaller than 0.5 but larger than zero. If we choose a_0 and Δa in such a way, that $a_0 + \Delta a < a_c$ the first of the described in the previous subsection cases appears. In our simulation, we chose $a_0 = 3.8284$ and $\Delta a = 0.00001$. Initially, we set $k = 0.01$. The maximum laminar phase length for a_0 is equal to 75, so that it is comparable with the inverse of k. In such a case, a broadening of the right peak is obtained — and it can be seen in our simulations. Figure 1a) presents the laminar phase length distribution for this case — a histogram of two million laminar phase lengths. A broadening of the right peak is clearly seen.

If we now decrease the value of k to 0.00001, we observe quite a different laminar phase length distribution, presented in Figure 1b). In this case, the splitting of the right peak is clearly seen in good agreement with the results of the previous subsection.

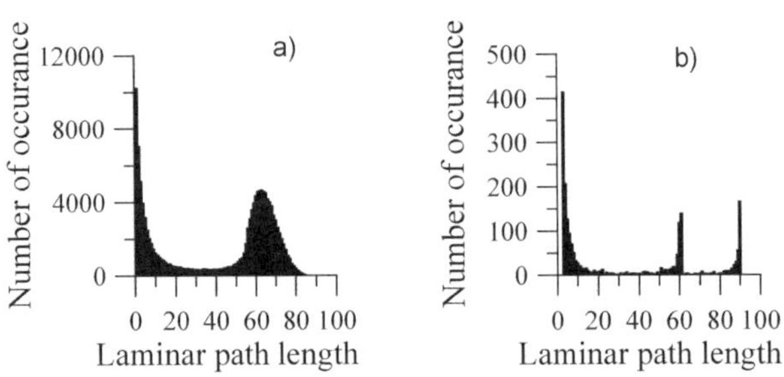

Fig. 1. Laminar phase length distribution for the case of logistic map with with a paramteric DMN for a)$k = 0.1$ and b) $k = 0.00001$. See explanation in text.

The similar effect is observed in continous time sytems. We have made some simulation of Rössler system[9] with parametric DMN and have obtained the same effects, as described above: right peak broadenning and splitting, depending on the value of transition rate k.

3. Effect of Dichotomous Markov Noise on the Properties of a Boundary Crisis

3.1. *General remarks*

In the case of boundary crisis, we are typically interested in the value of mean resident time (RT) on the pre-crisis attractor and in the RT distribution. It is well known that the RT distribution is exponential:[1]

$$P(t) = \frac{1}{\tau} \exp\left(-\frac{t}{\tau}\right) \tag{10}$$

where τ is the mean RT. This allows us to model the behavior of the system with a boundary crisis by means of a Markov process. The probability of of the system remaining on pre-crisis attractor is given as the conditional probability:

$$p(\Delta t) = \frac{P(t + \Delta t)}{P(t)} = \exp\left(-\frac{\Delta t}{\tau}\right) \tag{11}$$

If we consider a discrete-time system, we set $\Delta t = 1$. If we also assume that the τ is large enough, we obtain:

$$p = p(\Delta t = 1) = \exp\left(-\frac{1}{\tau}\right) \cong 1 - \frac{1}{\tau} \tag{12}$$

and the probability of an escape from the attractor is given as:

$$q = 1 - p \cong \frac{1}{\tau} \tag{13}$$

Assume again that one of the parameters described the system a is driven by DMN. Moreover, the mean value of a and the amplitude of DMN are chosen in such a way as to obtain $a_1 = a_0 + \Delta a > a_c$ and $a_2 = a_0 - \Delta a < a_c$. a_c is the critical value of the parameter a at which the attractor is destroyed as an effect of its collision with a hyperbolic saddle residing on the border of the basin of attraction. Of course, if $a = a_2$, the escape of the trajectory of the system from the pre-crisis attractor is not possible. The transition rate (or the probability of a change of the state for a discrete-time system) is set equal to k. Below we will take into account only the discrete-time systems. A generalization to continuous-time systems is obvious.

Taking all these consideration into account we may model our system by means of a three-state Markov process:

- State 1: $a = a_2$. In this state the trajectory is on the pre-crisis attractor.

- State 2: $a = a_1$ and the trajectory is also on the pre-crisis attractor.
- State 3: $a = a_1$ and the trajectory has left pre-crisis attractor.

In the language of Markov chain models, state 3 is called the absorbing state because no transition neither to state 2 nor to state 1 is possible. The other probabilities of transition between states are equal:

$$p_{1\to 1} = 1 - k \tag{14}$$

$$p_{1\to 2} = kp = k\left(1 - \frac{1}{\tau}\right) \tag{15}$$

$$p_{1\to 3} = kq = \frac{k}{\tau} \tag{16}$$

$$p_{2\to 1} = k \tag{17}$$

$$p_{2\to 2} = (1-k)\,p = (1-k)\left(1 - \frac{1}{\tau}\right) \tag{18}$$

$$p_{2\to 3} = (1-k)\,q = \frac{1-k}{\tau} \tag{19}$$

In these expression we have assumed that the change of the value of DMN and the escape of the system from attractor are independent and used the equations (12) and (13). Let us define two parameters τ_1 and τ_2, the mean time of reaching (MTR) of state 3 through the system, while starting from state 1 or 2, respectively. From the theory of Markov chain it is known, that these two parameters are solutions of the following system of equations:[10]

$$\tau_i = 1 + \sum_{j\neq 3} (p_{i\to j}\tau_j) \tag{20}$$

In our case, it leads to the system of two equations:

$$\tau_1 - \frac{\tau - 1}{\tau}\tau_2 = \frac{1}{k} \tag{21}$$

$$-\tau_1 + \left(\frac{\tau - 1}{\tau} + \frac{1}{k\tau}\right)\tau_2 = \frac{1}{k} \tag{22}$$

If k has is nonzero, the above system has a solution:

$$\tau_1 = \frac{1}{k} + 2\,(\tau - 1) \tag{23}$$

$$\tau_2 = 2\tau \tag{24}$$

So, the average time τ_1 is tends to the infinity while k decreases to zero and it is the result is intuitively correct. Of course, for $k = 0$ (stationary case) the trajectory may not leave the pre-crisis attractor. More interesting is the second result — the time τ_2 is constant and independent of the value of k.

3.2. *An example: The logistic map*

It is well-known that the logistic map (8) undergoes the boundary crisis when the parameter a is getting larger than 4. Let us now set the value of $a_0 = 4.0$ and $\Delta a = 0.02$. MTR from state 1 and 2 as a function of k is depicted in Figure 2. We can easily recognize the functions given by the equations (23) and (24). The value of τ_2 is not constant exactly for all the values of k but remains at the same value through threc dccades. The differences between its behavior for k close to zero and the theoretical prediction (24) arises from the simplification used in (12). We also see that this nearly constant value of τ_2 is close to 38, while the value of τ is equal to 19.62, so it is about twice smaller as compared to the Markov chain model prediction.

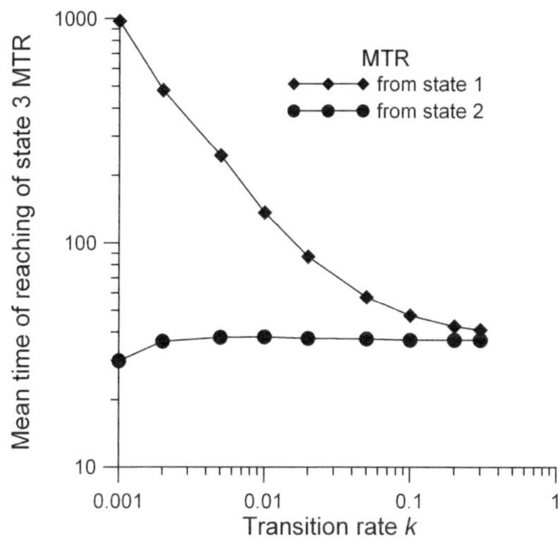

Fig. 2. Mean time of reaching of state 3 for the logistic map with a parametric DMN near the boundary crisis, while starting from state 1. and state 2. (see text for explanations)

4. Conclusion

We investigated the effect of parametric DMN on the properties of the dynamical systems exhibiting two kinds of chaotic transitions: type-I intermittency and boundary crisis.

In the first case, this kind of noise modifies the shape of the distribution of laminar phase lengths. Dependending on the rate transition, we may obtain a broadening of right peak or its splitting into two peaks. Similar phenomena are observed in real systems, e.g. authors of[11] report the case of intermittent chaos observed in human heart rate series. The laminar phase length distribution has a characteristic split right peak. This phenomenon was explained in[12] as a result of periodic parameter change. However, examples of the occurance of DMN in several chemical and biochemical systems are known[13–15] — so it is possible that some biochemical processes present in the human cardiovascular system are responsible for this effect.

In the case of systems with boundary crisis, we have obtained analytically the mean residence time of the trajectory on pre-crisis attractor in dependence on the properties of the parametric DMN. In our considerations, we have used a Markov chain model approximation. The results are in good agreement with numerical simulations.

References

1. E. Ott, Chaos in dynamical systems (Cambridge University Press, New York, 1993).
2. P. Manneville and Y. Pomeau, *Physics Letters A* **75**, p. 1 (1980).
3. D. S. J.E. Hirsch, B.A. Huberman, *Physical Review A* **25**, p. 519 (1982).
4. E. L. S. Choi, *Physics Letters A* **205**, p. 173 (1995).
5. I. Bena, *Journal of Modern Physics B* **20**, p. 2825 (2006).
6. W. Horsthemke and R. Lefever, Noise-induced transitions (Springer-Verlag, Berlin, 1984).
7. A. I. J.M. Gutierrez and M. Rodriguez, *Physical Review E* **48**, p. 2507 (1993).
8. C. V. den Broeck, *Journal of Statistical Physics* **31**, p. 467 (1983).
9. O. Roessler, *Zeitschrift fur Naturforschung A* **31a**, p. 259 (1976).
10. J. Jakubowski and R. Sztencel, Wstep do teorii prawdopodbieństwa (an introduction to the probability theory), in polish (SCRIPT, Warszawa, 2004).
11. J. Żebrowski and R. Baranowski, *Pramana* **64**, p. 543 (2005).
12. J. Gac and J. Żebrowski, *Physical Review E* **73**, p. 066203 (2006).
13. J. C. Giddings and H. Eyring, *Journal of Chemicla Physics* **59**, p. 416 (1955).
14. K. J. Mysels, *Journal of Chemicla Physics* **24**, p. 371 (1956).
15. G. I. Taylor, *Proceedings of Royal Society London* **219**, p. 186 (1953).

A Seven Mode Truncation of the Kolmogorov Flow with Drag: Analysis and Control

G. Gambino, M. C. Lombardo and M. Sammartino

Department of Mathematics
University of Palermo
Via Archirafi, 34
90123 Palermo, Italy

The transition from laminar to chaotic motions in a viscous fluid flow is investigated by analyzing a seven dimensional dynamical system obtained by a truncation of the Fourier modes for the Kolmogorov flow with drag friction. Analytical expressions of the Hopf bifurcation curves are obtained and a sequence of period doubling bifurcations are numerically observed as the Reynolds number is increased for fixed values of the drag parameter. An adaptive stabilization of the system trajectories to an equilibrium point or to a periodic orbit is obtained through a model reference approach which makes the control global. Finally, the effectiveness of this control strategy is numerically illustrated.

Keywords: Kolmogorov flow; finite dimensional approximation; adaptive control.

1. Introduction

In this work we analyze a finite-dimensional approximation of the so-called Kolmogorov flow with the inclusion of a drag friction effect. The introduction of the drag term into the original Kolmogorov model has been proposed to correct some disagreement of the experimental results with the theoretical results.

In the same setting as [1], we consider a seven mode truncation of the Kolmogorov flow with drag friction. The resulting dynamical system contains two parameters: the Reynolds number R and the drag coefficient μ. For fixed values of the drag coefficient we show that when R is increased secondary self-oscillating solutions bifurcate from the stationary solution. We give the analytical expressions of the bifurcation curves (in the μ-R plane) at whose crossings the Hopf bifurcations arise. An estimate of the period of the limit cycles shows that it is a decreasing function of the drag parameter. A further increase of the Reynolds number results in a period doubling cascade leading to transition to chaos. Moreover, following the approach proposed in [2] and [3], we design an adaptive controller which drives the trajectories of the system to an equilibrium point or to a periodic orbit. The adopted strategy consists in synchronizing two equal systems one of which stabilizes into a globally stable fixed point or a periodic orbit. The control is proved to be global by using Lyapunov stability analysis.[4]

2. Finite Dimensional Approximation

We consider the time dependent periodic Navier-Stokes flow excited by an external body force on the torus $T^2 = [0, 2\pi] \times [0, 2\pi]$:

$$\begin{gathered}\frac{\partial \triangle \psi}{\partial t} - \triangle^2 \psi + R\frac{\partial(\triangle\psi, \psi)}{\partial(x,y)} + \mu\triangle\psi = 64\cos(4y) \\ \psi(t, x+2\pi, y) = \psi(t, x, y+2\pi) = \psi(t, x, y)\end{gathered} \tag{1}$$

where ψ is the stream function, R is the Reynolds number and μ is the drag friction coefficient.

The stationary solution of (1) is $\psi_0 = -4/(16+\mu)\cos(4y)$, which depends on the drag friction parameter μ.

The Hopf bifurcation plays a central role in the investigation of the chaotic behavior of a viscous fluid as it often arises at the initial stage of the transition from laminar to turbulent flow.[5] Due to the infinite number of degrees of freedom of the model (1) it is difficult to obtain the analytical proof of the existence of Hopf bifurcations. Then, in the same spirit as [1] we obtain a seven mode approximation of equation (1) by imposing that the stream function has the following expression, derived from the truncation of the Fourier expansion:

$$\psi(t,x,y) = \psi_{0,1}(t)\cos(4y) + \sum_{m=1}^{2}\sum_{n=-1}^{1}\psi_{m,n}(t)\cos(mx + my + 4ny) \tag{2}$$

The resulting seven dimensional nonlinear dynamical system is:

$$\begin{cases} \dot{x}_1 = -(16+\mu)x_1 + R(-x_2x_3 + 3x_3x_4 + 8x_6x_7) - 4 \\ \dot{x}_2 = -(10+\mu)x_2 + R\left(\frac{14}{5}x_1x_3 - \frac{6}{5}x_3x_5 + \frac{36}{5}x_4x_6\right) \\ \dot{x}_3 = -(2+\mu)x_3 \quad - R(6x_1x_2 + 10x_1x_4 + 2x_2x_5 + 14x_4x_7) \\ \dot{x}_4 = -(26+\mu)x_4 + R\left(-\frac{14}{13}x_1x_3 - \frac{4}{13}x_2x_6 + \frac{38}{13}x_3x_7\right) \\ \dot{x}_5 = -(8+\mu)x_5 \quad + R(4x_1x_6 + 2x_2x_3) \\ \dot{x}_6 = -(8+\mu)x_6 \quad - R(4x_1x_5 + 12x_1x_7 + 8x_2x_4) \\ \dot{x}_7 = -(40+\mu)x_7 - R\left(\frac{4}{5}x_1x_6 + \frac{6}{5}x_3x_4\right) \end{cases} \tag{3}$$

where the Fourier coefficients have been relabeled as $x_1 = \psi_{0,1}$, $x_2 = \psi_{1,-1}$, $x_3 = \psi_{1,0}$, $x_4 = \psi_{1,1}$, $x_5 = \psi_{2,-1}$, $x_6 = \psi_{2,0}$ and $x_7 = \psi_{2,1}$. The system (3) admits the equilibrium point $E_0 \equiv (-4/(16+\mu), 0, 0, 0, 0, 0, 0)$ corresponding to the stationary solution ψ_0.

3. Analysis of the Bifurcations

To study the formation of the Hopf bifurcation as the Reynolds number is varied, we analyze the behavior of the solutions in a neighborhood of the equilibrium point E_0. The jacobian matrix evaluated at E_0 is a block diagonal matrix:

$$J = \begin{pmatrix} -(\mu+16) & 0 & 0 \\ 0 & A & 0 \\ 0 & 0 & B \end{pmatrix} \tag{4}$$

where the entries of A are the derivatives with respect to the variables x_2, x_3 and x_4 evaluated at E_0, and the entries of B are the derivatives with respect to x_5, x_6 and x_7 evaluated at E_0:

$$A = \begin{pmatrix} -\mu-10 & -\frac{56}{5(16+\mu)}R & 0 \\ \frac{24}{16+\mu}R & -\mu-2 & \frac{40}{16+\mu}R \\ 0 & \frac{56}{13(\mu+16)}R & -\mu-26 \end{pmatrix}, \quad B = \begin{pmatrix} -\mu-8 & -\frac{16}{16+\mu}R & 0 \\ \frac{16}{16+\mu}R & -\mu-8 & \frac{48}{16+\mu}R \\ 0 & \frac{16}{5(\mu+16)}R & -\mu-40 \end{pmatrix}$$

The negative eigenvalue $\lambda_1 = -16-\mu$ is easily obtained and is independent on the value of R. By applying the Routh-Hurwitz method to the characteristic polynomial of A one gets that at $R = \bar{R}(\mu)$ two eigenvalues cross the imaginary axis and a Hopf bifurcation occurs. We find that for $\mu = 0$, $\bar{R} = 4\sqrt{3510/29} \simeq 44$ corresponds to the value of the Hopf bifurcation estimated numerically in [1]. Using the Hopf theorem we estimate, up to second order in $R - \bar{R}$, the period of the limit cycle around the equilibrium point E_0:

$$T \approx 2\pi \sqrt{\frac{58 - 7\mu}{7\mu^3 + 706\mu^2 + 14516\mu + 61592}} \tag{5}$$

The above expression shows that this period is a decreasing function of μ.

Similar results are obtained by performing the linear analysis for B and the explicit expression of the bifurcation curve $R = \tilde{R}(\mu)$ is found.

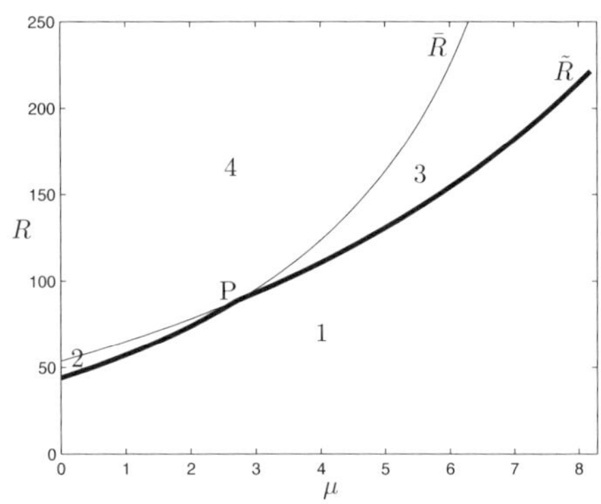

Fig. 1. Bifurcation curves $R = \bar{R}(\mu)$ and $R = \tilde{R}(\mu)$. The intersection point is $P \approx (2.72, 88.88)$.

The two bifurcation curves divide the plane into four regions in which the dynamical behavior of the system can be classified according to the sign of the eigenvalues. In region 1 all the eigenvalues have negative real part, then the system variables asymptotically converge to E_0. Since both $\bar{R}$ and $\tilde{R}$ are increasing functions of μ, the linear term modelling the drag friction acts as a stabilizer of the chaotic behavior.

In region 2, two eigenvalues of the matrix A have positive real part and all the other eigenvalues of the jacobian matrix have negative real part. On the left of Fig. 2 the numerical simulations show the limit cycles evolving from $E_0^+ \equiv (-4/(16+\mu), 0, 0.1, 0, 0, 0, 0)$ and $E_0^- \equiv (-4/(16+\mu), 0, -0.1, 0, 0, 0, 0)$, respectively. The initial conditions E_0^+ and E_0^- are chosen to describe small perturbations around the steady state solution in the direction of the unstable mode $\psi_\pm(0, x, y) = \psi_0 \pm 0.1\cos(x + y)$.[6] We have chosen to start from the two different initial conditions E_0^+ and E_0^- to capture the two limit cycles. For the parameter value $R = 48$ the trajectories evolve toward a unique stable limit cycle, independently from the initial conditions E_0^+ or E_0^-. However the limit cycle is approached counter-clockwise when one starts from E_0^+ and clockwise when one starts from E_0^-.

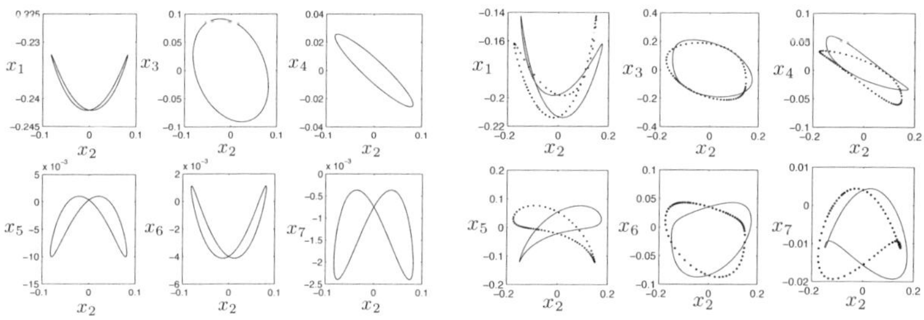

Fig. 2. Limit trajectories, evolving from E_0^+ and E_0^-, into the (x_2, x_i) planes. *Left*: The parameters $\mu = 0.2$ and $R = 48$. *Right*: The parameters $\mu = 0.2$ and $R = 58$.

For a higher value of the parameter R one can observe the behavior shown on the right of Fig. 2, where two symmetric (with respect to the transformation $(x_2, x_3, x_4) \to (-x_2, -x_3, -x_4)$) limit cycles are clearly visible. As the Reynolds number increases, each of these two limit cycles experiences a period doubling bifurcation (as shown on the left of Fig. 3) which in turn bifurcates into four–doubling and again into eight–doubling cycles. The cascade ends in the chaotic behavior shown on the right of Fig.3. The results of the numerical analysis are quite similar to the results of [1] where the authors considered the case $\mu = 0$. The same qualitative behavior can be observed for any $0 \leq \mu < 2.72$.

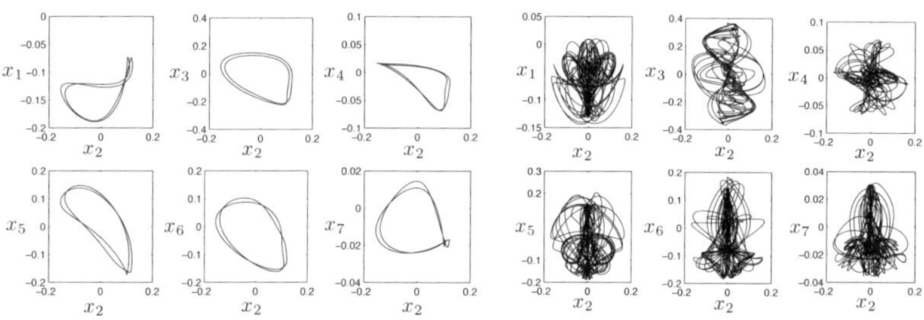

Fig. 3. *Left*: Two-doubling limit cycles in the (x_2, x_i) planes. The parameters $\mu = 0.2$ and $R = 69.5$. *Right*: The chaotic regime. The parameters $\mu = 0.2$ and $R = 160$.

Considering values of the drag coefficients $\mu > 2.72$ so that the first Hopf bifurcation will arise as the parameter R crosses the curve $R = \tilde{R}(\mu)$, one passes, in the $\mu - R$ space, from region 1 to region 3 (see Fig. 1). The qualitative behavior is here different because the oscillatory behavior is observed for the variable (x_5, x_6, x_7) (we recall that in this case the transition is due to the fact that the real parts of the eigenvalues of the matrix B become positive). The variables (x_2, x_3, x_4) are rapidly dampened to zero, while x_1 oscillates with twice the frequency of (x_5, x_6, x_7) as these variables appears in the dynamics of x_1 as a quadratic forcing term. The same route to chaos observed for $0 \le \mu < 2.72$ (for increasing R, a cascade of period-doubling with a final transition to a chaotic regime) can be observed by choosing μ to the right of the point P in Fig. 1. In these cases the time of convergence to the attractors is sensitively bigger than the relaxation time found in the cases with $\mu < 2.72$.

4. The Control Strategy

The method of adaptive controller used here has been introduced in [2] to synchronize and control dynamical systems. Here we give a brief outline of the method and refer the interested reader to [2] and [3]. Suppose one has two systems:

$$\dot{x} = f(x,t) + Bu \tag{6}$$

$$\dot{y} = g(y,t) \tag{7}$$

with $x \in \mathbb{R}^n$, $y \in \mathbb{R}^n$, $u \in \mathbb{R}^m$, and $B \in \mathbb{R}^{n \times m}$. In what follows the matrix B is chosen such that $\|B\| \le 1$.

The solutions of the so-called reference system (7) evolve on an attractor (e.g. an equilibrium point, or a limit cycle, or a chaotic attractor). The goal is to design an appropriate feedback controller $u = u(t)$ such that the solutions of the so-called plant system (6) asymptotically converge to the solutions of the reference system. Said differently one seeks a controller u such that the origin is a global attractor for the error equation $\dot{e}(t) = f(x,t) - g(y,t) + Bu$. Let us assume that the error dynamics can be reformulated as follows:

$$\dot{e}(t) = Le + m(x,t) - n(y,t) + Bu. \tag{8}$$

where the nonlinear functions $m(x,t), n(x,t) \in C^1$.

Assumption 1 Every trajectory of the reference system converges monotonously (in the sense that the distance is a decreasing function of time) toward a closed bounded attractor $\mathcal{A}$.

Assumption 2 A gain matrix $K \in \mathbb{R}^n$ exists such that $\hat{L} \equiv L - BK$ is a Hurwitz matrix, i.e. all its eigenvalues have a strictly negative real part.

Theorem 4.1. *Let $P \in \mathbb{R}^{n\times n}$ be a symmetric positive definite solution of the Lyapunov equation:*

$$P\hat{L} + \hat{L}^T P + I = 0, \tag{9}$$

(whose existence is guaranteed by the lemma in [7] , p. 210). Under the Assumptions 1 and 2, the controller u defined by

$$u(t) = -Ke(t) + k(t) \left\| B^T Pe \right\|^{-1} B^T Pe \,, \tag{10}$$

where k solves $\dot{k}(t) = -\|Pe\|$, guarantees that for every initial condition $(e(0), k(0)) = (e^0, k^0)$: $\lim_{t\to\infty} k(t) = k^* < \infty$ and $\lim_{t\to\infty} e(t) = 0$.

For the proof see [4]. Notice that the control law (10) requires $m(x,t)$ and $n(x,t)$ to be regular but, differently from [2], [3] and [8], it does not need any specific structure for the nonlinearities.

For the purpose of the adaptive control of the system (3), we introduce the following change of variables in such a way that the equilibrium state E_0 is translated to the origin:

$$y_1 = x_1 - \frac{4}{16+\mu}; \;\; y_i = x_i \text{ for } i = 2, \ldots, 7. \tag{11}$$

Now we consider our system of equations:

$$\dot{Y}^{(1)} = J_1 Y^{(1)} + V_1 + Bu \tag{12}$$

$$\dot{Y}^{(2)} = J_2 Y^{(2)} + V_2 \tag{13}$$

where $Y^{(i)} \equiv \left(y_1^{(i)}, \ldots, y_7^{(i)}\right)$, $i = 1, 2$, J_1 and J_2 are the linear part of the systems; they both have the same structure given by the diagonal block matrix J as in (4), and they differ for the parameters values μ_i and R_i.

The V_i contain the nonlinear terms and they have the same structure for the reference system and the plant system, the only difference being the values of the parameters R_i:

$$V_i = R_i \begin{pmatrix} -y_2^{(i)}y_3^{(i)} + 3y_3^{(i)}y_4^{(i)} + 8y_6^{(i)}y_7^{(i)} \\ \frac{14}{5}y_1^{(i)}y_3^{(i)} - \frac{6}{5}y_3^{(i)}y_5^{(i)} + \frac{36}{5}y_4^{(i)}y_6^{(i)} \\ -6y_1^{(i)}y_2^{(i)} - 10y_1^{(i)}y_4^{(i)} - 2y_2^{(i)}y_5^{(i)} - 14y_4^{(i)}y_7^{(i)} \\ -\frac{14}{13}y_1^{(i)}y_3^{(i)} - \frac{4}{13}y_2^{(i)}y_6^{(i)} + \frac{38}{13}y_3^{(i)}y_7^{(i)} \\ 4y_1^{(i)}y_6^{(i)} + 2y_2^{(i)}y_3^{(i)} \\ -4y_1^{(i)}y_5^{(i)} - 12y_1^{(i)}y_7^{(i)} - 8y_2^{(i)}y_4^{(i)} \\ -\frac{4}{5}y_1^{(i)}y_6^{(i)} - \frac{6}{5}y_3^{(i)}y_4^{(i)} \end{pmatrix} \tag{14}$$

The values μ_2 and R_2 for the reference system will be chosen so that the reference system will evolve toward an equilibrium point or toward a stable limit cycle. Our problem is to find a suitable adaptive controller u such that the chaotic system (12) is forced to behave like (13). The error equation in this case is given by:

$$\dot{e} = Le + V + Bu, \tag{15}$$

where $V \equiv V_1 - V_2$, $L \equiv J_1 - J_2$ and $e \equiv (e_1, \ldots, e_7)$, with $e_i = y_i^{(1)} - y_i^{(2)}$.

We choose to put the control input only in two equations of the given error system, in particular we define $B = (b_{ij}) \in \mathbb{R}^{7\times 2}$ with $b_{41} = b_{72} = 1$ and all the other entries equal to zero. We then find the gain matrix $K = (k_{ij}) \in \mathbb{R}^{2\times 7}$ such that the only nonzero entries are $k_{13} = 56R/(13(\mu+16))$ and $k_{26} = 16R/(5(\mu+16))$. It is straightforward to verify that the resulting matrix $\hat{L} = L - BK$ is a Hurwitz matrix.

Since the hypotheses of the theorem are satisfied, we design the control:

$$\begin{pmatrix} u_1 \\ u_2 \end{pmatrix} = \begin{pmatrix} -\frac{56R}{13(\mu+16)}e_3 + k(t) \\ -\frac{16R}{5(\mu+16)}e_6 + k(t) \end{pmatrix} \tag{16}$$

with $\dot{k} = -\|Pe\|$. The entries of matrix P are obtained by solving the Lyapunov equations (9). By the choice of the matrix B it follows that u_1 is inserted in the fourth equation of the error system (15) and u_2 into the seventh equation. Using this control, the error asymptotically decays to zero and $k(t)$ tends to a bounded value.

On the left of Fig. 4 we show how the global control (16) drags the trajectories of the plant system to the origin, when the reference model evolves towards the steady state. The control input decays to zero after only a few switchings. The right of Fig. 4 shows the effectiveness of the control (10) when the evolution of the reference system is periodic. In performing the numerical simulations we have chosen to use a more regular form of the control by substituting the norm in the evolution equation for $k(t)$ with its square, namely $\dot{k}(t) = -\|Pe\|^2$.

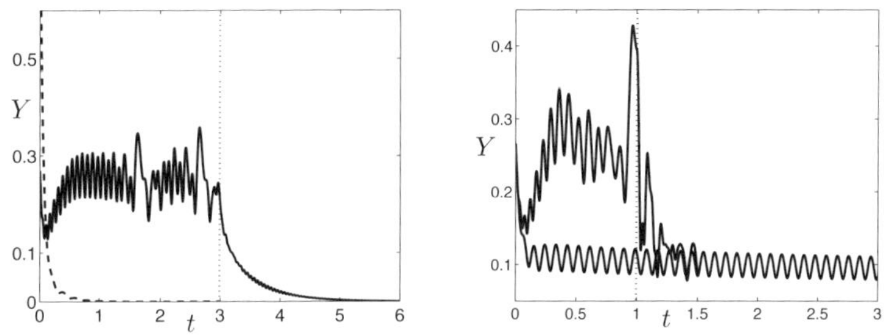

Fig. 4. Here $Y \equiv \sqrt{\sum_{i=1}^{7} \|y_i\|^2}$. *Left*: Adaptive control to a steady state. The controller starts to work at $t = 3$. *Right*: Adaptive control to a periodic evolution. The controller starts to work at $t = 1$.

References

1. Z. Chen and W. Price, *Chaos Solitons Fractals* **26**(4), 1195 (2005).
2. M. D. Bernardo, *Phys. Lett. A* **214**(4), 139 (1996).
3. M. D. Bernardo, *Internat. J. Bifur. Chaos Appl. Sci.* **6**(6), 557 (1996).
4. G. Gambino, M. C. Lombardo and M. Sammartino, *in press Chaos Solitons Fractals* **doi:10.1016/j.chaos.2007.11.003** (2008).
5. L. Landau, *C. R. Acad. Sci. USSR* **44**, 311 (1944).
6. Z. Chen, *Comm. Math. Phys.* **201**(1), 117 (1999).
7. S. Sastry, *Nonlinear Systems. Analysis, Stability and Control* (Springer–Verlag, New York, 1999).
8. D. Wagg, *Meccanica* **38**(2), 227 (2003).

Asymptotic Expansions of Transition Densities of the Superconducting Phase Differences in d. c. SQUID in Presence of White Noise

V. A. Gasanenko

Institute of Mathematics, National Academy of Sciences of Ukraine, Kyiv, Ukraine

R. De Luca and A. Fedullo

DMI, Universita degli Studi di Salerno, 84084 Fisciano (SA), Italy

We investigate the system of stochastic differential equation which describe the time phase evolution of a two-junction quantum interferometer in the presence of white noises. The effective algorithms of expansion of special Fokker-Plank equation with small and large parameters are proved. The estimations of approximations are obtained.

Keywords: Two Josephson junctions; current-phase relations; Brownian motion; two dimensional Fokker-Plank equation; small and large parameters; asymptotic expansion of solution.

1. Introduction

Recently many authors have been on solution of stochastic differential equations connected with the properties of d. c. SQUIDs (Superconductive QUantum Interference Devices); see ref. [2] and the references therein, for example. SQUID is superconducting loops interrrupted by two Josephson junctions (JJ). In the present work we describe the dynamical properties of d. c. SQUIDs in presence of stochastic action by means of the knowledge of the dynamics of the superconducting phases differences φ_1, φ_2 across two JJs, in analogous way as in ref. [1]. The basic equations with added white noises have the form

$$\frac{\Phi_0}{2\pi R}\frac{d\varphi_k}{dt} + I_j \sin\varphi_k = I_k + \sqrt{\gamma_1}dw_k, \quad k = 1, 2 \tag{1}$$

where R and I_j are the resistive parameter and the maximum Josephson current, respectively, I_k, $k = 1, 2$ are currents in two loop branches, Φ_0 is the elementary flux quantum , $\gamma_1 = (2k_BT)/R$, and w_k, $k = 1,2$ are independent Brownian motions. By introducing new processes $\xi_1(\tau) = \frac{\varphi_1(\tau)+\varphi_2(\tau)}{2}$ and $\xi_2(\tau) = \frac{\varphi_1(\tau)-\varphi_2(\tau)}{2}$ at normalized time $\tau = (2\pi RI_j/\Phi_0)t$ we can rewrite (1) in the following form

$$d\xi_k = -\frac{\partial W}{\partial x_k}(\xi_1, \xi_2)d\tau + \sqrt{\gamma}dw_k(\tau),\ k = 1,2 \tag{2}$$

where

$$W(x_1, x_2) = (1 - \cos x_2 \cos x_1) - \frac{i_B}{2}x_1 + \frac{1}{4\pi\beta}(x_2 + \pi\psi_{ex} - n\pi)^2,$$

$\gamma = \frac{k_BT}{E_j}$, $E_j = \frac{I_j\Phi_0}{2\pi}$, n is integer and i_B, β, ψ_{ex} are normalized current, inductance and applied magnetic flux respectively(see [1]). It is assumed that $\psi_{ex}(\tau) \in C^1$.

The main purpose of our investigation of (1), (2) is to obtain the procedure of computation of function $\frac{d}{d\tau}\langle\xi_1(\tau)\rangle_\beta$, where $\langle\eta\rangle$ denotes the expectation of random value η. This function defines the voltage (see [1]) across the d. c. SQUID.

The system of stochastic differential equations (2) with initial condition $(\xi_1(0), \xi_2(0)) = (x_{10}, x_{20}) =: x_0$ defines the following two dimensional Fokker-Plank equation for the probability density $p(\tau, x) := p(\tau, x_1, x_2)$ of distribution function $P(\xi_1(\tau) < x_1, \xi_2(\tau) < x_2)$

$$\frac{dp}{d\tau} = \nabla \cdot (p(\tau, x)\nabla W) + \frac{\gamma}{2}\triangle p(\tau, x),\ p(0, x) = \delta(x - x_0), \tag{3}$$

where $\nabla = \left(\frac{\partial}{\partial x_1}, \frac{\partial}{\partial x_2}\right)$ and $\triangle$ denotes the Laplace operator.

Sections 2 and 3 are devoted to the study of asymptotic expansions of solutions of problem (3) under $\beta \to 0$ and $\beta \to \infty$ respectively.

2. Asymptotic Expansion under Small β

We introduce the new variables and rescaled distribution:

$$\tau = t\beta^2 2\gamma^{-1},\ x_1 = z_1\beta,\ x_2 = z_2\beta, q(t, z_1, z_2) = p(t\beta^2 2\gamma^{-1}, z_1\beta, z_2\beta).$$

We may notice that the initial vector z_0 is equal to $(x_{01}\beta^{-1}, x_{20}\beta^{-1})$.

After such change we can write the problem (3) in the following form

$$\frac{\partial}{\partial t}q = L_0 q + \beta L_1 q(t, z) + \beta^2 L_2 q(t, z) + \beta^3 L_3 q(t, z),\ q(0, z) = \delta(z - z_0). \tag{4}$$

The operator L_0 is defined by the following way

$$L_0 : C^{(1,2,2)}([0,\infty)\times R\times R)\to C^{(1,2,2)}([0,\infty)\times R\times R),$$
$$L_0 f(t,z_1,z_2) = \left(\Delta - \gamma^{-1}\psi_e(t\beta^2 2\gamma^{-1})\frac{\partial}{\partial z_2}\right) f,$$

here

$$\Delta := \frac{\partial^2}{\partial z_1^2} + \frac{\partial^2}{\partial z_2^2}; \quad \psi_e(s) = \psi_{ex}(s) - n.$$

The operators with higher index are defined by the following way

$$L_i : C^1([0,\infty)\times R\times R)\to C^1([0,\infty)\times R\times R), i = 1,2,3.$$
$$L_1 f(t,z_1,z_2) = \gamma^{-1}\left(z^2\pi^{-1}\frac{\partial}{\partial z_2} - i_B\frac{\partial}{\partial z_1} + \pi^{-1}\right) f;$$
$$L_2 f(t,z_1,z_2) = \left(4\gamma^{-1}\cos(\beta z_1)\cos(\beta z_2)\right) f.$$
$$L_3 f(t,z_1,z_2) = 2\gamma^{-1}\left(\cos z_2\beta\frac{\sin z_1\beta}{\beta}\frac{\partial}{\partial z_1} + \cos z_1\beta\frac{\sin z_2\beta}{\beta}\frac{\partial}{\partial z_2}\right) f.$$

We note that, by the above definition, the following desired formulas are true

$$\begin{aligned}\langle\xi_1(t2\gamma^{-1}\beta^2)\rangle_\beta &= \int\limits_R x_1 \int\limits_R p(\tau,x_1,x_2)dx_1dx_2\\ &= \beta^3\int\limits_R z_1\int\limits_R q(t,z_1,z_2)dz_1dz_2\end{aligned}$$

and

$$\frac{d}{d\tau}\langle\xi_1(\tau)\rangle_\beta = \frac{\beta\gamma}{2}\frac{d}{dt}\left(\int\limits_R z_1\int\limits_R q(t,z_1,z_2)dz_1dz_2\right). \qquad (5)$$

Let $E(z,y,t,s)$, $t\geq s$ be the fundamental solution of the problem

$$\frac{\partial E}{\partial t}q = L_0^{z,t}E;\ E = \delta(z-y),\ \text{if } t = s,$$

here z,y are two dimensional vectors.

Let functions $q_k(t,z)$ are defined by the following formulas recurrently

$$q_0(t,z) = E(z, z_0, t, 0)$$

$$q_1(t,z) = \int_0^t ds \int_{R\times R} E(z,y,t,s) L_1 q_0(s,y) dyds$$

$$\vdots \tag{6}$$

$$q_k(t,z) = \int_0^t \int_{R\times R} E(z,y,t,s)$$

$$\times \left(L_1 q_{k-1}(s,y) + L_2 q_{k-2}(s,y) + L_3 q_{k-3}(s,y)\right) dyds, \quad k \geq 3.$$

Theorem 2.1. *Let $q(t,z)$ be solution of problem (4), then the following inequality take place at $\beta \to 0$*

$$\sup_{t,z} |q(t,z) - \sum_{k=0}^{n} \beta^k q_k(t,z)| \leq R\beta^{n+1},\ R < \infty.$$

Proof. We will be seek the function $q(t,z)$ in following form

$$q(t,z) = \sum_{k\geq 0} \beta^k q_k(t,z). \tag{7}$$

The functions q_k satisfies the normalizing conditions

$$q_0(z) = \delta(z - z_0); \quad q_k(0,z) = 0, \quad \sup_{t,z} |q_k(t,z)| < \infty, k \geq 1.$$

The above expression follow from the normalizing conditions for $q(t,z)$.
Substituting (7) in (6) and considering the above normalizing, conditions we obtain the system by which determination of the of the functions q_k is possible:

$$\frac{\partial}{\partial t} q_0 = L_0 q_0, \text{ under conditions } q_0(0,z) = \delta(z - z_0), \int_{R\times R} q_0(0,z)dz = 1;$$

$$\frac{\partial}{\partial t} q_1 = L_0 q_1 + L_1 q_0(t,z); \quad \frac{\partial}{\partial t} q_2 = L_0 q_2 + L_1 q_1(t,z) + L_2 q_0(t,z);$$

$$\frac{\partial}{\partial t} q_k = L_0 q_k + L_1 q_{k-1}(t,z) + L_2 q_{k-2}(t,z) + L_3 q_{k-3}, \quad k \geq 3 \tag{8}$$

under conditions $q_k(0,z) = 0$, $\sup_{t,z} |q_k(t,z)| < \infty$, $k \geq 1$.

It is well known, from theory of parabolic equations, that from (6) unique solutions of system (8) can be found for the functions $q_k(t,z)$.

Now we can consider the time evolution of this approximation. Let $R_n(t,z)$ denote the difference $R_n(t,z) = q(t,z) - \sum_{k=1}^{n} q_k(t,z)$, where $q_k(t,z)$ are defined by (6). Let H denote the operator of the right part of (4): $H := L_0 + \beta L_1 + \beta^2 L_2 + \beta^3 L_3$. By construction, the function $R_n(t,z)$ is the solution of the following problem

$$\begin{aligned} \frac{\partial}{\partial t} R_n(t,z) &= HR_n(t,z) \\ &-\beta^{n+1}(L_1 q_n + L_2 q_{n-1} + L_3 q_{n-2}) \\ &-\beta^{n+2}(L_2 q_n + L_3 q_{n-1}) - \beta^{n+3} L_3 q_n \\ =: HR_n(t,z) &+ \beta^{n+1} K_n(t,z);\ R_n(0,z) = 0,\ \sup_{t,z} |R_n(t,z)| < \infty. \end{aligned} \tag{9}$$

Let us denote by $F(z,y,t,s)$ the fundamental solution of problem (4). The results of Section IV of monograph [3] allow us to claim that solution of (9) is unique and has the following form

$$R_n(t,z) = \beta^{n+1} \int_0^t ds \int_{R\times R} F(z,y,t,s) K_n(s,y) dy.$$

Since coefficients of (9) are smooth, we can use the following estimate of a fundamental solutions $F(x,\xi,t,\tau)$ from the same monograph [3] (Section IV).

$$|D_t^r D_x^s F(x,\xi,t,\tau) \le c(t-\tau)^{-\frac{n+2r+s}{2}} \exp\left(-C\frac{|x-\xi|^2}{t-\tau}\right), \tag{10}$$

where $2r+s \le 2$, $t > \tau$. From latter inequality, from construction of the function K_n and given that $F(z,y,t,s)$ obeys of (10), we have the following upper bound $\sup_{t,z} |R_n(t,z)| \le \beta^{n+1} R$, $R < \infty$. Proof is completed. □

3. Asymptotic Expansion under Large β

In this case the parameter β is large, we make the following changes of variables

$$\beta = \epsilon^{-1},\ x_1 = z_1\epsilon,\ x_2 = z_2\epsilon,\ \tau = \gamma 2^{-1}\epsilon^2 t,\ p(\gamma 2^{-1}\epsilon^2 t, z_1\epsilon, z_2\epsilon) = h(t,z_1,z_2).$$

It is assumed that $\beta \to \infty$, which gives $\epsilon \to 0$.

Let z_0 denotes the vector $(x_{10}\beta, x_{20}\beta)$.

After these changes are made, the problem (3) transforms into the equivalent problem

$$\frac{\partial}{\partial t} h = \Delta h + \epsilon V_1 h + \epsilon^2 V_2 h + \epsilon^3 V_3 h; \quad h(0,z_1,z_2) = \delta(z - z_0). \tag{11}$$

We shall define operators V_1, V_2, V_3 in a similar way as done with operators L_i in Section 2.

$$V_i : C^{1,1,1}([0,\infty) \times R \times R) \to C^{1,1,1}([0,\infty) \times R \times R), i = 1,2,3.$$

$$V_1 f(t, z_1, z_2) = -\gamma^{-1} i_B \frac{\partial}{\partial z_1} f,$$

$$V_2 f(t, z_1, z_2) = 2\gamma^{-1} \left(\cos(\epsilon z_1)\cos(\epsilon z_2) + 2^{-1}\psi_e(\gamma 2^{-1}\epsilon^2 t)\right) f,$$

$$V_3 f(t, z_1, z_2) = 2\gamma^{-1} z_2 \frac{\partial f}{\partial z_2} + (\gamma\pi)^{-1} f$$

$$+ 2\gamma^{-1} \left(\cos z_2 \epsilon \frac{\sin z_1 \epsilon}{\epsilon} \frac{\partial}{\partial z_1} + \cos z_1 \epsilon \frac{\sin z_2 \epsilon}{\epsilon} \frac{\partial}{\partial z_2} \right) f.$$

Set now $z = (z_1, z_2), y = (y_1, y_2)$ and introduce the following function

$$\Gamma(t, z) = \frac{1}{4\pi t} \exp\left(-\frac{z^2}{4t}\right) \text{ if } t > 0 \quad \text{and} = 0 \quad \text{if } t \le 0, \text{ here } z^2 = \sum_{i=1}^{2} z_i^2.$$

Let functions $h_k(t, z)$ are defined by the following formulas recurrently

$$h_0(t, z) = \Gamma(t, z - z_0),$$

$$h_1(t, z) = \int_0^t ds \int_{R\times R} \Gamma(t - s, z - y) V_1 h_0(s, y) dy,$$

$$h_2(t, z) = \int_0^t ds \int_{R\times R} \Gamma(t - s, z - y) \left(V_2 h_0(s, y) + V_1 h_1(s, y)\right) dy,$$

$$h_k(t, z) =$$

$$\int_0^t ds \int_{R\times R} \Gamma(t - s, z - y) \left(V_3 h_{k-3} + V_2 h_{k-2} + V_1 h_{k-1}\right)(s, y) dy, \ k \ge 3.$$

Theorem 3.1. *Let $h(t, z)$ be the solution of problem (11), then the following inequality take place at $\beta \to \infty$*

$$\sup_{t,z} |q(t, z) - \sum_{k=0}^{n} \beta^{-k} h_k(t, z)| \le R\beta^{-(n+1)}, \ R < \infty.$$

This theorem can be proved in much the same manner as Theorem 1. For this case the fundamental solution is presented by the function $\Gamma(t, z)$.

In complete analogy with (5), we may obtain the following formulas

$$\langle \xi_1(\gamma 2^{-1}\epsilon^2 t)\rangle_\beta = \int\limits_R x_1 \int\limits_R p(\tau, x_1, x_2)dx_1dx_2$$
$$= \beta^{-3}\int\limits_R z_1 \int\limits_R h(t, z_1, z_2)dz_1dz_2.$$

and

$$\frac{d}{d\tau}\langle \xi_1(\tau)\rangle_\beta = \frac{\beta^{-1}\gamma}{2}\frac{d}{dt}\left(\int\limits_R z_1 \int\limits_R h(t, z_1, z_2)dz_1dz_2\right)$$

Let us consider the fundamental solution E from the Section 2 more detail. First of all, note that we may face a special situation when differential operator $L_{t,z}$ is given as sum of two operators

$$L_0^{t,z} f = L_0^{t,z_1} f + L_0^{t,z_2} f$$

where every operator depends only on one space variable and on time, so that

$$L_0^{t,z_1} g_1(z_1, t) = \frac{\partial^2}{\partial z_1^2} g_1; \quad L_0^{t,z_2} g_2(z_2, t) = \left(\frac{\partial^2}{\partial z_2^2} + \gamma^{-1}\psi_e(t2\gamma^{-1}\beta^2)\frac{\partial}{\partial z_2}\right) g_2.$$

The fundamental solution $E(z, y, t, s)$ for the operator $L_0^{z,t}$ is presented by the following product $E(z, y, t, s) = E_1(z_1, y_1, t, s)E_2(z_2, y_2, t, s)$, where $E_i(z_i, y_i, t, s)$ is fundamental solution, satisfying the one-dimensional equation

$$\frac{\partial}{\partial t}E_i(z_i, y_i, t, s) - L_0^{t,z_i}E_i(z_i, y_i, t, s) = 0; \ E_i(z_i, y_i, t, s) = \delta(z_i - y_i), \ t = s.$$

It is clear that $E_1(z_1, y_1, t, s) = \frac{1}{2\sqrt{\pi(t-s)}} \times \exp\left(-\frac{(z_1-y_1)^2}{4(t-s)}\right)$.

For construction of fundamental solution E_2 it may use algorithm from monograph [3]. It should be observed here that Theorems 1, 2 ensure the general approach to expansion of solution of the problems (4) and (11). We can obtain more detail schemes of expansions if we shall consider the expansions of functions-coefficients of operators L_i and V_i by small parameters.

4. Conclusions

We have determined Fokker-Plank equations for the probability density function of the superconducting phase differences of a two junction quantum interferometer in presence of white noises. We showed that opportune choice of scales for time and space variables allows to construct effective algorithm of approximation of solutions of these equations in the cases of small or large inductance values.

References

1. F. Romeo and R. De Luca, Effictive non-sinosoidal curren-phase dependence in conventional d.c. SQUID. *Physica Leters A*, 328: 330-334, 2004.
2. Ya. S. Greenberg, Theory of the voltage-current characterictic of high T_C DC SQUIDS. *Physica C*, 371: 156-172, 2002.
3. O. A. Ladyijenskai, V. A. Solonnikov and N. N. Uralceva, Linear and quasi-linear equations of parabolic type. Nauka, Moskow, 1967.

Chaos in the Brain: Novel Methodologies for Epilepsy Diagnosis and Seizure Detection

S. Ghosh-Dastidar* and H. Adeli†

The Ohio State University, Columbus, OH 43210 USA
*Emails: *ghosh-dastidar.3@osu.edu, †adeli.1@osu.edu*

This article is a summary of the authors' research on the application of the chaos theory along with wavelets and neural networks for automated EEG-based epilepsy diagnosis and seizure detection. The authors developed a novel wavelet-chaos-neural network methodology for classification of EEGs into healthy, ictal, and interictal EEGs based on the premise that EEG sub-bands may yield more accurate information compared with EEGs about constituent neuronal activities underlying the EEG. Certain changes in the EEGs that are not evident in the original full-spectrum EEG may be amplified when each sub-band is analyzed separately. Wavelet analysis is used to decompose the EEG into *delta*, *theta*, *alpha*, *beta*, and *gamma* sub-bands. Three parameters are employed for EEG representation: standard deviation (quantifying the signal variance), correlation dimension, and largest Lyapunov exponent (quantifying the non-linear chaotic dynamics of the signal). It was discovered that a particular mixed-band feature space consisting of nine parameters and a Levenberg-Marquardt Backpropagation Neural Network (LMBPNN) resulted in a high classification accuracy of 96.7%. In addition, a novel principal component analysis-enhanced (PCA) cosine radial basis function neural network classifier has been developed that yields comparable classification accuracy but is more robust with a lower standard deviation to changes in training data compared to LMBPNN. These statistics are especially remarkable because even the most highly trained neurologists do not appear to be able to detect interictal EEGs more than 80% of the times.

1. Introduction

Epilepsy is a common disorder affecting approximately 1% of the population in the United States and is commonly accompanied by intermittent abnormal firing of neurons in the brain leading to recurrent and spontaneous seizures (NINDS, [1]). Brain activity during a seizure (ictal state) is different from the activity in the normal state with respect to frequency and pattern of neuronal firing. In contrast to a neuronal firing rate of 80 times per second in a normal state, neurons may fire as fast as 500 times per second during a seizure. Moreover, the spatio-temporal pattern of neuronal firing gradually evolves from a normal state, first to a preictal (interictal) state and then to an ictal state.[2,3] Another characteristic of the interictal state is that neurons in multiple parts of the brain start firing in synchronization with those in the epileptogenic focus which leads to the ictal state.

At present, epileptic seizure detection and epilepsy diagnosis are performed manually based on visual examinations of electroencephalograms (EEGs) by highly trained neurologists. In epilepsy monitoring units, seizure detection is also performed by automated computer models. However, such models require close human supervision due to frequent false alarms and missed detections. Epilepsy diagnosis is more complicated due to excessive myogenic artifacts, interference, overlapping symptomatology with other neurological disorders, and low understanding of the precise mechanism responsible for epilepsy and seizure propagation.

Although neurologists routinely employ visible EEG markers such as interictal spikes to identify epileptic/interictal EEG, these spikes have subtly variable shapes that are not detected consistently by existing spike detection algorithms.[4–6] Recently, the problem has also been studied using an approach that does not rely on visible markers. This approach is based on the invisible spatio-temporal evolution of the EEG signals as the system moves from a seizure-free to a preictal and then to an ictal state.[2,7–9]

In our research, the clinical epilepsy and seizure detection problem is modeled as a three-group classification problem. The three subject groups are: a) healthy subjects (normal EEG), b) epileptic subjects during a seizure-free interval (interictal EEG), and c) epileptic subjects during a seizure (ictal EEG). Epilepsy diagnosis is modeled as the classification of normal EEGs and interictal EEGs. Seizure detection is modeled as the classification of interictal and ictal EEGs.[8,9]

An effective automated model to solve this classification problem requires a thorough investigation of two aspects. Firstly, an appropriate *classification feature space* (i.e., classification parameter or combinations of parameters) has to be identified that can facilitate the desired classification with high accuracy and robustness. Secondly, a *classifier* has to be designed that can accurately model the classification problem based on the identified feature space. These two aspects of the research are not easily separable because, usually, some feature spaces are more appropriate for certain classifiers. As a result both aspects should be investigated simultaneously. This article is a summary of the authors' research on an integrated multi-paradigm approach for analysis of EEGs through ingenious integration of chaos theory, wavelets, and neural networks.

2. Wavelet-Chaos Methodology for EEG Sub-Band Analysis

Wavelet-Based EEG Decomposition

In recent years, the idea that the EEG is a chaotic system that represents brain dynamics has grown considerably. It has been postulated that changes in these EEG dynamics can be used to identify various mental states. Iasemidis and Sackellares[10] studied these nonlinear dynamics in patients with temporal lobe epilepsy and in their subsequent studies concluded that the chaos in the brain was reduced in the preictal phase.[11,12] Similar findings confirming the reduction in complexity of neuronal firing during the preictal phase were also reported by Elger and Lehnertz.[13,14] Lopes da Silva et al.[3] reported the synchronization of neuronal firing in different parts of the brain during a seizure.

Our research challenged the belief that the EEG represents the dynamics of the entire brain as a unified system and has to be treated as a whole. In reality, an EEG represents the effect of the superimposition of diverse processes in the brain. Very little research has been done to separately study the effects of these individual processes. Generally, each EEG is decomposed into five EEG sub-bands: *delta* (0–4 Hz), *theta* (4–8 Hz), *alpha* (8–12 Hz), *beta* (13–30 Hz), and *gamma* (30–60 Hz). A premise of our research is that there is no good reason for the entire EEG to be more representative of brain dynamics than the individual frequency sub-bands. In fact, the sub-bands may yield more accurate information about constituent neuronal activities underlying the EEG and consequently, certain changes in the EEGs that are not evident in the original full-spectrum EEG may be amplified when each sub-band is analyzed separately.[7]

In order to extract the individual EEG sub-bands a wavelet filter is employed. Wavelet transform uses a variable window size over the length of the signal, which allows the wavelet to be stretched or compressed depending on the frequency of the signal.[4,15,16] This results in excellent feature extraction from nonstationary signals such as EEGs. In this work, the discrete wavelet transform (DWT) based on dyadic (powers of 2) scales and positions is used to make the algorithm computationally very efficient without compromising accuracy. The EEG is decomposed into progressively finer details by means of multi-resolution analysis as shown in Fig. 1. After four levels of decomposition five separate EEG sub-bands are obtained. Subsequently, the Correlation Dimension (CD, representing the complexity of the system) and Largest Lyapunov Exponent (LLE, representing the chaoticity

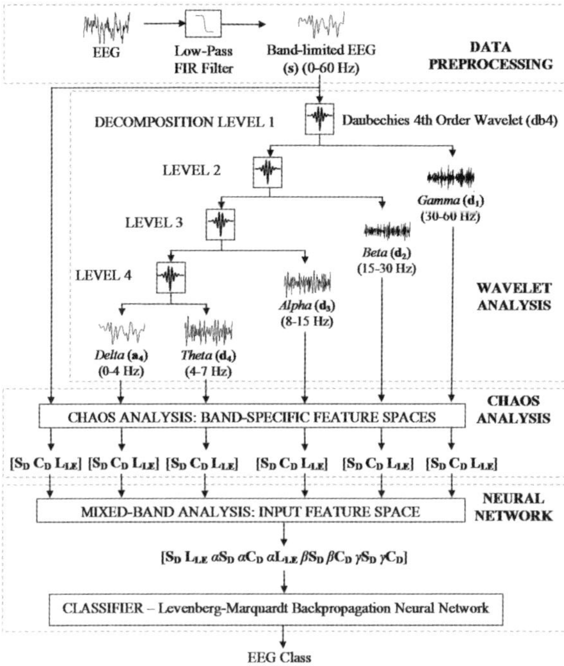

Fig. 1. Overview of the wavelet-chaos-neural network methodology for the 3-class EEG classification problem

of the system) are employed to characterize chaotic dynamics in the EEG and EEG sub-bands.[7] In addition, the standard deviation (STD, characterizing the signal variance) of the EEGs and EEG sub-bands is selected to investigate the subsequent classification accuracy.[8]

Chaos Analysis

The EEG (or EEG sub-band) is represented as a time series vector, $\mathbf{X} = \{x_1, x_2, \ldots, x_N\}$ (comprised of N voltage readings by an electrode) where the subscript indicates the time instant. In order to embed the EEG in an appropriate phase-space representation for chaos analysis, two parameters must be identified: optimum lag and minimum embedding dimension (MED). To find the optimum lag, the EEG is tested for mutual information which is a measure of the extent of overlap between the information contained in $\mathbf{X}_T$ (EEG at time T) and $\mathbf{X}_{T+m}$ (EEG at time $T + m$; where

m is the time lag). Varying the value of m yields mutual information coefficients (I_m) for different values of the time lag. The optimum lag, m_0, is the first local minimum on the graph of I_m versus m. MED (d_M) for the EEG is estimated using Cao's method.[17] Next, the i-th time-delay vector is reconstructed from the EEG time series vector, $\mathbf{X}$, using the optimum lag and the estimated MED. The reconstructed time delay vector in the lagged phase space has the form $\mathbf{Y}_i(d) = \{x_i, x_{i+m_0}, \ldots, x_{i+m_0(d-1)}\}$ where $i = 1, 2, \ldots, N - m_0(d-1)$, d is the embedding dimension, and x_i is the i-th data point in the EEG.

For a finite signal consisting of $N_C = N - m_0 d_M$ points, represented by $\mathbf{Y}_1(d_M), \ldots, \mathbf{Y}_{N_C}(d_M)$, CD is computed using the Takens method[18,19] as:

$$v = -\left[\frac{2}{N_C(N_C-1)}\sum_{i=1}^{N_C}\sum_{j=i+1}^{N_C}\log\left(\frac{\|Y_i(d_M) - Y_j(d_M)\|}{\epsilon}\right)\right]^{-1} \tag{1}$$

where $Y_i(d_M)$ and $Y_j(d_M)$ are the lagged phase space locations of the i-th and j-th points respectively for the selected embedding dimension d_M, and ϵ is the radius of the measuring unit found by trial and error to be 10% of the maximum attractor size.[17] The average trajectory divergence, D_T, of the attractor for a given prediction length, T, is expressed mathematically according to Wolf's method[20] as:

$$D_T = \frac{1}{N_S}\sum_{i=1}^{N_S}\left|\frac{Y_{i+T}(d) - Y'_{i+T}(d)}{Y_i(d) - Y'_i(d)_i}\right| \tag{2}$$

where $Y_i(d)$ and $Y_j(d)$ are neighboring points on separate trajectories in the phase space, and $Y_{i+T}(d)$ is the location of the point that evolved from $Y_i(d)$ along the trajectory. The prediction length, T, is measured in increments of time used for the EEG. The LLE ($\lambda_{\max}$) is then computed as the slope of the graph of the natural logarithm of trajectory divergence, D_T, versus the prediction length, T (see [7]). The methodology for chaos analysis of EEG and EEG sub-bands is detailed in Fig. 2.

3. Classification Feature Space: Preliminary Results

The primary purpose of the wavelet-chaos analysis is to perform feature enhancement to improve the subsequent classification accuracy. Additionally, it reduces a huge input space into a more compact and therefore manageable feature space. Prior to the wavelet-chaos analysis the dimensions of the input space was $N \times 6 \times 3K$ (N data points per EEG or EEG sub-band, 6 EEGs and EEG sub-bands and 3 groups of K EEGs). After the

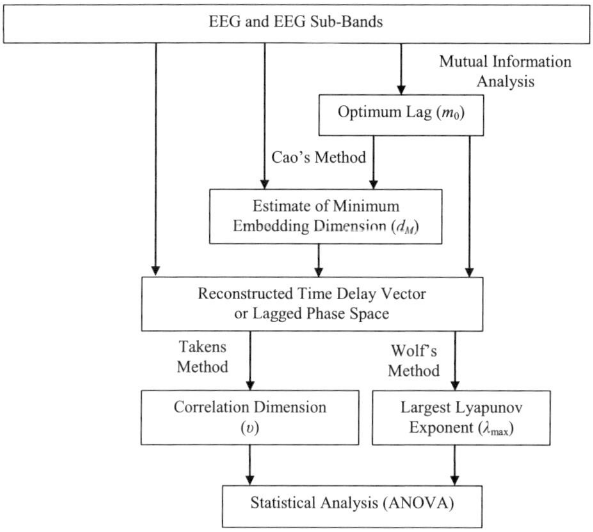

Fig. 2. Chaos quantification of EEG and EEG sub-bands

wavelet-chaos analysis, each EEG was represented as a P-dimensional data point in the reduced input space, F, dubbed *feature space*. The dimensions of the feature space are $P \times 3K$, where $P \in 1, 2, \ldots, 18$ is selected out of the three parameters (STD, CD, and LLE represented by the vectors $\mathbf{S_D}$, $\mathbf{C_D}$, and $\mathbf{L_{LE}}$ of size $3K$) obtained from the band-limited EEG as well as its five sub-bands (total of $6 \times 3 = 18$).

The wavelet-chaos methodology was investigated with three groups of EEG data: normal, interictal, and ictal, made available online by Dr. Ralph Andrzejak of the Epilepsy Center at the University of Bonn, Germany (`www.meb.uni-bonn.de/epileptologie/science/physik/eegdata.html`). The diagnosis was temporal lobe epilepsy and the hippocampal formation was identified as the epileptogenic focus). Each group contains $K = 100$ single channel EEG segments of 23.6 sec duration each sampled at 173.61 Hz (see [21]). Each segment contains $N = 4097$ data points recorded at a sampling rate of 173.61 Hz. Each EEG segment is considered as a separate EEG signal resulting in a total of 300 EEGs.

After extensive statistical analysis we observed that only the LLE, but not the CD, computed from the full-spectrum EEG showed significant differences in the three groups. However, when the CD and LLE were computed from certain specific EEG sub-bands (but not all), the resulting differences in parameter values among the three groups were statistically significant at the 99% confidence level.[7]

4. Classifier Design: Artificial Neural Networks

Band-Specific and Mixed-Band Analysis

Investigating all combinations of all three parameters, STD, CD, and LLE, for all EEGs and EEG sub-bands would have been extremely computationally expensive. Therefore, to reduce the computing time and output analysis to a more manageable one, the investigation was performed in two phases: a) band-specific analysis and b) mixed-band analysis, as shown in Fig. 1. In the first phase, the six types of signals (EEG and the five EEG sub-bands) were considered one by one. For each type of signal, the EEGs were classified based on STD, CD, and LLE. Consequently, each signal was represented by a point in a 1- to 3-dimensional feature space. This is called *band-specific* analysis in this article and the corresponding feature space is referred to as the *band-specific* feature space. The band-specific analysis was performed with multiple types of classifiers. One objective of band-specific analysis was to identify the classifiers that yield accurate classification and to eliminate the less accurate ones. Another objective was to identify specific combinations of parameters that may increase the classification accuracy of the selected classifiers.

In the second phase, the research was continued with the more promising classifiers and combinations of eighteen parameters selected from all six types of EEG and EEG sub-bands in the first step. As a result, each EEG was represented by a point in a 2- to 18-dimensional feature space. This is called *mixed-band* analysis in this research and the corresponding feature space is referred to as the *mixed-band* feature space.[8] Although many classifiers were investigated, only the two classifiers that yielded highest classification accuracies and robustness are described in this section.

Levenberg-Marquardt Backpropagation Neural Network (LMBPNN)

The LMBPNN classifier employs a standard feedforward neural network with one hidden layer in addition to the input and output layers. The difference lies in the training of LMBPNN which uses a modified form of the traditional backpropagation algorithm in order to increase convergence speed and reduce entrapment in local minima.[22] The design, training, and testing of the LMBPNN classifier are described in Ghosh-Dastidar et al.[9]

Band-specific feature spaces based on single parameters yielded low to very low classification accuracies ($< 65.0\%$) except for STD computed from the *gamma* sub-band which yielded a classification accuracy of 86.4%. Band-specific feature spaces based on combinations of two and three parameters led to more accurate classification (in the range of 65 – 90%). In general, it was observed from the statistical analysis in the previous section and the band-specific analysis, that the parameter combinations or feature spaces that showed higher classification accuracy were $[\mathbf{S_D\ C_D}]$ when computed from the *beta* and *gamma* sub bands and $[\mathbf{S_D\ L_{LE}}]$ when computed from the band-limited EEG and the *alpha* sub-band. Much lower classification accuracy was obtained from the *delta* and *theta* sub-bands. The two highest values of classification accuracy were obtained when the $[\mathbf{S_D\ L_{LE}}]$ feature space was computed from the band-limited EEG (89.9%) and the $[\mathbf{S_D\ C_D}]$ feature space was computed from the *gamma* sub-band (87.3%).

Overall, the following nine parameters were selected for further investigation as they yielded higher classification accuracies: STD computed from the band-limited EEG and*alpha*, *beta*, and *gamma* sub-bands; CD computed from *alpha*, *beta*, and *gamma* sub-bands; and LLE computed from the band-limited EEG and *alpha* sub-band.[8] In phase two of this research, it was discovered that a mixed-band feature space consisting of all nine aforementioned parameters $[\mathbf{S_D L_{LE}}\ \alpha\mathbf{S_D}\ \alpha\mathbf{C_D}\ \alpha\mathbf{L_{LE}}\ \beta\mathbf{S_D}\ \beta\mathbf{C_D}\ \gamma\mathbf{S_D}\ \gamma\mathbf{C_D}]$ resulted in the highest classification accuracy of 96.7% (with a standard deviation of 2.9). In this notation, the parameter prefix denotes the EEG sub-band from which the parameter is computed. Absence of a prefix indicates that the parameter is computed from the band-limited EEG.[8]

Principal Component Analysis (PCA)-Enhanced Cosine Radial Basis Function Neural Network (RBFNN)

Next, a new accurate two-stage classifier was developed to model the EEG classification problem more accurately and increase the robustness of EEG classification (with respect to EEG variations across subjects and various mental states) based on the 9-parameter mixed-band feature space (Fig. 3). In the first stage, principal component analysis (PCA) was employed for feature enhancement by transforming the 9-parameter feature space into a new feature space more amenable to subsequent classification. PCA is usually used for dimensionality reduction to reduce the computational requirement, and in some cases, for denoising.[23,24] Although, dimensionality reduction was also achieved in this work, that was not the primary objective. Rather, the rearrangement of the input space along the principal components of the data was used to improve the classification accuracy of the cosine radial basis function neural network (RBFNN) employed in the second stage.[9]

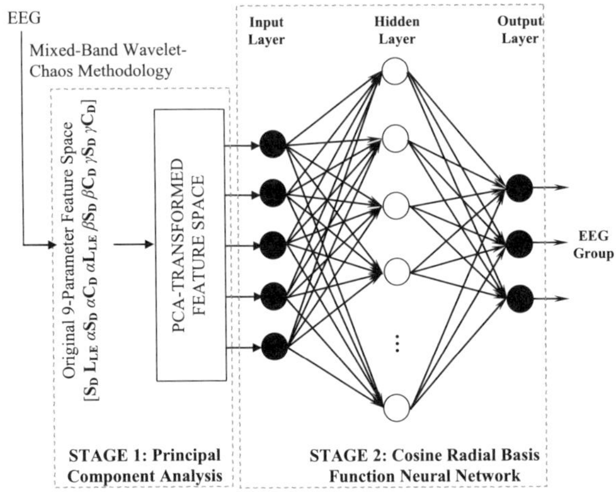

Fig. 3. Overview of the wavelet-chaos-neural network methodology for the 3-class EEG classification problem and architecture of the two-stage PCA-enhanced cosine RBFNN classifier with one hidden layer

PCA is a statistical method used primarily to transform the input space into a new lower dimensional space. The coordinate system is rotated using a linear transformation. The axes (or components) of the new coordinate system are linear combinations of the original axes. The primary axis or principal component is selected to represent the direction of maximum variation in the data. The secondary axis, orthogonal to the primary axis, represents the direction of the next largest variation in the data and so on. In the reoriented space, most of the variation in the data is concentrated in the first few components. Consequently, the components that account for most of the variability are retained whereas the remaining components are ignored. As a result, dimensionality is reduced without compromising the accuracy of data representation in any significant way.[24]

The training input was subjected to PCA to determine the axes of the new reoriented input space. Eigenvectors corresponding to the lowest eigenvalues are discarded. The reorientation of the feature space based on the remaining eigenvectors increases the angular separation between individual data points and leads to improved cosine RBFNN classification accuracy. The details of classifier design (including the selection of parameters), training, and testing are described in detail in Ghosh-Dastidar et al.[9]

The PCA-enhanced cosine RBFNN classifier yielded a classification accuracy of 96.6%. The standard deviation was a low 1.4% over 100 repetitions (with randomly selected training data) which implies a high degree of robustness to changes in training data. Overall, the classification accuracy of the PCA-enhanced RBFNN is comparable to the LMBPNN classifier but the sensitivity to the choice of training data is reduced to half.[9]

5. Concluding Remarks

Based on the wavelet-chaos-neural network methodology created in this work, the 9-parameter mixed-band feature space

$$[\mathbf{S_D}\ \mathbf{L_{LE}}\ \alpha\mathbf{S_D}\ \alpha\mathbf{C_D}\ \alpha\mathbf{L_{LE}}\ \beta\mathbf{S_D}\ \beta\mathbf{C_D}\ \gamma\mathbf{S_D}\ \gamma\mathbf{C_D}]$$

yielded the most accurate classification results. When this feature space was used as input to the LMBPNN and PCA-Enhanced cosine RBFNN classifiers, maximum classification accuracies of 96.7% and 96.6% are achieved, respectively. The results of this research and extensive parametric analysis led to the conclusion that all three key components of the wavelet-chaos-neural network methodology are important for improving the accuracy of EEG classification. The EEG was decomposed into sub-bands using wavelet transforms which is instrumental in creating the mixed-band features spaces with an improved accuracy over band-specific feature spaces. The parameters used in these feature spaces are obtained by means of statistical analysis (STD) and chaos analysis (CD and LLE). Although, these parameters individually were unable to differentiate between the three EEG groups, their combinations improved the classification accuracy significantly especially when multiple-parameter mixed-band feature spaces were employed. Finally, the LMBPNN and PCA-enhanced cosine RBFNN classifiers classified the EEGs into the three groups with a high degree of accuracy and robustness.

References

1. National Institute of Neurological Disorders and Stroke (NINDS), Seizures and epilepsy: Hope through research, `http://www.ninds.nih.gov/health_and_medical/pubs/seizures_and_epilepsy_htr.htm` (2004).
2. L. D. Iasemidis, L. D. Olson, J. C. Sackellares and R. Savit, *Epilepsy Res.* **17**, 81 (1994).
3. F. H. L. da Silva, J. P. Pijn, and W. J. Wadman, *Prog. Brain Res.* **102**, 359 (1994).

4. H. Adeli, Z. Zhou and N. Dadmehr, *J. Neurosci. Methods* **123**(1), 69 (2003).
5. J. Gotman, *Journal of Clinical Neurophysiology* **16**(2), 130 (1999).
6. P. J. Durka, *BioMedical Engineering OnLine* **2**(1), 1 (2003).
7. H. Adeli, S. Ghosh-Dastidar and N. Dadmehr, *IEEE Transactions on Biomedical Engineering* **54**(2), 205 (2007).
8. S. Ghosh-Dastidar, H. Adeli and N. Dadmehr, *IEEE Transactions on Biomedical Engineering* **54**(9), 1545 (2007).
9. S. Ghosh-Dastidar, H. Adeli and N. Dadmehr, *IEEE Transactions on Biomedical Engineering* **55**(2), 512 (2008).
10. L. D. Iasemidis and J. C. Sackellares, The temporal evolution of the largest Lyapunov exponent on the human epileptic cortex, in *Measuring chaos in the human brain*, eds. D. Duke and W. Pritchard (World Scientific, Singapore, 1991) pp. 49–82.
11. L. D. Iasemidis, D. S. Shiau, J. C. Sackellares and P. Pardalos, *Transition to epileptic seizures: Optimization* (American Mathematical Society, 2000).
12. L. D. Iasemidis, D. S. Shiau, W. Chaovalitwongse, J. C. Sackellares, P. M. Pardalos, J. C. Principe, P. R. Carney, A. Prasad, B. Veeramani and K. Tsakalis, *IEEE Trans. Biomed. Eng.* **50**(5), 616 (2003).
13. C. Elger and K. Lehnertz, Ictogenesis and chaos, in *Epileptic seizures and syndromes*, ed. P. Wolf (Libbey, London, U. K., 1994) pp. 547–552.
14. C. E. Elger and K. Lehnertz, *Eur. J. Neurosci.* **10**(2), 786 (1998).
15. S. G. Mallat, *IEEE Trans. Pattern Anal. Mach. Intell.* **11**(7), 674 (1989).
16. X. Jiang and H. Adeli, *Integrated Computer-Aided Engineering* **10**(3), 287 (2003).
17. L. Cao, *Physica D* **110**(1-2), 43 (1997).
18. G. Williams, *Chaos theory tamed* (National Academy Press, Washington D.C., 1997).
19. S. Borovkova, R. Burton and H. Dehling, *Ann. Appl. Probab.* **9**(2), 376 (1999).
20. J. B. Wolf, H. L. Swift, Swinney and J. A. Vastano, *Physica D* **16**, 285 (1985).
21. R. G. Andrzejak, K. Lehnertz, C. Rieke, F. Mormann, P. David and C. E. Elger, *Phys. Rev. E* **64**(6), (061907)1 (2001).
22. M. T. Hagan, H. B. Demuth and M. Beale, *Neural Network Design* (PWS Publishing Company, Boston, MA, 1996).
23. L. Zhukov, D. Weinstein and C. Johnson, *Engineering in Medicine and Biology Magazine* **19**(3), 87 (2000).
24. H. Lee and S. Choi, PCA+HMM+SVM for EEG pattern classification, in *Proceedings of Seventh International Symposium on Signal Processing and Its Applications*, 1, 541(2003).

Digital Data Encryption Based on Discrete Additive Systems

V. Grigoras* and C. Grigoras†

*'Gh. Asachi' Technical University, Iasi, Romania

†'Gr. T. Popa' University of Medicine and Pharmacy, Iasi, Romania

and

Romanian Academy, Iasi Branch, Institute for Computer Science, Romania

Emails: *grigoras@etc.tuiasi.ro, †crmn_g@yahoo.com

The present contribution deals with digital data encryption, by means of the inverse system approach chaos synchronization, applied to additive, discrete-time, chaotic systems. Due to their digital filter type architecture, additive chaotic systems allow general and flexible designs of both chaotic emitter and synchronizing receiver, regardless of the system order. The proposed approach is also known to have higher resistance to known attacks. The resulting encryption/decryption setup is similar to classical recursive encryption systems; it can be included in the class of private key encryption algorithms, thus benefiting from standard key management procedures, but can be designed to use far larger key lengths compared to classical systems.

In the following, we present discrete-time additive systems basics, the design method for the encryption and decryption systems, the conditions imposed to the Lyapounov exponents of the encryption system and the selection of the channel quantization step, in order to achieve best performance. Special attention is given to the *statistical properties* of the resulting random processes, under fixed and floating point state representation, and their influence on the encryption key space dimension.

Keywords: Discrete additive systems; chaos synchronization; chaotic encryption.

1. Introduction

Chaos synchronization seemed an attractive approach for secure communication since its discovery.[1,2] Due to inherent sensitivity, chaos modulation is less attractive than chaos encryption. The latter has to perform well in several aspects: design flexibility, good statistical properties, large encryption key space and sufficient resistance to known attacks.[3]

The present contribution concentrates on the discrete-time additive systems approach[4,5] for digital data encryption, by means of chaos synchronization.[6,7] To achieve chaos synchronization, in the chosen environment, the inverse system approach is applied, leading to floating point encryption and decryption algorithms that have high resistance to known attacks.[8,9] Conditions for the filter coefficients, Lyapounov exponents and channel quantization step are derived.

The first section of this paper deals with chaos synchronization and modulation, with special emphasis on discrete time additive systems. Application of the proposed approach to data encryption is shortly highlighted. The next section presents the possibility of quantizing the modulating and transmitted signals, to model plaintext and cipher text, preserving floating point implementation of the chaotic emitter and synchronizing receiver. The fourth section is devoted to the statistical aspects of the proposed approach, mainly the limit cycle length of the encryption system, under state quantization conditions. The concluding section includes some further research suggestions.

2. Discrete-Time Chaotic Additive Systems

The proposed encryption method starts from a chaos synchronization setup, composed of a discrete time nonlinear emitter system, with chaotic behavior, and a globally stable receiver, designed to synchronize with the emitter. When synchronization is achieved, in order to make use of this connection, the aspects regarding chaos modulation must be addressed. Thus we take the inverse system approach and apply the information-bearing, or modulating, signal, $m[k]$, to the emitter input, transmit the chaotic output, $y[k]$, and recover an approximation of the modulating signal, $\tilde{m}[k]$, at the receiver output, as suggested in Fig. 1.

In the following, we concentrate on the particular case of discrete-time additive systems of the Lure type. They are built in a feedback loop, using the residue algebraic nonlinear function, suggested by the two's complement overflow characteristic in digital filters[3] depicted in Fig. 2 and given by the following equation:

$$r(x) = x - \text{round}(x)$$

Due to the additivity property of the algebraic nonlinear function, and use only of additional linear elements, the resulting systems are additive too. This result, introduced in [4,5], simplifies the analysis of the nonlinear system response calculation and, as a consequence, its design. The emitter is built starting from the topology of an all pole recursive digital filter, including the residue algebraic nonlinear function at the summing node, as presented in Figure 3, leading to the equation:

$$y[k] = r\left(m[k] - \sum_{n=1}^{N} h[n] \cdot y[k-n]\right) \tag{1}$$

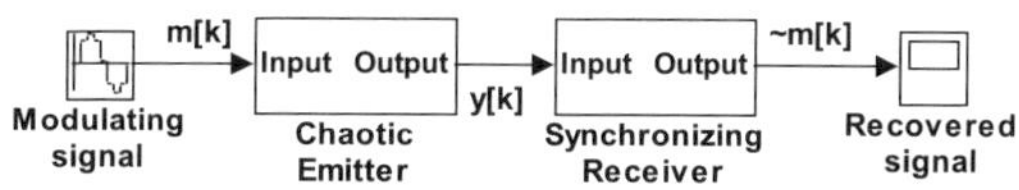

Fig. 1. The general chaos encryption/decryption setup

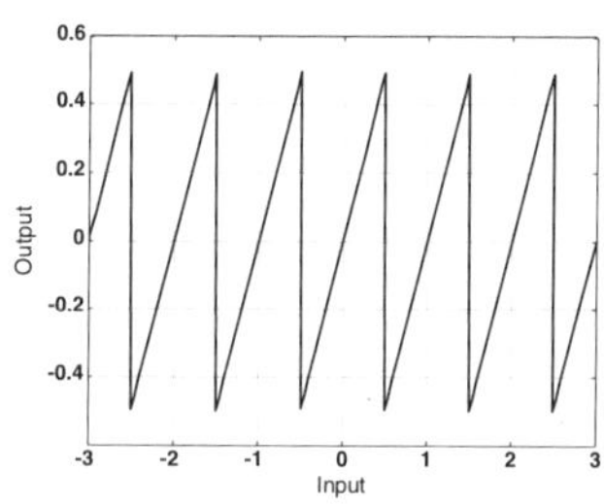

Fig. 2. The residue algebraic nonlinear function used in the chaotic emitter

In (1) the coefficients $h[n]$, $n = 1, \ldots, N$, represent the impulse response samples of the FIR filter in the feedback loop, that become the denominator coefficients in the IIR linear prototype transfer function:

$$H(z) = \frac{1}{1 + \sum_{n=1}^{N} h[n] \cdot z^{-n}}$$

As demonstrated in [4], a discrete-time emitter will always behave chaotically, by using any unstable linear prototype with the residue nonlinear function, $r(x)$.

By using the inverse system approach, and taking into account the additivity property of the symbolic residue algebraic function defining the nonlinear behavior of the emitter, the receiver system is a finite memory one, described by the difference equation:

$$\tilde{m}[k] = r\left(\tilde{y}[k] + \sum_{n=1}^{N} h[n] \cdot \tilde{y}[k-n]\right) \tag{2}$$

The corresponding synchronizing receiver, which implements (2), has the feed forward topology presented in Fig. 4.

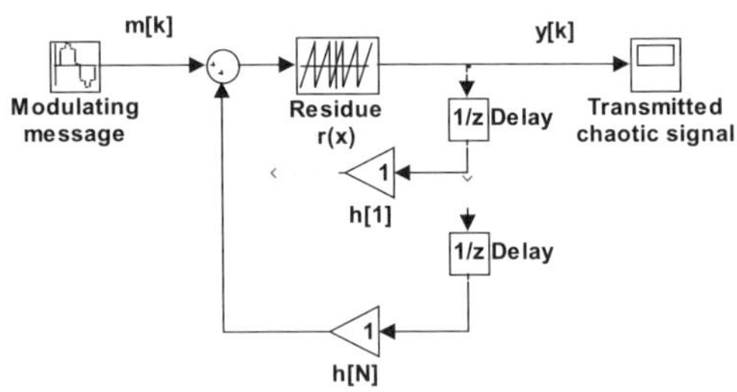

Fig. 3. Generic topology of the additive chaotic emitter

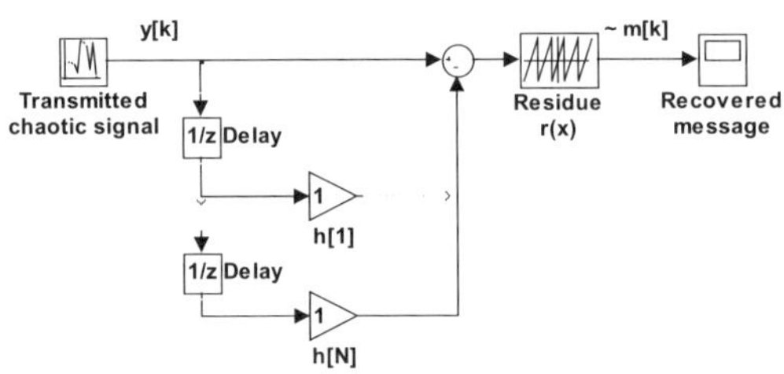

Fig. 4. Block diagram of the synchronizing receiver

As demonstrated in [7], if the emitter and receiver share the same set of coefficients, no matter the differences in initial conditions, the receiver synchronizes on the emitter in finite time. The synchronization transient lasts for a time interval equal to the order of the two systems, N. After that the modulating and recovered signals will be identical, as suggested in the example of simulation in Fig. 5(a). An unauthorized receiver, with different coefficient values, will never synchronize, as can be easily seen in Fig. 5(b).

3. Input and Output Signal Quantization

In the following, the previous results are used for encryption purposes. We model the input alphabet as a finite set of integer values, obtainable by quantizing the finite range M, input signal, with a specified quantization step q, leading to an alphabet with $C = M/q$ characters, which can be represented in fixed point on $B = \log_2(C)$ bits. The encrypted text is also obtained by quantization, in this case of the emitter output, and performed on B_1 bits. This way, the channel noise, that affects analog chaos modulation, no longer affects the system. Instead, the quantization noise, that has known limited amplitude, has to be taken into account.

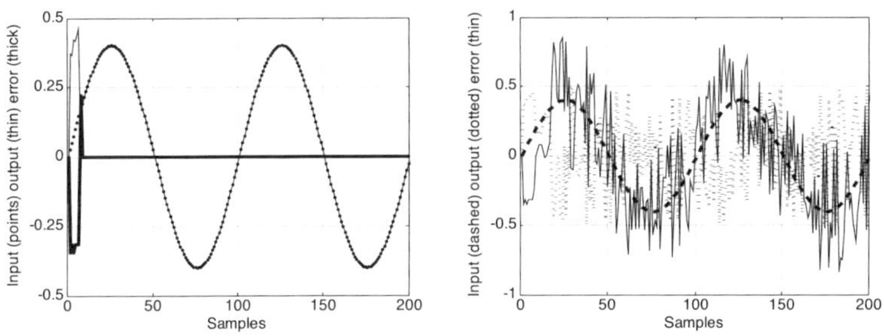

(a) Time evolution (b) The lack of synchronization case

Fig. 5. Graphical representation of the synchronization process time evolution

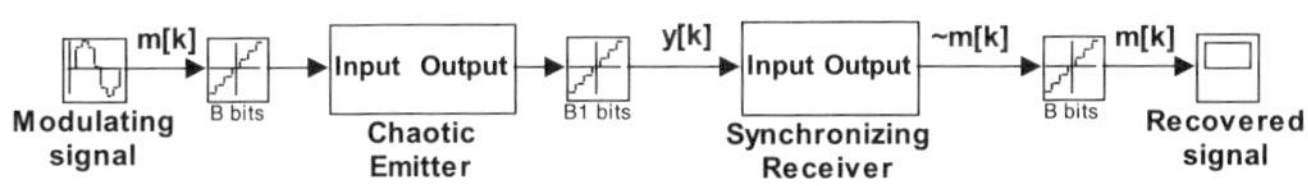

Fig. 6. Block diagram of the digital signal synchronization setup, achieving encryption/decryption

The resulting block diagram, for the proposed encryption/decryption system is presented in Fig. 6.

In [6] it is demonstrated that, if the channel noise is given by the quantization noise coming from a digital transmission channel, the number of bits used for its quantization gives the maximum value of the perturbation signal:

$$q = M \cdot 2^{-B_C}$$

To ensure zero error recovery of the modulating signal, we must quantize this signal too, to the number of bits restricted by:

$$2^{-B_C} \cdot \sum_{n=1}^{N} |h[n]| < 2^{-B_M - 1}$$

This result leads to the logarithmic inequality:

$$B_C - \log_2 \left(\sum_{n=1}^{N} |h[n]| \right) > B_M + 1 \tag{3}$$

Relation (1) shows that, even when integer (quantized) values are sent over the communication channel, the bit rate is somewhat increased for encrypted signals compared to their direct transmission. The increase in bit number results as:

$$\Delta B = B_C - B_M > \log_2 \left(2 \cdot \sum_{n=1}^{N} |h[n]| \right)$$

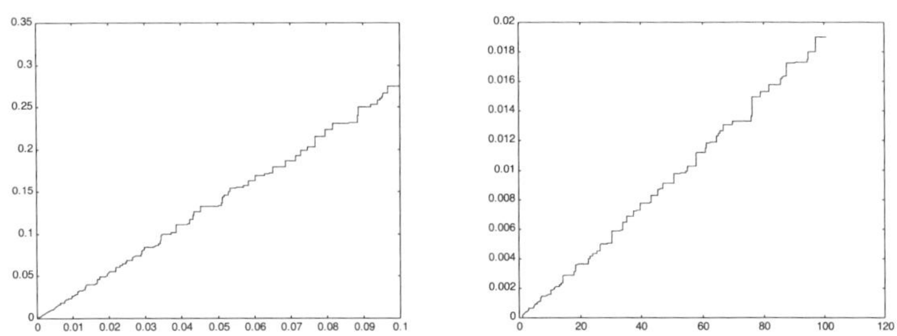

(a) Error dependence on the quantization noise maximum value

(b) Error dependence on the sum of the absolute value of the filter coefficients

Fig. 7. Maximum (l_1) synchronization error dependence

This increase is larger when the receiver coefficients are larger too. Thus the extra data volume to be transmitted in case of encryption is growing with the values of the receiver coefficients, but so are the Lyapounov exponents of the chaotic emitter, and thus the degree of data privacy. The graphs depicted in Fig. 7 highlight these results.

4. State Quantization Effects

If only input and output signals are quantized, the resulting encryption/decryption system has lower speed, due to the higher complexity of the floating point calculations in both encryption and decryption algorithms. Choosing to quantize the state of the two systems eliminates this drawback, making the proposed approach competitive to classical encryption algorithms in terms of speed and complexity. The main resulting drawback consists of the fact that the emitter (encryption algorithm) becomes a finite automaton, having periodic orbits as the most complex nonlinear dynamics achievable. The aim in such a system design is to obtain the longest period length possible for an order N system, with state variables quantized on B bits:

$$T = 2^{N \cdot B} - 1$$

We describe the additive chaotic emitter by the state equations:

$$\begin{cases} x[k+1] = f\big(x[k], m[k]\big) \\ \quad\;\; y[k] = g\big(x[k], m[k]\big) \end{cases}$$

where $m[k]$, $x[k]$ and $y[k]$ denote the input, state and output signals, while $f(\cdot)$ represents the state transition nonlinear vector function and $g(\cdot)$ denotes the output nonlinear scalar function. The state transition nonlinear vector function depends on the linear combination of state variables, the residue function, $r(\cdot)$ and the state quantization function $QS(\cdot)$. For instance, if the simple recursive topology in Fig. 3 is chosen for the emitter, the following state transition function results:

$$Q_S\left(r\left(\begin{pmatrix} h[1] & h[2] & \cdots & h[N] \\ 0 & 1 & \cdots & 0 \\ \vdots & \vdots & \ddots & \vdots \\ 0 & 0 & \cdots & 1 \end{pmatrix} \cdot \begin{pmatrix} x_1[k] \\ x_2[k] \\ \vdots \\ x_N[k] \end{pmatrix}\right)\right)$$

To identify a limit cycle of length L, the following nonlinear system of algebraic equations must be solved, in the conditions of null input:

$$\Big\{ x[k] = f^{(L)}\left(x[k], 0\right)$$

where $f^{(L)}(\cdot)$ denotes the L-times iterative composition of the state transition function with itself. The problem of determining the length of the emitter system limit cycles, resumes in iteratively applying the state transition function to the initial condition, for a number of $2^L - 1$ times at most. The previous equation gives the loop stop condition. If we choose to quantize both coefficients and state-variables on B bits, the complexity of the problem is:

$$C_p \simeq 2^{B\cdot(N+2)} \tag{4}$$

As seen from equation (4), the problem complexity is exponential on both system order and number of bits, making its implementation a time consuming task. Algorithms that test for already found states and groups of coefficients, avoiding redundancy, could slightly reduce the complexity, but not its exponential order.

Some preliminary simulation results, based on a brute force implementation of the presented algorithm, are shown in Fig. 8. Simulations were made on a first order, discrete-time additive system, with state variables quantized on 6 and 8 bits. Simulation results show that an important fraction of the coefficient range lead to the longest limit cycle length, while the remaining coefficient set ensure shorter limit cycles and fixed points.

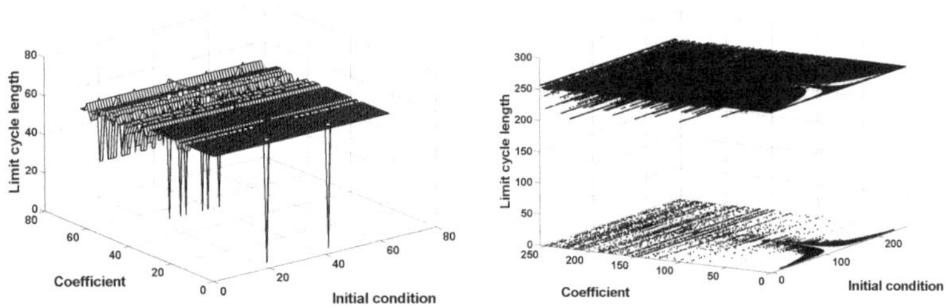

(a) 6-bit state quantization (b) 8-bit state quantization

Fig. 8. Dependence of the limit cycle length on the values of the feedback coefficient and the initial condition of a first order system

5. Conclusions

The present paper gives quantitative estimations on the performance of encryption systems based chaos synchronization in discrete-time additive systems. The proposed approach is based on the inverse system type of chaos synchronization, which has proved to be the least sensitive to attacks. The recursive structure of the chaotic encryption system gives a further security advantage, due to a dynamic relation between the cipher text and plaintext. Both theoretical and experimental results suggest that large

Lyapounov exponents ensure better security. The large number of possible encryption keys, high computational complexity necessary to break a chaotic code and the good resistance to known types of attacks can be further enhanced by periodic key modification. The proposed system is easily cascadable with existing encryption/compression/error-correction systems to increase the overall system performance.

Further research is needed to design hardware accelerators for floating-point additive encryption systems, within reasonable cost. For designs using a fixed point approach, further refinement of the limit-cycle length calculation algorithm is necessary, in order to reduce the computing time, imposed by the exponential complexity of the brute-force approach.

References

1. L. M. Pecorra and T. L. Carrol, *Physical Review Letters* **64**, 821 (February 1990).
2. L. M. Pecorra and T. L. Carrol, *IEEE Transactions on Circuits and Systems* **CAS 38**, 453 (April 1991).
3. L. Kocarev, *IEEE Circuits and Systems Magazine* **1**(3), 6 (2001).
4. V. Grigoras and L. Goras, Additive nonhomogeneous discrete systems, in *Proceedings of Nonlinear Dynamics in Electronic Circuits, NDES'94*, (Krakow 29–30 July, 1994).
5. V. Grigoras and L. Goras, A Z-transform approach to additive nonlinear discrete systems analysis, in *Proceedings of European Conference on Circuit Theory and Design, ECCTD'95*, (Istanbul 29–31 August, 1995).
6. V. Grigoras, *International Journal on Chaos Theory and Applications* **4**, 15 (December 1999).
7. A. Leuciuc and V. Grigoras, *International Journal of Chaos Theory and Applications* **2**, 3 (May 1997).
8. V. Grigoras and C. Grigoras, Chaos encryption method based on large signal modulation, in additive nonlinear discrete-time systems, in *Proceedings of WSEAS NOLASC '06*, (Bucharest October 16–18, 2006).
9. V. Grigoras and C. Grigoras, *WSEAS Transactions on Circuits and Systems* **5**, 1608 (November 2006).

Local Lyapunov Exponents: A New Way to Predict Chaotic Systems

D. Guégan

PSE, CES-MSE, Université Paris-I Panthéon Sorbonne, 106 boulevard de l'Hôpital, 75013 Paris, France
Email: Dominique.Guegan@univ-paris1.fr

J. Leroux

Institute for Applied Economics, HEC Montréal and CIRPÉE 3000 chemin de la Côte-Ste-Catherine, Montréal, QC H3T 2A7, Canada
Email: justin.leroux@hec.ca
Fax: (1) 514-340-6469

We propose a novel methodology for forecasting chaotic systems which is based on exploiting the information conveyed by the local Lyapunov exponent of a system. We show how our methodology can improve forecasting within the attractor and illustrate our results on the Lorenz system.

1. Introduction

It is well known that some deterministic chaotic systems can be difficult to forecast accurately. Indeed, their extreme sensitivity to initial conditions amplifies slight deviations from a trajectory in the state space into dramatic changes in future behavior.

In this chapter, we propose a novel methodology for forecasting deterministic series which corrects for the inevitable bias of most non-parametric predictors (such as the ones based on kernels, radial functions, neural nets, wavelets, etc.; see [1] and [2]) by incorporating additional information on the local chaoticity of the system via the so-called local Lyapunov exponent (LLE). To the best of our knowledge, while several works exist on the forecasting of chaotic systems (see, e.g., [3–6]), none exploit the information conveyed by the LLE. The general intuition of the methodology we propose in this chapter can be viewed as a complement to existing forecasting methods, and can be extended to chaotic time series. For illustrative purposes, we describe how our methodology can be used to improve upon the well-known nearest-neighbor predictor on the Lorenz system.

The nearest-neighbor predictor has proved to be a simple yet useful tool for forecasting chaotic systems (see [7,8]). In the case of a one-neighbor predictor, it takes the observation in the past which most resembles today's state and returns that observation's successor as a predictor of tomorrow's state. The rationale behind this nearest-neighbor predictor is quite simple: given that the system is assumed to be deterministic and ergodic, one obtains a sensible prediction of the variable's future by looking back at its evolution from a similar, past situation. For predictions more than one step ahead, the procedure is iterated by successively merging the predicted values with the observed data.

We show that we can improve predictions in a chaotic system by incorporating information carried by the system's LLE. The LLE (see [9,10]) represents the local dispersion rate of a system at a given point: a positive value meaning that two nearby points in the state space tend to grow apart over time, while a negative value indicates that nearby points will come closer together in the near future (but may diverge later on). In other words, the LLE is a measure of local chaoticity of a system, as sensitivity to initial conditions is characteristic of chaotic systems.

By definition, the LLE tells us precisely by how much the distance between the current state and its nearest neighbor will expand (or contract) over time, so that we can easily obtain the distance between the nearest-neighbor predictor (i.e., the neighbor's successor) and the future we are predicting (tomorrow's state). Thus, we know exactly by how much to correct the prediction of the nearest-neighbor predictor. We use this fact to develop a new methodology.

The rest of the chapter is organized as follows. In Section 2, we develop our methodology by first pointing out why the nearest-neighbor predictor is biased and then suggesting how to correct this bias using information carried by the system's LLE. In Section 3, we present simulations carried out on the well-known Lorenz system to illustrate the extent of the substantial potential accuracy gains our methodology yields. In Section 4, we present how one can achieve these potential gains almost perfectly thanks to a simple rule of thumb to solve the selection problem arising from our methodology. Section 5 concludes by pointing to directions in future work in order to refine our selection process and our methodology in general.

2. Methodology

Consider a one-dimensional series of T observations from a chaotic system, $(x_1, ...x_T)$, whose future values we are trying to forecast, d is the embedding dimension used to detect the attractor. A possible embedding method, involving building a d-dimensional orbit, (X_t), with $X_t = (x_t, x_{t-1}, ..., x_{t-(d-1)})$, is described in [11].

By definition, the local Lyapunov exponent (or LLE) of a dynamic system characterizes the rate of separation of infinitesimally close points of an orbit. Quantitatively, two neighboring points in phase space with initial separation δX_0 are separated, t periods later, by the distance:

$$|\delta X| \approx |\delta X_0| e^{\lambda_0 t},$$

where $|\cdot|$ represents the modulus of the considered vectors and λ_0 is the local Lyapunov exponent of the system in the vicinity of the initial points. Typically, this local rate of divergence (or convergence, if $\lambda_0 < 0$) depends on the orientation of the initial vector δX_0. A dynamic system is considered to be (locally) chaotic if $\lambda_0 > 0$, and (locally) stable if $\lambda_0 < 0$. (see, e.g., [12]).

Our goal is to exploit the local information carried by the LLE to improve upon existing methods of reconstruction and prediction. We propose a methodology which builds upon the classical nearest-neighbor predictor, which we now recall.

Consider an orbit $(X_1, ..., X_T)$ whose one-step-ahead future, X_{T+1}, we are trying to predict. The nearest-neighbor predictor returns $\hat{X}_{T+1} = X_{i+1}$, where X_i is the element of the orbit with minimal distance to X_T. Because the dynamic system at hand is aperiodic (or else, forecasting would not be an issue), the nearest-neighbor predictor is inevitably biased. Indeed, because $|X_T - X_i| > 0$, it must also be the case that:

$$|\hat{X}_{T+1} - X_{i+1}| \approx |X_T - X_i| e^{\lambda_i} > 0, \tag{1}$$

where λ_i can be approximated in practice by the following expression:

$$\hat{\lambda}_i = \ln \frac{|X_{i+1} - X_{j+1}|}{|X_i - X_j|} \quad \text{with} \quad X_j = \arg \min_{t \neq i, T} |X_t - X_i| \tag{2}$$

It follows from Expression (1) that knowing the distance between the predictee and the nearest neighbor as well as the LLE at the nearest neighbor allows us to predict the distance of the predictee's image to the neighbor's image. Note that this is true regardless of the sign of λ_i; i.e., regardless of whether the system is locally chaotic or locally stable. Moreover, because the orbit considered results from the embedding of a one-dimensional series, we also know all but the first coordinate of $X_{T+1} = (x_{T+1}, x_T, ..., x_{T-d+2})$. Hence, X_{T+1} lies at the intersection of the sphere of radius $|X_T - X_i|e^{\hat{\lambda}_i}$ centered on X_T and the line defined by $\{(z, x_T, ..., x_{T-d+2})|z \in \mathbb{R}\}$ which, in the Euclidean space, amounts to solving the following polynomial for $z \in \mathbb{R}$:

$$\sqrt{(z-x_{i+1})^2+(x_T-x_i)^2+...+(x_{T-d+2}-x_{i-d+2})^2}-|X_T-X_i|e^{\hat{\lambda}_i}=0 \quad (3)$$

Typically, two candidates emerge, $\hat{x}^-_{T+1}$ and $\hat{x}^+_{T+1}$, respectively underestimating and overestimating the true value of observation x_{T+1} (see Figure 1)[a].

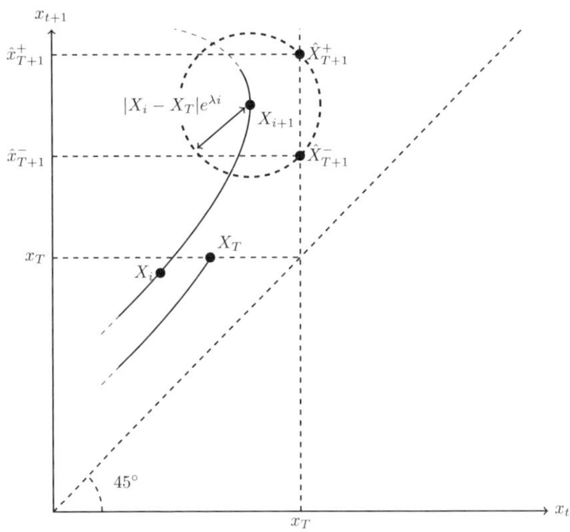

Fig. 1.

One difficulty lies in determining when the nearest-neighbor predictor overestimates or underestimates the true value to be predicted. Being able to discriminate accurately between $\hat{x}^-_{T+1}$ and $\hat{x}^+_{T+1}$ may significantly improve the accuracy of the nearest-neighbor predictor, as we next illustrate. We discuss this point in Section 4.

[a]The situation whereby Expression (3) has no real solution would only arise if λ_i had been greatly underestimated, which never occurred to us in practice using Expression (2).

3. Simulations

We illustrate our point by simulating a well-known chaotic system: the Lorenz system,[13] which is characterized by the following system of differential equations:

$$\frac{dx}{dt} = \sigma(y - x)$$
$$\frac{dy}{dt} = x(R - z) - y$$
$$\frac{dz}{dt} = xy - bz$$

We simulated this system with values $\sigma = 16$, $R = 45.92$ and $b = 4$, initial values $x_0 = -10$, $y_0 = -10$ and $z_0 = 30$, and a step size of 0.01. Taking 5,000 observations, deleting the first 1,000 ensure that we are working within the attractor and considering the values on the x-coordinate as its own series, we successively predicted the last 1,000 in-sample observations. Each prediction was carried out with the full—and true—information set leading up to it, each time using the best of the two candidates, $\hat{x}^-_{T+1}$ and $\hat{x}^+_{T+1}$(measured in distance to the—known—successor). We obtain results which are always better than with the nearest-neighbor predictor and a mean-squared error which is roughly two-thirds that of the nearest-neighbor predictor: 0.1552 for the raw nearest-neighbor predictor, versus 0.1056 after LLE-correction.

This example shows the interest of the method and the great potential in improving upon the accuracy of the nearest-neighbor predictor by incorporating the information contained in local Lyapunov exponents as in Expression (1). Another example based on the logistic map is available in [14].

4. Solving the Selection Problem

Several aspects of the implementation are still to be refined. Consistently discriminating between the two candidates, $\hat{x}^-_{T+1}$ and $\hat{x}^+_{T+1}$, can prove to be a difficult task due to the inherent chaotic nature of the systems at hand. As a first guess, one can select the candidate which maximizes the colinearity between the $X_{i+1} - X_T$ vector and the vector $\hat{X}_{T+1} - X_T$ (with $\hat{X}_{T+1}$ standing for $\hat{X}^-_{T+1}$ or $\hat{X}^+_{T+1}$). With the simulation of the Lorenz system described above, we achieve 94.3% accuracy over 1000 predictions, suggesting that this method of discrimination might be reasonable. Results are summarized in Table 1 below.

Notice that the LLE ranges from −1.4353 to 1.4580 on the portion studied. In other words, the attractor contains stable and chaotic regions. Our predictions are more accurate on 1000 predictions than on 100, which is likely due to the fact that the corresponding latter portion of the attractor is more chaotic, which is confirmed by the fact that the LLE is larger on average in that region (0.2756 versus 0.1940).

Table 1. Lorenz system

Number of predictions	100	1000
Success rate	87%	94.30%
Average error	0.0738	0.0390
NNP average error	0.1075	0.0548
Best candidate average error	0.0689	0.0372
otherSE	0.2041	0.1036
mean LLE (min;max)	0.2756(-0.9861;1.3639)	0.1940 (-1.4353;1.4580)
mean LLE on mistakes (min;max)	0.4685(-0.4020;1.3639)	0.4354 (-0.5142;1.3639)

Table 2 provides the size of errors on small ranges of LLE on 1000 predictions.

Table 2. Average errors and success rates

	Average errors			Success rates		
LLE	success	failure	raw 1-NN	success	failure	success rate
-1.5,1.3	0.0077	-	0.0077	1	0	100%
-1.3,-1.1	-	-	-	-	-	-
-1.1,-0.9	0.0335	-	0.0335	3	0	100%
-0.9,-0.7	0.0286	-	0.0286	3	0	100%
-0.7,-0.5	0.0451	0.0169	0.0479	67	1	98.51%
-0.5,-0.3	0.0541	0.005	0.0594	92	1	98.91%
-0.3,-0.1	0.0505	0.0899	0.0626	98	2	97.96%
-0.1,0	0.0477	0.0347	0.0580	47	4	91.49%
0,0.1	0.0249	0.027391	0.0365	61	4	93.44%
0.1,0.3	0.0342	0.0876	0.0539	109	7	93.58%
0.3,0.5	0.0290	0.0594	0.0626	195	8	95.90%
0.5,0.7	0.0325	0.0847	0.0560	223	22	90.13%
0.7,0.9	0.0236	0.0562	0.0267	18	2	88.89%
0.9,1.1	0.0152	0.396	0.0273	13	3	76.92%
1.1,1.3	0.0433	0.0632	0.0333	9	2	77.88%
1.3,1.5	0.1594	0.0432	0.0339	4	1	75%
Overall	0.0373	0.0667	0.0548	943	57	94.3%

We observe that errors can be small even when we make mistakes, but can also be relatively large with accurate selection. However, the size of errors is relatively stable over the range of LLEs when selection is successful. This seems to indicate that our method accurately corrects for the dispersion of neighboring trajectories as measured by the value of the LLE. If this were not the case, one would expect the size of errors to be larger for larger values of LLEs. In fact, errors become large only for values of the LLE near the upper end of their range. A possible reason for this sudden increase may be that our estimator for the value of the LLE is not sufficiently robust in regions of high chaoticity. We expect that a more sophisticated estimation method for the LLE may solve this issue.

Our method selects perfectly for very low values of the LLE. Selection mistakes start to appear past a value of LLE of -0.5, but the success rate does not seem to fall significantly until the value of the LLE increases up to 0.7. The success rate falls as the LLE grows very large, however, which is in line with the common intuition that the system is then (locally) more chaotic and, thus, more difficult to predict.

5. Concluding Comments

We find that our methodology may lead to substantial improvements to existing non-parametric predictors. Despite the candidate selection problem it poses, we were able to reap most of the potential benefits thanks to a simple rule of thumb. However, more sophisticated selection procedures should be considered. For instance, it may be sensible to condition the selection process on the size of the LLE, acting on the intuition that trajectories are likely to be more stable when the LLE is small, and more erratic when it is large. More specifically, the bottom row of Table 1 suggests that the current rule of thumb makes mistakes where the value of the LLE is large. We plan on investigating this line of reasoning in future work.

Such increased precision in short-run predictions may translate into accuracy gains for medium-run predictions, which is currently unsatisfactory with existing techniques. In addition, the general intuition behind the proposed method readily applies to other non-parametric predictors. Next steps include enhancing predictions via better estimations of the LLE, either by using more neighbors, or neural network methods (see [15]). Naturally, our ultimate goal is to evaluate how our method holds up when confronted to real data, and particularly to intra-day financial and economic time series.

References

1. M. Shintani and O. Linton, *Journal of Econometrics* **120**, 1 (2004).
2. D. Guégan and L. Mercier, Stochastic and chaotic dynamics in high-frequency financial data, in *Signal Processing and Prediction*, 1998 pp. 365–372.
3. J. D. Farmer and J. J. Sidorowich, *Physical Review Letters* **59**, 845 (1987).
4. R. Doerner, B. Hübinger, W. Martienssen, S. Grossman and S. Thomae, *Chaos, Solitons and Fractals* **1**, 553 (1991).
5. S. Rinaldi, R. Casagrandi and A. Gragnani, *Chaos, Solitons and Fractals* **12**, 313 (2001).
6. F. A. Guerra and L. S. Coelho, *Chaos, Solitons and Fractals* **35**, 967 (2009).
7. S. Yakowitz, *Journal of Time Series Analysis* **8**, 235 (1987).
8. D. Guégan, *Les Chaos en Finance: Approche Statistique* (Economica Eds, 2003).
9. H. D. I. Abarbanel, Local and global lyapunov exponents on a strange attractor, in *Nonlinear Modeling and Forecasting, SFI Studies in the Science of Complexity*, eds. M. Casdagli and S. Eubank (Addison-Wesley, 1992) pp. 229–247.
10. R. C. L. Wolff, *Journal of the Royal Statistical Society B* **54**, 353 (1992).
11. F. Takens, *Progress in Nonlinear Differential Equations and their Applications* **19**, 405 (1996).
12. B. Bailey, Local lyapunov exponents: predictability depends on where you are, in *Nonlinear Dynamics and Economics*, ed. K. et al 1997
13. E. N. Lorenz, *Journal of Atmospheric Science* **20**, 130 (1963).
14. D. Guégan and J. Leroux, *Chaos, Solitons and Fractals* (2009).
15. R. Gençay and T. Liu, *Physica D* **108**, 119 (1997).

On the Complete Chaotic Maps that Preserve Prescribed Absolutely Continuous Invariant Densities

W. Huang

Nanyang Technological University
Nanyang Avenue, Singapore 639798
Email: awhhuang@ntu.edu.sg

A nonsingular chaotic map $F : [0,1] \to [0,1]$ is said to be complete if there exists a partition $0 = a_0 < a_1 < \cdots < a_{r-1} < a_r = 1$ such that the restriction τ_i of τ to the open interval (a_{i-1}, a_i) is a monotonically continuous function which can be extended to the closed interval $[a_{i-1}, a_i]$ with $F_i(a_{i-1}) \in \{0,1\}$, $F_i(a_i) \in \{0,1\}$ and $F_i(a_{i-1}) + F_i(a_i) = 1$, for $i = 1, 2, ..., r$. A most general formulation for complete chaotic maps that preserve a prescribed invariant densities are provided.

Keywords: Complete maps; invariant densities; IFPP; chaos.

1. Introduction

A set universal formulas for the construction of two-segmental complete chaotic maps preserving specified invariant densities was introduced by the current author.[1] The current research intends to generalize the relevant results to multi-segmental complete chaotic maps. Our formulations do not only help to understand the probabilistic characteristics of nonsingular chaotic maps but also provide a practical approach to construct such chaotic maps with specified probabilistic as well as analytical characteristics.

2. Complete Chaotic Maps

Let $J = [0,1]$ be the unit-interval. Let $F : J \to J$ be a nonsingular chaotic map. "Nonsingularity" means that $m(F^{-1}(X)) = 0$ whenever $m(X) = 0$ for a measurable set $X \subset J$, where m denotes the Lebesgue measure.

Definition 2.1. A nonsingular map $F : J \to J$ is said to be *complete* if

i) there exists a partition $A = \{a_i\}_{i=0}^{r}$, with $0 = a_0 < a_1 < \cdots <$

$a_{r-1} < a_r = 1$, of the unit interval J such that for each integer i ($i = 1, 2, ..., r$), the restriction F_i of F to the open interval (a_{i-1}, a_i) is a monotonically continuous function that can be extended to the closed interval $J_i = [a_{i-1}, a_i]$;

ii) $F_i(a_{i-1}) \in \{0, 1\}$, $F_i(a_i) \in \{0, 1\}$ and $F_i(a_{i-1}) + F_i(a_i) = 1$, for $i = 1, 2, ..., r$.

We shall call F_i extended to J_i as i-th branch of F, $s_i \doteq F_i(a_i) - F_i(a_{i-1})$ as its sign index to indicate the sloping direction of F on i-th subinterval J_i, $i = 1, ..., r$, and $S = \{s_i\}_{i=1}^r$ as the sign pattern of $F = \{F_i\}_{i=1}^r$. In particular, we shall let $I_+ \doteq \{i|s_i > 0\}$ and $I_- \doteq \{i|s_i < 0\}$ be the positive and negative index sets that indicate the set of branches with positive and negative slopes, respectively. For instance, the chaotic map defined by

$$F(x) = \begin{cases} F_1(x) = 3\sqrt{x}, & x \in [0, 1/3), \\ F_2(x) = \sqrt{4/3 - 3x^2} & x \in [1/3, 2/3), \\ F_3(x) = \sqrt{(9x^2 - 4)/5}, & x \in [2/3, 1], \end{cases}$$

is complete with $A = \{0, 1/3, 2/3, 1\}$, $S = \{1, -1, 1\}$, $I_+ = \{1, 3\}$ and $I_- = \{2\}$.

Definition 2.2. A complete map F is said to be complete chaotic (fully chaotic) if it is

i) ergodic with respect to the Lebesgue measure, and
ii) chaotic in the probabilistic sense, that is, an absolutely continuous invariant density φ is preserved. In other words, for

$$\mu(x) = \int_0^x \varphi(z)\, dz, x \in J.$$

we have

$$\mu(A) = \mu\left(F^{-1}(A)\right) \tag{1}$$

for any set of $A \subset J$ with $m(A) \neq 0$, where m denotes the Lebesgue measure.

For the convenience of exposition, we need to define a special class of monotonic increasing one-to-one function on the unit-interval.

Definition 2.3. A C^1 function $g : J \to J$ is said to be a *generating function* if g meets the following requirements: i) $g(0) = 0$ and $g(1) = 1$; ii) g is continuous; and iii) $g'(x) > 0$ for almost all $x \in J$.

Define $\mathcal{G}$ as the set of generating functions.

Corollary 2.1. *A complete chaotic map $F = \{F_i\}_{i=1}^r$ preserves a normalized absolutely continuous invariant measure μ if and only if there exists a set of generating functions $G = \{g_i\}_{i=1}^r$, with $g_1(x) = x$ and $g_i \in \mathcal{G}$, for $i = 2, 3, ..., r$, so that all its branches can be expressed as*

$$F_1(x) = \begin{cases} \mu^{-1}\Big(\sum_{j\in I_+} [\mu(a_{j-1} + (a_j - a_{j-1})\, g_j(x/a_1)) - \mu(a_{j-1})] \\ \quad + \sum_{j\in I_-} [\mu(a_j) - \mu(a_j - (a_j - a_{j-1})\, g_j(x/a_1))] \Big), & \text{if } F_1' > 0, \\ \mu^{-1}\Big(\sum_{j\in I_+} [\mu(a_j - (a_j - a_{j-1})\, g_j(x/a_1)) - \mu(a_{j-1})] \\ \quad + \sum_{j\in I_-} [\mu(a_j) - \mu(a_{j-1} + (a_j - a_{j-1})\, g_j(x/a_1))] \Big), & \text{if } F_1' < 0, \end{cases} \tag{2}$$

$$F_i(x) = \begin{cases} F_1\Big(a_1 g_i^{-1}\Big(\dfrac{x - a_{i-1}}{a_i - a_{i-1}}\Big)\Big), & \text{if } F_i' F_1' > 0, \\ F_1\Big(a_1 g_i^{-1}\Big(\dfrac{a_i - x}{a_i - a_{i-1}}\Big)\Big), & \text{if } F_i' F_1' < 0, \end{cases} \quad \text{for } i = 2, 3, ..., r.$$

Proof. We prove the "only if" part, then the "if" part follows straightforwardly.

For any complete chaotic map F defined on $J = \cup_{i=1}^r J_i$, for each $x \in J$, F has r preimages $F_i^{-1}(x)$, $i = 1, 2, ..., r$. Therefore, for any $x \in J_1$, $F_1(x)$ has r preimages in each subinterval J_i, which are linked with x through the branching functions $\omega_i : J_1 \to J_i$, defined by

$$\omega_i(x) \doteq F_i^{-1}(F_1(x)), \ \forall x \in J_1, \ i = 1, 2, ..., r. \tag{3}$$

Depending on the sign indices of F_1 and F_i, the i-th branching function ω_i is a monotonic function that satisfies the following:

i) if $F_i' F_1' > 0$, then $\omega_i'(x) > 0$ for almost all $x \in J_1$ with $\omega_i(a_0) = a_{i-1}$ and $\omega_i(a_1) = a_i$;
ii) if $F_i' F_1' < 0$, then $\omega_i'(x) < 0$ for almost all $x \in J_1$ with $\omega_i(a_0) = a_i$ and $\omega_i(a_1) = a_{i-1}$.

In particular, we have $\omega_1 = x$.

Let $X = [0, F_1(x)]$, $x \in J_1$, then Frobenius-Perron equation (1) simplifies to

$$F_1(x) = \mu^{-1}\left(\sum_{j\in I_+} \Big[\mu(\omega_j(x)) - \mu(a_{j-1})\Big] + \sum_{j\in I_-} \Big[\mu(a_j) - \mu(\omega_j(x))\Big] \right). \tag{4}$$

Secondly, for any branching function $\omega_i{:}J_1 \to J_i$, $i = 1, 2, ..., r$, there exists a unique $g_i \in \mathcal{G}$ such that

$$\omega_i(x) = \begin{cases} a_{i-1} + (a_i - a_{i-1})\, g_i(x/a_1), & \text{if } F_i' F_1' > 0, \\ a_i - (a_i - a_{i-1})\, g_i(x/a_1), & \text{if } F_i' F_1' < 0. \end{cases} \tag{5}$$

In particular, for $g_1(x) = x$, we have $\omega_1(x) = x$. Substituting (5) into (4) yields the expression for F_1 in (2).

Finally, it follows directly from the definition of branching functions (3) that $F_i(x) = F_1\left(\omega_i^{-1}(x)\right)$, for $i = 1, 2, ..., r$. Substituting $\omega_i^{-1}(x)$ with

$$\omega_i^{-1}(x) = \begin{cases} a_1 g_i^{-1}\left(\dfrac{x - a_{i-1}}{a_i - a_{i-1}}\right), & \text{if } F_i' F_1' > 0, \\ a_1 g_i^{-1}\left(\dfrac{a_i - x}{a_i - a_{i-1}}\right), & \text{if } F_i' F_1' < 0, \end{cases}$$

leads to the conclusion. □

The set of monotonically increasing function $G = \{g_i\}_{i=1}^r$, where $g_1 = x$ and $g_i \in \mathcal{G}$, for $i \neq 1$, is referred to as the set of generating functions for F. Utilizing the generating functions (which all share a common domain J) overcomes the difficulty in specifying branching functions as in (4) (which map J_1 to other subintervals) and hence has advantage in simplifying the general formulation.

3. A Unified Formulation

The general formulation provided by Corollary 2.1 suggests that, for any given sign pattern S, there exist infinitively many complete chaotic maps that preserve a specified invariant measure μ. A more compact formulation can be obtained through the concept of topological conjugation.

Theorem 3.1. *A complete chaotic map $F = \{F_i\}_{i=1}^r$ preserves a normalized absolutely continuous invariant measure μ if and only if there exists a set of generating functions $G = \{g_i\}_{i=1}^r$, with $g_1(x) = x$ and $g_i \in \mathcal{G}$, for $i = 2, 3, ..., r$, so that each and every branch of F can be expressed as*

$$F_i(x) = \begin{cases} \mu^{-1}\left(\sum\limits_{j=1}^{r} \beta_j g_j\left(g_i^{-1}\left(\dfrac{\mu(x) - \alpha_{i-1}}{\beta_i}\right)\right)\right), & \textit{if } F_1' > 0 \textit{ and } F_i' > 0, \\ \mu^{-1}\left(\sum\limits_{j=1}^{r} \beta_j\, g_j\left(g_i^{-1}\left(\dfrac{\alpha_i - \mu(x)}{\beta_i}\right)\right)\right), & \textit{if } F_1' > 0 \textit{ and } F_i' < 0, \\ \mu^{-1}\left(1 - \sum\limits_{j=1}^{r} \beta_j g_j\left(g_i^{-1}\left(\dfrac{\alpha_i - \mu(x)}{\beta_i}\right)\right)\right), & \textit{if } F_1' < 0 \textit{ and } F_i' > 0, \\ \mu^{-1}\left(1 - \sum\limits_{j=1}^{r} \beta_j g_j\left(g_i^{-1}\left(\dfrac{\mu(x) - \alpha_{i-1}}{\beta_i}\right)\right)\right), & \textit{if } F_1' < 0 \textit{ and } F_i' < 0, \end{cases} \tag{6}$$

where $\alpha_i \doteq \mu(a_i)$ *and* $\beta_i \doteq \alpha_i - \alpha_{i-1}$, *for* $i = 1, 2, ..., r$.

Proof. The family of complete chaotic maps that preserve the Lebesgue measure (uniform invariant density) can be obtained through substituting into (2) with $\mu(x) = \mu^{-1}(x) = x$. To distinguish with the general chaotic maps preserving arbitrary invariant measure, we replace F and $A = \{a_i\}_{i=0}^r$ with $\tilde{F}$ and $B = \{\alpha_i\}_{i=0}^r$, respectively. It is easy to verify that (2) can be simplified into an unified expression:

$$\tilde{F}_i(x) = \begin{cases} \sum_{j=1}^{r} (\alpha_j - \alpha_{j-1})\, g_j\left(g_i^{-1}\left(\dfrac{x-\alpha_{i-1}}{\alpha_i - \alpha_{i-1}}\right)\right), & \text{if } F_1' > 0 \text{ and } F_i' > 0, \\ \sum_{j=1}^{r} (\alpha_j - \alpha_{j-1})\, g_j\left(g_i^{-1}\left(\dfrac{\alpha_i - x}{\alpha_i - \alpha_{i-1}}\right)\right), & \text{if } F_1' > 0 \text{ and } F_i' < 0, \\ 1 - \sum_{j=1}^{r} (\alpha_j - \alpha_{j-1})\, g_j\left(g_i^{-1}\left(\dfrac{\alpha_i - x}{\alpha_i - \alpha_{i-1}}\right)\right), & \text{if } F_1' < 0 \text{ and } F_i' > 0, \\ 1 - \sum_{j=1}^{r} (\alpha_j - \alpha_{j-1})\, g_j\left(g_i^{-1}\left(\dfrac{x-\alpha_{i-1}}{\alpha_i - \alpha_{i-1}}\right)\right), & \text{if } F_1' < 0 \text{ and } F_i' < 0. \end{cases} \tag{7}$$

By conjugation law, any complete chaotic map F, which preserves an invariant measure μ, can be expressed as a topologically conjugate of F given by $F = \mu^{-1} \circ \tilde{F} \circ \mu$. The monotonically increasing property of μ guarantees that F and $\tilde{F}$ have identical sign patterns. Moreover, two set of partitioning points are related through $\alpha_i \doteq \mu(a_i)$. Substituting $\tilde{F}$ with (7) yields (6). □

Apparently, given a sign pattern, $S = \{F_i'\}_{i=1}^r$, and a set of generating functions, $G = \{g_i\}_{i=1}^r$, the set of complete chaotic maps defined by (6) is usually different from the ones defined by (2). It can be verified that the former is identical to the latter if and only if $\mu(x) = x$.

Example 3.1. Consider a 3-to-1 complete chaotic map with $\mu(x) = x^2$ ($\varphi(x) = 2x$) $A = \{0, 1/3, 2/3, 1\}$ and $G = \{x, x^2, x\}$, Eq. (6) yield two sets of complete chaotic maps:

i) Corresponding to $s_1 > 0$, we have

$$\bar{F}_1(x) = x\sqrt{3(2+9x^2)}$$

$$\bar{F}_2(x) = \begin{cases} \sqrt{9x^2 - 1 + 2\sqrt{3(9x^2-1)}/3}, & \text{for } \bar{F}_2'(x) > 0, \\ \sqrt{4 - 9x^2 + 2\sqrt{3(4-9x)}/3}, & \text{for } \bar{F}_2'(x) < 0. \end{cases}$$

$$\bar{F}_3(x) = \begin{cases} \sqrt{270x^2 - 120 + 15\sqrt{5(9x^2-4)}/15}, & \text{for } \bar{F}_3'(x) > 0, \\ \sqrt{3(1-x^2)(19-9x^2)/5}, & \text{for } \bar{F}_3'(x) < 0. \end{cases}$$

Fig. 1(c) illustrates the relevant functional forms.

ii) Corresponding to $s_1 < 0$, we have

$$\tilde{F}_1(x) = \sqrt{(1-9x^2)(3x^2+1)}$$

$$\tilde{F}_2(x) = \begin{cases} \sqrt{5+9x^2-2\sqrt{12-27x^2}}/3, & \text{for } \tilde{F}_2'(x) > 0, \\ \sqrt{10-9x^2-2\sqrt{3(9x^2-1)}}/3, & \text{for } \tilde{F}_2'(x) < 0. \end{cases}$$

$$\tilde{F}_3(x) = \begin{cases} \sqrt{(4-9x^2)(3x^2-8)}/5, & \text{for } \tilde{F}_3'(x) > 0, \\ \sqrt{3(1-x^2)(9x^2+11)}/5, & \text{for } \tilde{F}_3'(x) < 0. \end{cases}$$

Fig. 1(d) illustrates the relevant functional forms. A typical computer simulation of realized frequency is provided in Fig. 1(b)

The beauty of general expression (6) lies in that it does not only provide an analytical solution to the standard inverse Frobenius-Perron problem, but also enables us to examine many analytical properties that are difficult to prove with the conventional approaches. For instance, consider a special case where $g_i(x) = x$, $i = 1, 2, ..., r$, it can be verified that, for all i, we have

$$\sum_{j=1}^{r} \beta_j (\mu(x) - \alpha_{i-1}) / \beta_i = 1 - \sum_{j=1}^{r} \beta_j (\alpha_i - \mu(x)) / \beta_i$$

so that (6) can be simplified into

$$F_i(x) = \begin{cases} F_i(x) = \mu^{-1}\left(\dfrac{\mu(x) - \mu(a_{i-1})}{\mu(a_i) - \mu(a_{i-1})}\right), & \text{if } F_i' > 0; \\ F_i(x) = \mu^{-1}\left(\dfrac{\mu(a_i) - \mu(x)}{\mu(a_i) - \mu(a_{i-1})}\right), & \text{if } F_i' < 0. \end{cases} \tag{8}$$

Therefore, for the special class with $g_i(x) = x$, for all i, at each predetermined interval, both upward and downward sloping branch are uniquely determined, i.e., their functional forms are independent of the derivative of F_1. On the other hand, the map given by (8) is nothing but a topological conjugate of complete piecewise linear chaotic map defined by

$$T_i(x) = \begin{cases} \dfrac{x - \alpha_{i-1}}{\alpha_i - \alpha_{i-1}}, & \text{if } T_i' > 0, \\ \dfrac{\alpha_i - x}{\alpha_i - \alpha_{i-1}}, & \text{if } T_i' < 0, \end{cases} \qquad i = 1, 2, ..., r.$$

It is well-known that $T = \{T_i\}_{i=1}^{r}$ preserves the Lebesgue measure.

This observation reveals the fact that *the class of complete chaotic maps that are conjugated to the complete piecewise linear chaotic maps consist of only a subset of all complete chaotic maps that preserve the same absolute continuous invariant measures.*

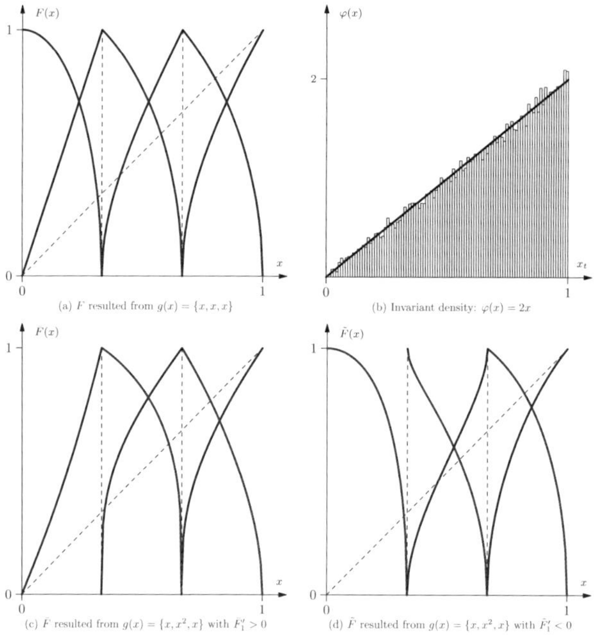

Fig. 1. Complete chaotic maps that preserve $\varphi(x) = 2x$.

Example 3.1 (Continued). *Fix* $\mu(x) = x^2$, $A = \{0, 1/3, 2/3, 1\}$ *but change* G *to* $\{x, x, x\}$, *Eqs (8) lead to*

$$F_1(x) = \begin{cases} 3x, & \text{if } F_1' > 0, \\ \sqrt{1 - 9x^2}, & \text{if } F_1' < 0, \end{cases}$$

$$F_2(x) = \begin{cases} \sqrt{3x^2 - 1/3}, & \text{if } F_2' > 0, \\ \sqrt{4/3 - 3x^2}, & \text{if } F_2' < 0, \end{cases}$$

$$F_3(x) = \begin{cases} \sqrt{(9x^2 - 4)/5}, & \text{if } F_3' > 0, \\ \sqrt{9(1 - x^2)/5}, & \text{if } F_3' < 0. \end{cases}$$

Fig. 1(a) illustrates the relevant functional forms.

Example 3.2. Consider a 3-to-1 complete chaotic map with $\mu(x) = 2\arcsin(\sqrt{x})/\pi$ ($\varphi(x) = 1/(\pi\sqrt{x(1-x)})$), $A = \{0, 1/4, 3/4, 1\}$ and $G =$

$\{x, x, x\}$, it can be verified that Eq. (8) results in

$$F_i(x) = \begin{cases} x(4x-3)^2, & \text{if } F_i' > 0, \\ (1-x)(1-4x)^2, & \text{if } F_i' < 0, \end{cases} \quad \text{for } i = 1, 3,$$

$$\text{and} \quad F_2(x) = \begin{cases} (1-x)(1-4x)^2, & \text{if } F_2' > 0, \\ x(4x-3)^2, & \text{if } F_2' < 0. \end{cases}$$

4. Conclusion

We have characterized the multi-segmental complete chaotic maps that preserve prescribed invariant densities and provide a set universal formulas for the construction of such maps. Future research can be directed to explore the mutual implication family, as have been done for two-segmental complete chaotic maps,[2] the maps preserving the densities with special characteristics such as the uniform density[1] and the symmetric densities.[3] The issues about the multi-segmental chaotic maps with special functional forms such as closed functional forms,[4] the Tent-map-like structure,[5] the reciprocal structure[6] and the linear branch[7] can be investigated.

References

1. W. Huang, *Chaos, Solitons & Fractals* **25**, 449 (2005).
2. W. Huang, *Nonlinearity* **18**, 1375 (2005).
3. W. Huang, *Chaos, Solitons & Fractals* **38**(4), 1065 (2008).
4. W. Huang, *Discrete Dynamics in Nature and Society* **2006**, 1 (2006).
5. W. Huang, *Discrete Dynamics in Nature and Society* **2005**(3), 357 (2005).
6. W. Huang, *Chaos, Solitons & Fractals* **24**, 287 (2005).
7. W. Huang and Y. Zhang, *Journal of Economic Behaviour and Organization* **62**(3), 389 (2007).

Chaotic Behavior of a Self-Replicating Robotic Population

G. Kaloutsakis
The Johns Hopkins University
Baltimore, MD, USA

A self-replicating modular robot has been developed that could operate with white noise input. The robot moves randomly around producing a Brownian style motion. We examine the influence of the initial conditions to the results of self-replication. We train a Feed-Forward Neural Network using back propagation to simulate the process results. We verify the approximation capabilities of the method using experimental data.

Keywords: Mobile robots; self-replication; robotic population; chaotic behavior.

1. Introduction

Self-replicating robots are self-reproducing machines. They are the physical example of self-automata population. A self-replicating mobile robot can move around, multiply by building functional replicas or die because of failure or termination of it's power apothems. Most of self-replicating robots that have been demonstrated in literature create copies of themselves by assembling modular replicas.[1–3] There still exists the challenge of creating a robot that can demonstrate manufacturing abilities and a robot that except of self-replication, will be demonstrating an extra useful operation i.e. energy production.

Our self-replicating model for un-structured environments is based on a probabilistic approach. A robot might not replicate for sure, but has a high chance of doing so. We are not also able to predict how many robots will be replicated at the end. We have experimentally determined that in a small bounded area about half of the modules to be assembled are compiling functional replicas. The system was tested for different initial conditions since we were interested on checking it's general ability to replicate.

The system has a highly non-linear behavior. That happens due to the Brownian type of motion and the infinitely high number of initial conditions that can be applied. Except of the random initial conditions the system is evolving randomly as well and that increases the complexity.

2. The Robot

Our self-replicating mobile robot consists of three identical modules. The modules get connected to each other with permanent magnets. They have a pizza slice shape and combining three of them gives us a round structure. Each module consists of a plastic frame, a servo motor, a Basic-Stamp controller from Parallax Inc., five AA batteries, two permanent magnets and an omnidirectional wheel (Figure 1). The operation of each module is independent to the others but the wheels will start moving only if three modules are form a robot. This happens because of metal inserts that are attached on the sides of the chassis and close the power circuit when being pushed by the neighbor module. In that way when a robot is not fully assembled, i.e. two modules only being attached to each other, the motor will not be powered and the wheels will not move.

Fig. 1. Three modules assembled to a robot. Each module consists of 1-Batteries, 2-Servo motor, 3-Controller, 4-Omnidirectional wheel and 5-Permanent magnets.

Because of the polarity of the magnets, if we put two modules in a box and shake it, there is only one way that they will get attached to each other. In the same way if we put three modules, first, two modules will get attached to each other and then there is only one way that the third module will get attached to them and this is the way of forming a full robot.

The controllers have been programmed to change the velocities of the wheels randomly. They send pulses to the servo motor with frequency that is generated by a normal distribution with mean zero (white noise). Thus the speed and the direction that the wheels move is random. The three omnidirectional wheels have been installed symmetrically and result a holonomic motion.[4]

By writing the robot's kinematics equations in the standard form and then solving the Fokker-Planck equations using the inverse Fourier transform properties,[5] we computed the distribution of the robot's position for a fixed time (Figure 2(a)). We can see that it is an isotropic gaussian distribution and thus we conclude that the robot will move randomly around. We can also see that the robot is performing a non-linear-chaotic motion from the trajectory plot generated from the kinematics equations (Figure 2(b)).

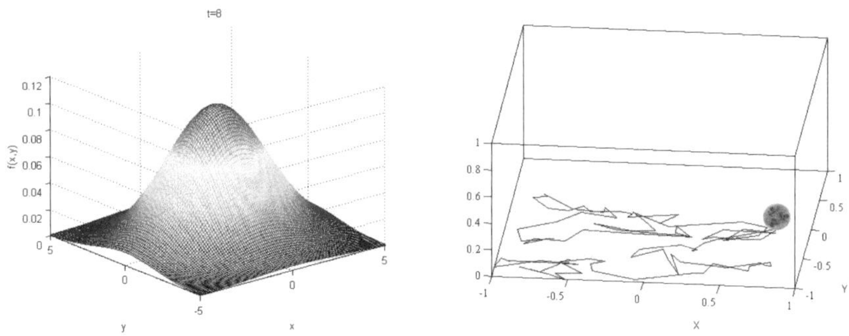

(a) Robot's position distribution for fixed time t=8 sec

(b) Trajectory generated with analytic simulation.

Fig. 2. Robot's position and trajectory.

3. The experiment

We place a robot and many modules inside a bounded planar surface. The robot moves around pushing the single modules. Eventually some of the modules come close to each other and because of the magnets they get assembled to double module formations or to new robots. Immediately after the replicas get assembled they start moving around as well, accelerating the rhythm that the system is changing.

For a one by one meter bounded area with twelve modules, one robot and random initial parts position and the orientation, generated from a uniform distribution, we experimentally computed that there is a small chance that all the modules will end up in double formations, while the average number of replicas assembled is two.[6] This result was measured over fifty completed runs.

3.1. *Initial conditions*

The initial position and orientation of the modules statistically affects the result of the self-replicating process. Since the position and orientation are continue measurements, there is infinitely high number of them that can be applied initially. If we convert their space from continue to discreet we will be able to choose from a certain number of initial conditions. We construct a square grid of 0.1 meters per line for positioning the modules in our arena. Regarding their orientation we divide the circle to eight arcs and choose the orientation of the modules to be equal to one of these eight discreet cases.

If we consider the modules to be distinguishable then we can have $2.145 \cdot 10^{117}$ different cases of initial conditions. If we consider that the twelve modules are not distinguishable then the number of applicable initial conditions is $4.49 \cdot 10^{108}$. This is still a quite high number and shows that we can't experimentally conclude how do the initial conditions effect the outcome of the experiment. We would like to be able to determine for the modules, certain initial positions and orientations that result higher probability for self-replication.

4. The Neural Network

Neural Networks (NN) can be used for approximating functions.[7,8] They can even approximate non-linear functions with higher accuracy than other classic methods, i.e. polynomial interpolation.

We use a Feed-Forward (FF) NN with thirty nine inputs, four hidden layers of twenty nodes each and one output. Inputs of the NN are the initial conditions, coordinates and orientations, while output is the outcome of the self-replication which is the number of total replicas produced. We have chosen a high number of layers and nodes per layer compared to other NNs in literature because the number of the high number of inputs. We use the Gradient Descent Learning (GDL) algorithm for training. GDL is an optimization process that produces a point $\mathbf{\Theta}$ in the NN weight space by minimizing a function $L(\mathbf{\Theta})$ (see [9]).

For training data we use the values measured from the fifty self-replication experiments.[6] We choose sigmoidal activation functions for the nodes of the hidden layers and a linear function for the output nodes. The NN successfully approximates the system's behavior with an error of 10^{25} (Figure 3). We tuned our training process for 100 epochs but it stopped at 30 epochs since the measured error was already quite small.

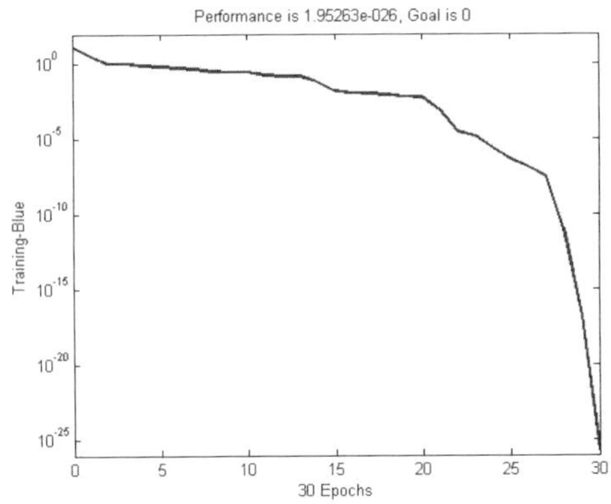

Fig. 3. The error of the NN's training process.

5. The Simulation

Using the NN we trained above, we can simulate the outcome of our self-replicating process for certain initial conditions. In this way we can instantly approximate how many replicas will be produced for given initial conditions. That result will not always be precise for single runs, but we expect that over many runs with the same initial conditions the predicted result will be the same to the real one.

To test our assumption we randomly generate a set of initial conditions. Inputting these values to the trained NN we get as an output three replicas. That means that we expect our self-replicating process, with these initial conditions, will produce three more copies of the robot.

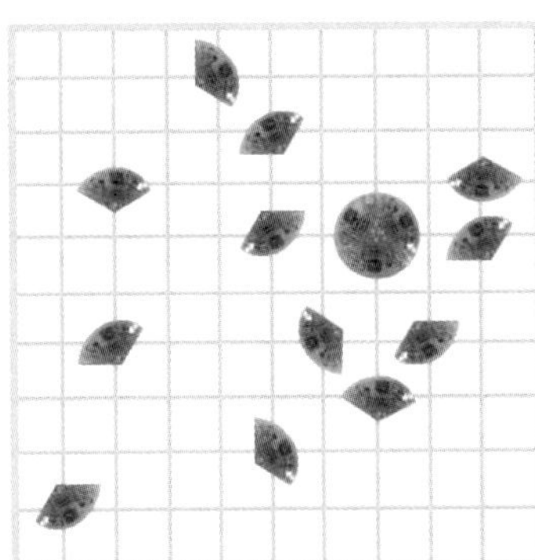

Fig. 4. Randomly generated initially conditions for testing the NN.

We set up the robots in the arena and have them complete ten self-replicating runs. In each time the robots are placed on the same positions and orientations, which were generated above (Figure 4). We record the number of robots assembled. In five cases the experiment produced three replicas, in three cases produced two replicas and in two cases produced four replicas. The average number of replicas produced is 2.9 when the predicted number from the NN was 3.

6. Conclusions and Future Work

We trained a NN to approximate the number of replicas produced in a self-replicating process for given initial conditions. The process is complex and non-linear. The robots used, are demonstrating complex trajectories that are similar to the random motion of particles in fluids (Brownian motion). The data we used was based on experimental measurements. For randomly generated initial conditions the NN approximated the average number of replicas produced by ten runs. Like that the NN can be used for predicting the success of the self-replicating process immediately after the initial conditions become known.

The NN can be also used for testing a large number of initial conditions that would require many months of real experiments. It can help us to come up with conclusion about which positions and orientations for the modules lead to more successful runs.

In order to strengthen our results more testing is needed. That is training a NN with larger data set and then testing the results in more than one cases of initial conditions.

References

1. K. Lee and G. S. Chirikjian, *IEEE Robotics and Automation Magazine* **14(4)**, 34 (2007).
2. A. C. J. Suthakorn and G. S. Chirikjian, *Proceedings of the IEEE/ASME Intl. Conf. on Advanced Intelligent Mechatronics* **1**, 137 (2003).
3. B. A. V. Zykov, E. Mytilinaios and H. Lipson, *Nature* **435**, 163.
4. G. Kaloutsakis and G. S. Chirikjian, *17th CISM-IFToMM Symposium on Robot Design, Dynamics, and Control* **submitted** (2007).
5. Y. Zhou and G. S. Chirikjian, *Proceedings of the IEEE ICRA* **2**, 1594.
6. G. Kaloutsakis and G. S. Chirikjian, *11th International Symposium on Experimental Robotics* **under submission** (2008).
7. A. Barron, *Machine Learning* **14**, 114.
8. K. Funahashi, *Neural Networks* **2**, 183.
9. D. Wilson and T. Martinez, *Neural Networks* **16**, 1429.

Chaos and Geodynamics in Nature Systems: Virtual Nature Systems Approach

V. I. Klenov

98-5-80 Profsouznaya St., 117485, Moscow, Russia
Email: klenovvaleriy@mail.ru

The objects of investigation are the most widespread Nature Systems on the Land - River Basins, their geodynamics and geoinformatics. For the purpose we worked out the Virtual Nature Systems (VNS) as computer doubles of River Basins and of their components in view to research their dynamics and evolution under continual influence of external factors, in analogy with the Actual Nature Systems (ANS).

Keywords: River Basin; Virtual Nature System; matrix; computer mapping; natural records; phase trajectory; Information Loss Law.

1. Introduction

River Basins belong to the Open Non-Equilibrium Systems - ONES,[1] which are always under continual pressure of external Exogenous (Climatic) and Endogenous (Tectonics) factors. Observing for river basins by geomorphology methods shows that Water and Sediments flows throughout the River Basin generate oscillations (meandering and branching) and thresholds (changes of structure and sharp hazardous flows), both in dynamics and evolution time scales. The problem for reconstruction of its history by observed tracks (sediments and river terraces) is in assessment of ability or these tracks to be natural records for the history has the obstacle in the Information Loss Law, as a result for own dynamics of the Systems. The discussed are: a) Information lacks and distorts in the Natural Records; b) properties and problems of the Virtual Nature Systems.

2. A 'Cost' for Ignorance the Information Loss Law in the ONES

This problem is a foundation for elaboration of computer doubles of Actual Nature Systems (ANS). The Virtual Nature System - River Basin (VNS) can reproduce dynamics and evolution of the Actual Nature Systems. For the purpose was used basic proposal for oscillations in evolution of nature systems. The Virtual River Basins exists under the same external influence as the ANS, by take it account major inner interactions and flows in the System. The first simplified computer model for the River Basins was worked out as several segments of single valley with estimation of water flows through the System. Increasing of water flows downstream and tectonic deformations for all or separate sections were provided by table functions. Change of riverbed's width during its cutting was generates by stochastic processes or as a table function for oscillation of river's discharge. Tectonic deforms of section increases upstream from a mouth. It is deterministic model with stochastic input of external factors. The Virtual Valley was experiences for many scenarios of both factors. For even section was estimated transport capacity of flows, which caused several responses: cutting of upper section in dependence of transport capacity of sediments, in dependence of difference between sections height. In dependence on value of transport capacity the response was by three cases: erosion, denudation, if simple transit of sediments. The riverbed width changes by widening of narrowing of the bed.[2] The general proposal was stochastic variation of riverbed oscillation in the time without trends by stochastic stationary process.

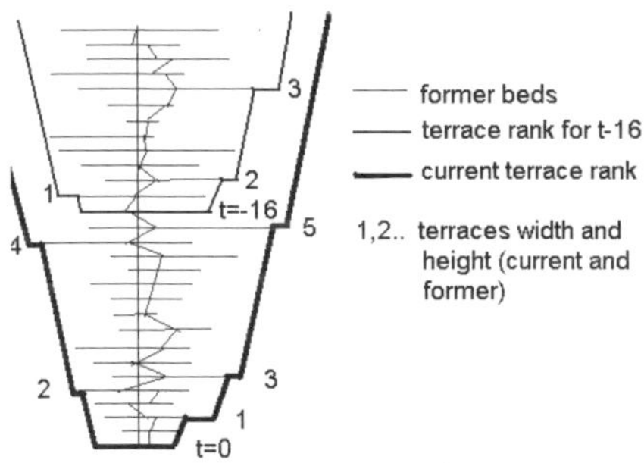

Fig. 1. Computed terrace complex (cross section) by proposal of multiple Riverbed oscillations

In the process of simulation was received effect of forming of river terraces (Figure 1). On the Figure the thick borders of cross section define the cross section at the moment $t = 0$ (present time). Terraces on both borders are usually interpreted as general stages of evolution during the Quaternary. Thin lines are the alternative hypothesis of riverbed's width oscillation in time. This process is accompanied by multiple forming and later removes of the levels. Number of level during the time varies by statistic oscillation of width usually between 2-5 levels. Both the first and second hypothesis explain terrace rank, but the second one (high frequency oscillations) is preferable because of it better explains variability of together observed and modeled terrace complexes along a valley (Figure 2). The comparison of observed and computed complexes[3,4] resulted (both by data of Observations and Statistic modeling) in several important conclusions as follows:

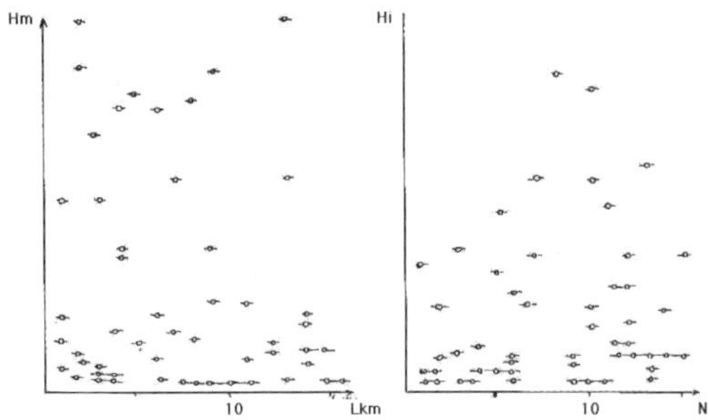

Fig. 2. Observed (left) and computed terrace complexes

(1) Stationary stochastic process of riverbed oscillation is fixed in present state of a valley as a Terrace Rank with increasing of a vertical distance between the neighbor terraces upslope;
(2) Most of terraces were removed in process of a valley deepening;
(3) The initially uninterrupted levels (along a valley) becomes fragmentary;
(4) Most of terraces are fragmentary initially;
(5) The heights of synchronic terrace levels (former riverbeds) oscillate during a valley, making levels hardly correlated.
(6) Terrace rank statistically is the Invariant (Figure 1) to a history of a river valley, because of it saves and reproduces not a history as a 'Natural Record' but in realty reflects own dynamics of a valley.

In the Geomorphology field practice, the levels for more or less equal height are assigned as simultaneous (because of absence of truthful dating of each fragmentary level). The existing methods of the dating make the hypothesis of stochastic oscillations preferable to explain observed facts.

These results in the following general conclusions: a) temporal blanks of a history between terraces increases upslope, making an illusion of trend for a climate/tectonics history, b) increasing for a frequency of terraces downslope makes illusion for increasing frequency of climate/tectonics oscillations from the Past. These illusions are still happens not only among non-specialists, but among geologists and geomorphologists, especially in explanation of the same trends in other natural records (ice, sediments).

The general conclusion is that the Natural Records cannot be used for any historic interpretations, because they cannot be objects of statistic and spectral analysis by making false trend for frequency and trends in the history and having increasing in the Past blanks in the records. Unfortunately, this simple model and many other statistic models of terrace ranks, terrace complexes, and sedimentation dynamics are until now ignored by geologists and geomorphologists. The false trends are written in accepted numerous Stratigraphy Schemes. Otherwise, the most part of the history stays blank, and in these blanks may be inserted any historical events, and not by a single way! These results are in coincidence with general properties of the ONES. Primary, just now interpretations of trends in Ice, Marine, Continental sediments, and in other Natural Records have not truthful foundations.

Research for Temporal Dynamics of the Nature System results in formulation of the Information Loss Law,[3] which is written by simple statistic (and probabilistic) equation:

$$P(h) = h^{-b}$$

where, P is the probability of terrace saving on a height above riverbed, h is the height above a riverbed, and b is a parameter, which decreases below 1 with lowering of valley's slopes sharpness ($b \leq 1$).

3. Benefits for the ONES

The VNS-RB was implied by the DOS/Fortran77 for a middle scale river basin (the Moscow River upstream), with grid's resolution 100m. The VNS-RB for the Moscow River upstream was implied and satisfactory validated by of meteorology records for 20 years.[5] The main purpose was validation of spring and summer floods by statistic and other criterions. The extended version of the VNS[6] obtains ability to recognize and follow up disastrous debris flows. It was done by stretching of a time step simultaneously with initiation of debris flows (in any part of a mountainous basin) by exceeding of geomorphology-based threshold. Then the VNS tracked the debris flow downstream until conditions of debris flows continued. The thresholds were calibrated by two years daily meteorology records (with and without debris flow events), and then the monitoring was prolonged for about 20 years with full coincidence of independently observed and computed accidents, with effects for debris flows meandering, branching, strong local sedimentation, forming/cutting of sediment jams, and others.

The VNS-RB was rewritten for the Visual Fortran language and was installed for the same mountainous basin, with a grid's spatial resolution 10 meters. It is satisfactory geomorphologic condition for estimation of debris flows widening, narrowing, and branching. Because of debris flows happen often inside a one day recording step then during the current step the tracks are set over current map, by overlapping.

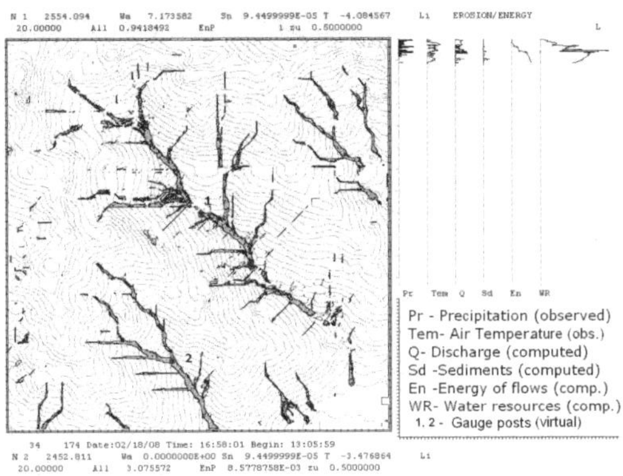

Fig. 3. Computer's map for Debris Flows Risk

Mapping of Debris Flows Risk (Figure 3) is set in separate Matrix layer. It is a foundation for Land Use planning, for Insurance, and for other aims. The map was evaluated by the PC Windows XP under scenario of Warming (by enlarged Air Temperature and Precipitation). The Mapping is provided regularly and automatically in all time of monitoring. The computer image (Figure 3) contains all tracks of debris flows for only 30 days. Size of the Matrix is 540×540 cells. It is the prototype of the System for regional real-time monitoring of Disasters for any vulnerable area. Evaluation scenarios for earthquakes, strong storms, and enlarging human' activity demonstrates that the ONES strongly sensitive to decreasing of soil resistance by tectonic faults and other reasons, sensitive to any deforming and incline of earth surface with response by catastrophic debris flows. The ONES does not generate storms and earthquakes, but it skills for quick response on these external impacts by sharp enlarge for disastrous processes.

The executing of the VNS-RB is in good similarity to earth processes: virtual water fills up all reservoirs, destroys dams after their overfilling, and loads sediments and pollutions, as actual water flows do it. Velocities of water flows are variable over are in accordance with riverbed incline and with other geomorphology conditions. All information on current processes may be read by location of cursor on any cell, by contour mapping of an area, and by corresponding records for observed and computed influences and processes on assigned points (virtual Gauge Posts). It provides functions for the GeoDynamics and GeoInformatics for the Nature Systems. The GeoDynamics of the ONES is a result for reforming of external (Meteorology) and internal (Tectonic) influences to non-equilibrium Environmental flows. The non-linear Geodynamics of the ONES is a main source and reason of Disasters (Floods, Debris Flows, Dam faulting, et al.). The GeoDynamics & Geoinformatics of the Nature Systems are indivisible.

The accuracy of the VNS is provided by continual comparison of computed events with actual ones[5,6] because they both have the single sources – actual external influences. May be, in the future will be available real-time connections with information sources.[7–9] Then the 2D satellite data should be immediately reformed by the Information machine of the VNS to assessments of all actual flows, with a recognition of all hazardous flows in any sites of an area. The efficiency for Warning-Response System depends on efficiency of interactions between: satellite–computer–end users. Locations and power of water related accidents might appear on display or even in the Internet through a few minutes after or even **earlier than** accidents initiation (but after impacts), in dependence of the accidents sites and distance to vulnerable sites. The sources and tracks of extreme flows are automatically recognized by the VNS and followed up through an area.

The external spatial-temporal dynamics of the Active Energy (Precipitation and Temperature) activates spatial-temporal Geodynamics of a Nature System. The corresponding Geoinformatics provides timely observing, assessment and urgent response (may be, automatic) on hazardous situations. Spatial-temporal stochastic dynamics of exterior factors (climatic), spatial-temporal stochastic dynamics of interior influences (tectonics) both provide spatial temporal stochastic GeoDynamics, as spatial-temporal structure and power of flows. This 'Wild Dance' of external reasons and System's responses (Geodynamics) is satisfactory recognized and followed by the Dynamic Mapping (GeoInformatics). This complex task is reliable for the VNS both in actual and scenario experiences, under condition of high-regular or real-time monitoring by the 2D information sources. Particularly, continual mapping for Debris-Flows Risk and for Floods Danger should be completed by 'Summary' or 'Current' (or both) regimes for mapping.

4. Problems for the ONES

The Open Non-Equilibrium Nature Systems (ONES) are major mechanics of non-equilibrium processes in most nature systems and major source of all water related disasters over the Earth. It becomes obvious in result of direct observing and modeling for response of the Nature Systems on all external powers. When non-linear pressure meets non-linear response, then disasters happen.

The major problem for the ONES and the VNS is that the systems are non-predicable because of deficiency for prediction of exterior reasons (Weather, Climate, Earthquakes, and other Tectonics). Resources of the Virtual Systems still exceed their provision by predicted data, by mechanics of processes and interactions in the external governing Systems. For example, precipitation pattern for earthquake's Power, Place, and Time are still non-predicable to be inserted in the VNS, but are immediately estimated by response of the Nature System on the events.

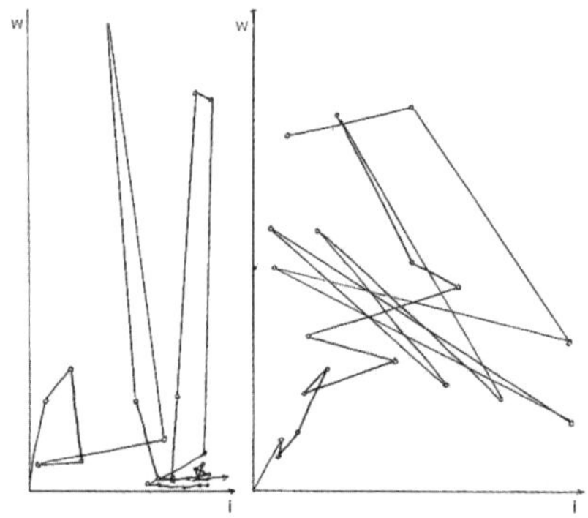

Fig. 4. Phase trajectories of a Basin's model, in axes: "Bed's Width" "Riverbed's Incline" under small (left) and large Precipitation

In other words, it is the strong necessity in acceptable methods for extraction of continual data not from non complete Natural Records, but from a memory of the Natural Systems. This is a property of the ONES' phase trajectories (Figure 4) to never coincide with non disturbed one, and to never return to not-disturbed trajectory after external influence. Always is a hope to extract the memory by moving 'Ahead to the Past' along the trajectory to restore all relevant external influences ('Natural records') and all history of the System's dynamics. The alternative is that the actual Past and the Future will be always unknown for use in practice.

In the beginning for the Nature Systems modeling was done attempt to restore a Map for the Past together with corresponding restoring for records of influences.[10] The attempt was to evaluate step by step the Past states and external regimes by moving back along the Phase trajectory (similar to Figure 2).

For the purpose was applied statistic method for step-by-step moving to the Past along the trajectory with multiply returning to previous state after non-successful step and with saving of satisfactory steps. Initially the small size Matrix and simple model (initial version of the VNS-RB) was computed on ancient computer for several steps ahead. Then the resulted last stage of 'direct' modeling was assigned as the initial for estimation of the Past. The first (and all current) step of the evaluation determines all following steps. Therefore, the task was restricted by the first step. The received before all steps of the direct modeling are not used in the 'Back Task', and are used only for estimation of exactness (what never happens in realty). Results of the numerical experiment are below:

Table 1. Assessment for the 'Back Task' decision

A – 'Initial' relief	4.770 3.758	2.898 2.286	1.321 1.046
B – 'Reconstructed' Past relief	4.961 3.856	3.009 2.344	1.360 1.070
C – 'True' Past relief	4.942 3.865	3.000 2.345	1.367 1.072

Difference between **B** and **C** is a measure of accuracy. Resulted is the meaningful trend to decision of the Back Task. The method for compare for evaluated **B** and a priory known **C** is not acceptable in actuality. For the purpose of actual cases it is proposed other method. It is a difference between the initial **A** and the estimated **A** by return 'Back to the Future' from the **B** to the **A**. The task was not finished exactly, but it seems, that a single way to complete GeoInformatics for the 'ONES' lies in use of completed Virtual Nature Systems. Common mistake for reconstruction of the Elevation Matrix was 8%. Mistake for the reconstructed Precipitation was 12%. Converge of decision by many iterations gives a hope to increase accuracy by powerful computing. However, it was applied only for a model.

5. Conclusions

Together with the VNS River Basin were worked out other models. The first is the VNS Coastal Zone, which was satisfactory applied for evaluation of bottom/shore abrasion and sedimentation under wind wave influence in several gulfs of the Japan Sea. The second is the coupled VNS which joins models for the River Basins and for the Coastal zone ('RIDEC'). Evaluated was a variety of scenarios for Dynamics and interactions between two major Systems of the Earth. The RIDEC uses separate complexes of external drivers and of inner processes inside a common Matrix. In the RIDEC processes of both separate systems are partially overlapped. The RIDEC was installed as the Virtual Double for the Rhine Basin and for the Coastal Zone of North Sea. It was used for scenarios of GeoDynamics under common and local Precipitation, for soft and sharp tectonics distorts of the vast area, trends, and oscillations of a sea level, and others, and for several years. This is convenient object for evaluation for traces of Cyclones simultaneously with sea level changes, for influence of Land Use policy on erosion, for influence of dams/reservoirs on floods mitigation, for evaluation of toxic pollutants spreading through rivernet and coastal zone, and other tasks.

The discussed above GeoDynamics of the ONES by their Virtual Doubles shows total deficiency of data for long natural records of external factors in view to divide problematic Natural climate warming from the Actual Warming due non-rational human's activity. The GeoInformatics and the GeoDynamics of the ONES is provided only by real-time and for directly observed meteorological and other records.

Otherwise, necessary records due the Information Loss Law in the Natural Records de facto do not provide the GeoInformatics for the Evolution of the ONES. It stays the governing problem for reconstruction the Past and Forecasting the Future of the ONES. The method for restoration the 'Memory' of the Natural Systems seems the evaluation and analysis for their Strange Attractors for the purpose to imply their properties to follow Back along a single string for the Evolution.

References

1. I. Prigogine and I. Stengers, *Order Out of Chaos* (Heinemann, London, 1984).
2. V. I. Klenov, *Geomorphologia* **1**, 32 (1986).
3. V. I. Klenov, River terraces and information losses, in *New methods for Geography explorations*, (MFGO, Moscow, 1981) pp. 54–62.
4. V. I. Klenov, *Geomorphologia* **4**, 39 (1985).
5. V. I. Klenov, Experience of nature systems simulation, in *Hydroinformatics 2000, Proceedings, Iowa, USA*, (2000).
6. V. I. Klenov, Debris-flow recognition using an extended version of the river basin simulation model, in *Debris-Flow hazards mitigation: Mechanics, Prediction, and assessment, Millpress, Rotterdam*, (2003).
7. H. Fukui, Gis next. digital earth, its' ideal, situation and future, in *From Digital Earth to Digital Asia*, 9, 11 (2003).
8. O. Altan, Use of photogrammetry, remote sensing, and spatial information technologies in disaster management, especially earthquakes, in *Geo-information for Disaster Management*, (Delft University of Technology, the Netherlands, 2004).
9. R. Laurini, S. Servigne and G. Noel, Real-time for disaster management, in *Geo-information for Disaster Management*, (Delft University of Technology, the Netherlands, 2005).
10. V. I. Klenov, *Geomorphologia* **1**, 51 (1989).
11. V. I. Klenov, The moving digital earth technology (mde) for monitoring of forthcoming disasters, in *Proceedings of the 3rd International ISCRAM Conference, Newark, USA*, (2006).

Self-Organization Processes in a Slow-Flowing Gravitational Compressible Cosmological Body

A. M. Krot

United Institute of Informatics Problems of National Academy of Sciences of Belarus
Minsk, Belarus
Email: alxkrot@newman.bas-net.by

This paper shows that strength of the gravitational field of a slowly contracting spheroidal body satisfies a second order differential equation of parabolic type for the case of unobservable velocities of particles. Therefore gravitational waves of a soliton type can be propagated in a slowly compressible gravitating spheroidal body. The equation for spherical autowaves of magnitudes of gravitational field strength is derived. This paper introduces the velocity of autowave front propagation for gravitational strength magnitude. The equation of movement of autowave front in a slowly compressible gravitating spheroidal body is found.

Keywords: Slowly compressible gravitating spheroidal body; strength of gravitational field; spherical autowaves of magnitudes of gravitational strength; equation of movement of autowave front propagation.

1. Introduction

The statistical theory for a cosmological body forming (so-called the spheroidal body model) has been proposed in the works [1–4]. Within the framework of this theory, gravitating bodies have fuzzy outlines, and they are represented by means of *spheroidal* forms. A slowly evolving in time process of a gravitational compression of a spheroidal body (close to an unstable mechanical equilibrium state) has been investigated in [2]. The proposed theory follows from the conception for forming a spheroidal body as a cosmological body (in particular, proto-Sun) inside a gas-dust nebula (protoplanetary cloud); it permits to derive the form of distribution functions for an immovable spheroidal body and rotating one (as well as their mass density, potential and strength of gravitational field)[1,2] and also to find the distribution function of specific angular momentum for a rotating spheroidal body.[3,4]

This work studies a slowly evolving process of gravitational condensation of a spheroidal body from an infinitely distributed gas-dust substance in space. The equation for an initial evolution of mass density function of a gas-dust cloud is considered here. It is found this equation coincides completely with the analogous equation for a slowly gravitational compressed spheroidal body.[2]

The first equation for spherical autowaves in a spheroidal body is obtained relative to a strength vector $\vec{a}$ of gravitational field. We suppose that a spheroidal body as a system in homogeneous limit case ($\partial\vec{a}/\partial r = 0; r = r_*$ is a point of the spheroidal body mass density bending)[1,2] has two stable solutions $\vec{a}^{(1)}(r,t)$, $\vec{a}^{(2)}(r,t)$ and one instable solution $\vec{a}^{(3)}(r,t)$. Analogously to [5] we seek solutions corresponding to the bigger magnitude of vector $\vec{a}^{(2)}$ in the left part ($r_2 = 0$) and the smaller one $\vec{a}^{(1)}$ in the right part ($r_1 = \infty$).There becomes a long transient domain (a autowave front) between the states $\vec{a}^{(1)}$and $\vec{a}^{(2)}$. A location R of autowave front can be determined (by analogy with [5]) as a value r when the strength magnitude corresponds a mean (instable) stationary magnitude of vector $\vec{a}^{(3)}$, i.e. $\vec{a}(R,t) = \vec{a}^{(3)}$. Then the velocity of autowave front propagation is to be found by differentiation of this obtained equation. Taking into account the first equation for spherical autowaves together with the definition of velocity of autowave front propagation we obtain the equation of autowave front motion in a slowly gravitational compressible spheroidal body under consideration.

2. Differential Equations for Physical Values Describing a Slow-Flowing Gravitational Compression under Unobservable Velocities of Particles

Let us find differential equations satisfied by physical values describing a slow-flowing process of gravitational compression of a spheroidal body under unobservable velocities of particles. Making use of the reasoning presented in the works [2,6,7], we assume similarly that $\alpha = \alpha(t)$ is a positively definite, monotonically increasing time function of gravitational compression. First of all, let us consider the mass density as the function of two variables r and t:

$$\rho = \rho(r, \alpha(t)) = M \left(\frac{\alpha}{2\pi}\right)^{3/2} e^{-\alpha r^2/2}, \tag{1}$$

where r is a distance from the mass center of spheroidal body, t is a time, M is a spheroidal body mass. Taking into account the spheroidal body's centrally-symmetrical distribution of mass density, in paper [2,7] it has been shown that

$$\frac{\partial \rho}{\partial t} = -\left(\frac{1}{2\alpha^2} \cdot \frac{d\alpha}{dt}\right) \nabla^2 \rho, \tag{2}$$

where ∇^2 is the Laplacian. Denoting through the*gravitational compression coefficient*:[2,2,7]

$$G(t) = \frac{1}{2\alpha^2} \cdot \frac{d\alpha}{dt}, \tag{3}$$

we write down the equation of the spheroidal body's gravitational compression in the form:[7]

$$\frac{\partial \rho}{\partial t} = -G(t)\nabla^2 \rho. \tag{4}$$

Since $\alpha(t)$ is a monotonically increasing function then $\mathrm{G}(t) > 0$. This means that, according to (4), the initial disturbance of mass density will not decrease (as in the diffusion equation[8]) but increase. In [2], two particular cases of equation (4) have been considered: that of quasi-equilibrium gravitational compression ($\alpha = t/2G_1$) and that of initial gravitational compression ($\alpha = t^2/G_2$) of a weakly gravitating spheroidal body.

Let us find the differential equation form satisfied by the gravitational field strength $a(r,\alpha(t))$ (see [1,9]) in a process of gravitational compression.[2,6,7] We shall begin with expressing $a(r,\alpha)$ through $\rho(r,\alpha)$, applying (1):

$$a(r,\alpha) = \sqrt{\frac{2}{\pi}} \frac{\gamma\, M\alpha^{3/2}}{r^2} \int_0^r x^2 e^{-\alpha x^2/2} dx = \frac{4\pi\gamma}{r^2} \int_0^r x^2 \rho(x,\alpha) dx, \tag{5}$$

where γ is the Newton' gravitational constant. It is not difficult to obtain from (5) that

$$\rho(r,\alpha) = \frac{1}{4\pi\gamma} \frac{1}{r^2} \frac{d}{dr} (r^2 a(r,\alpha)).$$

Substituting the last equation in (2) one obtains:

$$\frac{\partial}{\partial \alpha} \left[\frac{1}{r^2} \frac{\partial}{\partial r} (r^2 a(r,\alpha))\right] = -\left(\frac{1}{2\alpha^2}\right) \nabla^2 \left[\frac{1}{r^2} \frac{\partial}{\partial r} (r^2 a(r,\alpha))\right]. \tag{6}$$

Applying the expression for the Laplace's operator of a scalar value in spherical coordinates: $\nabla^2 [\cdot] = \left(\frac{1}{r^2}\right) \left[\frac{\partial}{\partial r} \left(r^2 \frac{\partial}{\partial r} [\cdot]\right)\right]$, the expression (6) will be written in the form:[7]

$$\frac{\partial a(r,\alpha)}{\partial \alpha} = -\frac{1}{2\alpha^2} \frac{\partial}{\partial r} \left[\frac{1}{r^2} \frac{\partial}{\partial r} (r^2 a(r,\alpha))\right] \tag{7}$$

On the other hand, taking into consideration that the potential (irrotational) gravitational field strength $\vec{a}(r,\alpha)$ is a vector value, its Laplacian in spherical coordinates has the form:

$$\nabla^2\vec{a}(r,\alpha) = \mathrm{grad}(\mathrm{div}\,\vec{a}(r,\alpha)) = \frac{\partial}{\partial r}\left(\frac{1}{r^2}\left[\frac{\partial}{\partial r}\left(r^2 a(r,\alpha)\right)\right]\right)\vec{e}_r, \qquad (8)$$

where $\vec{e}_r$ is a unit basis vector along coordinate r. Substituting the equation (8) in (7), the gravitational field strength equation is written in the vector form:[7]

$$\frac{\partial\vec{a}}{\partial t} = -\,G(t)\nabla^2\vec{a} \qquad (9)$$

As is seen from equation (9), the differential equation of gravitational field strength describes the space-time dynamics of gravitational field strength value for a spheroidal body. If eq. (9) is considered to be a gravitational wave equation, then its solutions (5) are solitary waves (or solitons).[7] For the case of a quasi-equilibrium compression ($\alpha = t/2G_1$) the gravitational soliton is described by $a(r,t) = \frac{1}{2\sqrt{\pi}}\frac{\gamma M}{r^2}\left(\frac{t}{G_1}\right)^{3/2}\int\limits_0^r x^2\exp\left[-x^2t/4G_1\right]dx$. In a process of its propagation, the soliton $a(r,t)$ is deformed in such way that its "center of gravity" (the point of strength maximum) is displaced towards the center of coordinates of a spheroidal body.

Relating to the "center of gravity", let us note the point of the mass density inflection

$$r_* = 1/\sqrt{\alpha} = r_{\max} \qquad (10)$$

is simultaneously the point of strength maximum of gravitational field for a spheroidal body.[9]

If $r_* \sim R$, where R is a radius of a gas-dust nebula (protoplanetary cloud), then $\alpha \sim 1/R^2$ is a very small positive parameter of gravitational compression.[1–4,6,7,9]

3. Velocity of Autowave Front Propagation in a Slowly Gravitational Compressible Spheroidal Body

According to (8), (9) the magnitude of a strength vector of gravitational field $a(r,t)$ satisfies the following equation:

$$\frac{\partial a(r,t)}{\partial t} = -G(t)\left[\frac{\partial^2 a(r,t)}{\partial r^2} + (2/r)\frac{\partial a(r,t)}{\partial r} - (2/r^2)a(r,t)\right] \qquad (11)$$

It is known the analogous equation (11) in the theory of dissipative structures.[5,10] Generally speaking, concentrations $n_i(\vec{r},\ t)$ should be considered as continuous functions of coordinates $\vec{r}(x, y, z)$ and time t within framework of this theory. In general case, the local concentrations satisfy the field equations containing flow and diffusion terms:[5,10]

$$\partial n_i(\vec{r},\ t)/\partial t + \vec{v}\,\mathrm{grad}\, n_i(\vec{r},\ t) - D_i\nabla^2 n_i(\vec{r},\ t) = f_i(n_1,\ ...,\ n_I), \qquad (12)$$

where $\vec{v}$ is a hydrodynamic velocity, D_i is a coefficient of diffusion for $i-$th substance, f_i is a term describing a chemical production of substance. If the hydrodynamic flows are absent in the continuous system with diffusion, i.e. $\vec{v} = 0$, then for a chemical system with I components, the closed system of field equations $(i = 1, 2, ..., I)$ can be written as follows:

$$\partial n_i(\vec{r},\ t)/\partial t = f_i(n_1,\ ...,\ n_I) + D_i\nabla^2 n_i(\vec{r},\ t), \qquad (13)$$

where f_i is a nonlinear function describing a rate for producing $i-$ component, and a diffusion coefficient D_i is supposed to be a constant only in that case when interactions between components can be neglected.[5]

Let us study the main statements of theory for plane, cylindrical and spherical autowave front propagations of concentration[5] having in mind the one-component system case with a diffusion:

$$\partial n(\vec{r},\ t)/\partial t = f(n) + D\nabla^2 n(\vec{r},\ t). \qquad (14)$$

On introducing the dimensionless variables:

$$X = n/n_0\ ,\quad \tau = t/t_0\ ,\quad \xi = r\,[D\ t_0]^{-1/2}\ , \qquad (15)$$

let us transform the equation (14) to the form in spherical coordinates:

$$\partial X(\xi,\ \tau)/\partial\tau = f(X) + (2/\xi)(\partial X/\partial\xi) + \partial^2 X/\partial\xi^2. \qquad (16)$$

By means of generalizing, we obtain the equation for plane $(\nu = 1)$, cylindrical $(\nu = 2)$ and spherical $(\nu = 3)$ autowaves:[5]

$$\partial X(\xi,\ \tau)/\partial\tau = f(X) + ((\nu - 1)/\xi)\partial X/\partial\xi + \partial^2 X/\partial\xi^2. \qquad (17)$$

Let us note that on introducing the dimensionless variables:

$$E = a/a_0,\quad \tau = -t/t_0\ ,\quad \xi = r/\sqrt{Gt_0}, \qquad (18)$$

the equation (11) can be reduced to equation of the kind:

$$\partial E/\partial\tau = \partial^2 E/\partial\xi^2 + (2/\xi)\partial E/\partial\xi - 2E/\xi^2. \qquad (19)$$

Comparing the equation (19) with (17) we can see that

$$\nu = 3, \qquad (20)$$

$$f(E) = -2E/\xi^2, \qquad (21)$$

$$\xi = r/\sqrt{G(t)t_0}. \tag{22}$$

According to (19), (20) spherical autowaves of magnitudes of gravitational field strength can be appeared in a slowly gravitational compressible spheroidal body.[11] We suppose that a spheroidal body as a system in homogeneous limit case ($\partial E/\partial\xi = 0; \xi = \xi_*$ is a point of the spheroidal body mass density inflection (10)) has two stable solutions $E^{(1)}(\xi, t)$, $E^{(2)}(\xi, t)$ and one instable solution $E^{(3)}(\xi, t)$. Analogously to the stated in [5] we seek solutions corresponding to the bigger dimensionless magnitude of a vector strength $E^{(2)}$ in the left part ($\xi_2 = 0$) and the smaller one $E^{(1)}$ in the right part ($\xi_1 = \infty$). There becomes a long transient domain as an *autowave front*[5] between the states $E^{(1)}$ and $E^{(2)}$. A location R of autowave front can be determined as a value ξ when the dimensionless strength magnitude corresponds to a mean (instable) *stationary* dimensionless magnitude $E^{(3)}$:

$$E(R,\ \tau) = E^{(3)}. \tag{23}$$

From the condition of stationary relating to time it follows that

$$dE^{(3)}/d\tau = dE(R,\ \tau)/d\tau = 0, \tag{24}$$

whence the velocity of autowave front propagation $\dot{R}$ is to be found by differentiation of the equation (23) under condition (24):

$$\partial E(R,\ \tau)/\partial\tau + (\partial E(R,\ \tau)/\partial R)\dot{R} = 0. \tag{25}$$

If follows from (25) directly that

$$\dot{R} = -(\partial E/\partial\tau)/(\partial E/\partial R). \tag{26}$$

Taking into account the designations (20), (21) now let us consider the equation (19). By multiplying it on $\partial E/\partial\xi$ and then on $d\xi$ we obtain:

$$\frac{\partial E}{\partial\tau}\frac{\partial E}{\partial\xi}d\xi = \frac{\partial^2 E}{\partial\xi^2}\frac{\partial E}{\partial\xi}d\xi + +((\nu-1)/\xi)\left(\frac{\partial E}{\partial\xi}\right)^2 d\xi + f(E)dE, \tag{27}$$

where $\nu = 3$ and $f(E) = -2E/\xi^2$ according to (20), (21). Bearing in mind that the first addition in the right part (27) can be represented by the following:

$$(\partial^2 E/\partial\xi^2)(\partial E/\partial\xi) = \frac{1}{2}\partial\left\{(\partial E/\partial\xi)^2\right\}/\partial\xi,$$

let us rewrite the equation (27) and then integrate it from ξ_2 to ξ_1:

$$\begin{aligned}\int_{\xi_2}^{\xi_1}(\partial E/\partial\tau)(\partial E/\partial\xi)d\xi &= (1/2)\int_{\xi_2}^{\xi_1} d\left\{(\partial E/\partial\xi)^2\right\}\\ &+(\nu-1)\int_{\xi_2}^{\xi_1}(1/\xi)(\partial E/\partial\xi)^2 d\xi + \int_{E^{(2)}}^{E^{(1)}} f(E)dE.\end{aligned} \tag{28}$$

Since the points ξ_2 and ξ_1 correspond to *stationary* states of field with the dimensionless strength magnitudes $E^{(2)}$ and $E^{(1)}$ respectively then

$$(1/2)\int_{\xi_2=0}^{\xi_1=\infty} d\left\{(\partial E/\partial\xi)^2\right\} = (1/2)(\partial E/\partial\xi)^2 \Big|_{\xi_2=0}^{\xi_1=\infty} \\ = (1/2)\left[(\partial E/\partial\xi)^2\Big|_{\xi=\xi_1=\infty} - (\partial E/\partial\xi)^2\Big|_{\xi=\xi_2=0}\right] = 0. \tag{29}$$

Taking into account (29) let us rewrite the equation (28) representing an integral of energy:

$$\int_{\xi_2}^{\infty} (\partial E/\partial\tau)(\partial E/\partial\xi)d\xi = \int_{E^{(2)}}^{E^{(1)}} f(E)dE + (\nu-1)\int_{\xi_2}^{\infty}(1/\xi)(\partial E/\partial\xi)^2 d\xi$$

or

$$\int_{\xi_2}^{\infty} \left[-(\partial E/\partial\tau)/(\partial E/\partial\xi) + (\nu-1)/\xi\right](\partial E/\partial\xi)^2 d\xi = \int_{E^{(1)}}^{E^{(2)}} f(E)dE. \tag{30}$$

The analogous equation (30) relating to a dimensionless concentration X has been obtained by W. Ebeling for the first time.[5] Since a dependence of concentration X from ξ has the form like (1) then the derivative $\partial X/\partial\xi$ has a sharp maximum near a point $\xi = \xi_* = R$ in a domain of sharp fall of the function $X(\xi)$. This fact permits to write the analogous equation (30) relating to X approximately in the form:[5]

$$[-(\partial X/\partial\tau)/(\partial X/\partial R) + (\nu-1)/R]\int_{\xi_2}^{\infty}(\partial X/\partial\xi)^2 d\xi = \int_{X^{(1)}}^{X^{(2)}} f(X)dX. \tag{31}$$

However, in the case of equation (30) relating to the dimensionless strength magnitude E the function $E(\xi)$ near point $\xi = \xi_* = R$ has a sharp maximum, therefore, $\partial E/\partial\xi|_{\xi=R} = 0$. In this connection, designating by

$$g(\xi) = -(\partial E/\partial\tau)/(\partial E/\partial\xi) + (\nu-1)/\xi; h(\xi) = (\partial E/\partial\xi)^2, \tag{32}$$

let us use the 2nd theorem on average:

$$\int_{\xi_2}^{\xi_1} g(\xi)h(\xi)d\xi = g(\theta)\int_{\xi_2}^{\xi_1} h(\xi)d\xi \qquad (\xi_2 \le \theta \le \xi_1)$$

under condition that g and h are continuous functions besides the function h does not change its sign (the last is evident from (32)).

Choosing $\theta = R$ based on the 2nd theorem on average, let us transform the left part of equation (30) to the form:

$$\begin{aligned}&\int_{\xi_2}^{\infty} [-(\partial E/\partial\tau)/(\partial E/\partial\xi) + (\nu - 1)/\xi] (\partial E/\partial\xi)^2 d\xi \\ &= \{-(\partial E/\partial\tau)/(\partial E/\partial R) + (\nu - 1)/R\} \int_{\xi_2}^{\infty} (\partial E/\partial\xi)^2 d\xi.\end{aligned} \tag{33}$$

Introducing the notation:

$$u = \int_{E^{(1)}}^{E^{(2)}} f(E)dE \Big/ \int_{\xi_2}^{\infty} (\partial E/\partial\xi)^2 d\xi \tag{34}$$

and using the equations (26),(30),(33) we obtain the *equation of movement of autowave front*:[5]

$$\dot{R} = u - (\nu - 1)/R. \tag{35}$$

Let us note that equation (35) is true only if the value R is large in comparison with a transient domain width Δ, i.e. $R \gg \Delta$. In the case of plane autowaves (when $\nu = 1$) this equation (35) becomes very simple one:

$$\dot{R} = u. \tag{36}$$

In other words, u is an *autowave velocity* whose value is defined by the formula (34). In the considered case when $\nu = 3$ (spherical autowaves) and $R = \xi_*$ the equation (35) has the form:

$$\dot{R} = u - 2/\xi_*. \tag{37}$$

According to (37) $\dot{R}$ is a velocity of spherical autowave front propagation while u is an autowave velocity (34) and $\xi_* = r_*/\sqrt{Gt_0}$ with regard to (22). Taking into account (10), (24) the formula (37) can be described as follows:

$$\dot{R} = u - 2\sqrt{\alpha G t_0}. \tag{38}$$

Let us estimate the autowave velocity (34) at long distance from the center of a slowly gravitational compressible spheroidal body, i. e. when $\xi \geq R$. A dimensionless strength magnitude of gravitational field in a remote zone is equal to

$$E^{(1)}(\xi, t) = \frac{4\pi\gamma\rho_0}{3a_0\alpha^{3/2}} \frac{1}{G(t)t_0\xi^2}. \tag{39}$$

Taking into account the formula (34) as well as the equation (39) it is not difficult to see that

$$u = -\frac{5}{6}\sqrt{\alpha G t_0}. \tag{40}$$

On substituting the formula (40) into equation (38), we obtain the velocity of of spherical autowave front propagation:

$$\dot{R} = -\frac{17}{6}\sqrt{\alpha G t_0}.$$

in a remote zone of a slowly gravitational compressible spheroidal body.[1–4,11]

References

1. A. Krot, *Achievement in Modern Radioelectronics* **8**, 66 (1996).
2. A. Krot, Statistical description of gravitational field: a new approach, in *Proc. SPIE's 14th Annual Intern. Symp. "AeroSense"*, 4038, 1318 (Orlando, Florida, USA, 2000).
3. A. Krot, The statistical model of rotating and gravitating spheroidal body with the point of view of general relativity, in *Proc.35th COSPAR Scientific Assembly*, A-00162, (Paris, France, 2004).
4. A. Krot, *Geophys. Res. Abstracts* **8**(A-00216) (2006).
5. W. Ebeling, *Origin of Structures at Irreversible Processes: Introduction in the Theory of Dissipative Structures* (Rostock, 1977).
6. A. Krot, The model of slow-gravitating spheroidal body based on statistical approach, in *CD-ROM Proc. 2nd World Conf. on Mathem. and Comput. in Physics*, (Vouliagmeni, Athens, Greece, 2000).
7. A. Krot, Gravidynamical equations for a weakly gravitating spheroidal body, in *Proc. SPIE's 15th Annual Intern. Symposium "AeroSense"*, 4394, 1271 (Orlando, Florida, USA, 2001).
8. L. Landau and E. Lifschits, *Fluid Mechanics* (Pergamon, Oxford, 1959).
9. A. Krot, Use of the statistical model of gravity for analysis of nonhomogeneity in earth surface, in *Proc. SPIE's 13th Annual Intern.Symposium "AeroSense"*, 3710, 1248 (Orlando, Florida, USA, 1999).
10. G. Nicolis and I. Prigogine, *Self-organization in Nonequilibrium Systems:From Dissipative Structures to Order through Fluctuation* (John Willey and Sons, New York, 1977).
11. A. Krot, *Chaos, Solitons and Fractals* (2008).

Disagreement between Newtonian and Relativistic Trajectories for a Low-Speed Kicked Dissipative System

B. L. Lan
School of Engineering, Monash University
Selangor, Malaysia
Email: lan.boon.leong@eng.monash.edu.my

It is shown that, contrary to conventional belief, the Newtonian and relativistic predicted trajectories for a low-speed delta-kicked dissipative system can eventually disagree completely. Similar results were found previously for two non-dissipative systems. These results imply that, for a slow-moving dynamical system, once the trajectory predicted by Newtonian mechanics is completely different from the empirically-correct trajectory predicted by special relativistic mechanics, Newtonian mechanics can no longer be used to study its dynamics. Special relativistic mechanics must be used instead.

Keywords: Dynamical systems; low speed motion; Newtonian dynamics; relativistic dynamics.

1. Introduction

It is conventionally believed,[1–4] that if the speed of a particle remains low, i.e., much less than the speed of light, then the trajectory predicted by special relativistic mechanics remains very well approximated by the trajectory predicted by Newtonian mechanics for the same parameter(s) and initial conditions. Einstein[1] wrote:

> Classical [Newtonian] mechanics required to be modified before it could come into line with the demands of the special theory of relativity. For the main part, however, this modification affects only the laws for rapid motions, in which the velocities of matter v are not very small as compared with the velocity of light. ...*for other motions [v very small compared with the velocity of light] the variations from the laws of classical mechanics are too small to make themselves evident in practice* (emphasis is mine).

Corben and Stehle[2] wrote:

> ...[Relativistic mechanics] when applied to problems in which $v \ll c$, leads to results experimentally indistinguishable from those obtained [using Newtonian mechanics] ...

Ford and Mantica[3] wrote:

> ...the predictions of special relativity must agree with those of Newtonian dynamics when particle speeds are small compared to the speed of light ...

More recently, McComb, in the preface to [4], wrote:

> ...for speeds $v \ll c$, Newton's laws are restored to us.

However, it was recently shown with two counterexample Hamiltonian (i.e. non-dissipative) systems[5,6] that although the particle speed is low, the Newtonian trajectory can eventually disagree completely with the relativistic trajectory, regardless of the type (chaotic or non-chaotic) of each trajectory. Another counterexample is given here for a delta-kicked dissipative system which is described by the dissipative standard map[7–9] from the Newtonian perspective. Implication of the previous[5,6] and present results for the study of dynamical systems is discussed.

2. Dissipative Dynamical Maps

The dissipative standard map **DSM**[7–9]

$$\begin{aligned} P_n &= bP_{n-1} - \frac{\kappa}{2\pi}\sin(2\pi X_{n-1}) \\ X_n &= (X_{n-1} + P_n) \bmod 1 \end{aligned} \tag{1}$$

is an important paradigm of nonlinear dynamics. The map, with dimensionless positive parameters b and κ(kappa), describes the Newtonian dynamics of a periodically-delta-kicked damped electron in a time-varying electric field in a plasma. Eqs.(1) map the dimensionless scaled position X and dimensionless scaled momentum P from just before the nth kick to just before the $(n+1)$th kick. The dissipative parameter b ranges between 0 (the infinite damping limit) and 1 (the Hamiltonian non-dissipative limit): larger b, smaller damping.

The relativistic dynamics of the same delta-kicked dissipative system is (Lan, document in preparation) described by the following map:

$$\begin{aligned} P_n &= bP_{n-1} - \frac{\kappa}{2\pi}\sin(2\pi X_{n-1}) \\ X_n &= \left(X_{n-1} + \bar{P}_n\right) \bmod 1 \end{aligned} \tag{2}$$

where:

$$\bar{P}_n = \frac{1}{\alpha(1-b)}\left[\ln\left(\alpha P_n + \sqrt{1+(\alpha P_n)^2}\right) - \ln\left(b\alpha P_n + \sqrt{1+(b\alpha P_n)^2}\right)\right]$$

The relativistic dissipative standard map **RDSM** above has, in addition to b and κ, another dimensionless positive parameter α. The ratio of the particle speed v to the speed of light c is given by:

$$\frac{v}{c} = \frac{b\alpha P}{\sqrt{1+(b\alpha P)^2}} \tag{3}$$

It is easily seen from Eq. (3) that $b\alpha P \ll 1$ implies $v \ll c$.

3. Results

In this example, the Newtonian trajectory is generated from the initial conditions $X_0 = 0.1$ and $P_0 = 1$ using the DSM with parameters $b = 0.1$ and $\kappa = 9.9$. The relativistic trajectory is generated from the RDSM using the same initial conditions and parameters b and κ, and $\alpha = 0.0001$. The magnitude of the relativistic momentum P generated by the RDSM is of the order of 1. Thus $b\alpha P \approx 10^{-5}$, which implies, according to Eq. (3), that $v \approx 10^{-5}c$, i.e., the speed v of the system is merely 0.001% of the speed of light c.

The position and momentum predicted by the two theories are plotted in Figures 1 and 2 respectively. The Newtonian and relativistic positions agree to certain extent for the first 10 kicks, but they do not agree at all from kick 11 onwards. Similarly, there is a complete disagreement between the Newtonian and relativistic momentums from kick 12 onwards.

The eventual complete breakdown of the agreement of the Newtonian trajectory with the relativistic trajectory at low speed was also found for other initial conditions and parameters. In all cases, the standard accuracy check[10] of comparing the double precision values to the corresponding values computed in quadruple precision verified that the breakdown of agreement is not a numerical artifact.

4. Conclusions

This paper shows that, contrary to conventional wisdom,[1–4] the Newtonian and relativistic predicted trajectories, from respectively the DSM (Eqs. (1)) and RDSM (Eqs. (2)), for a low-speed delta-kicked dissipative system can eventually disagree completely.

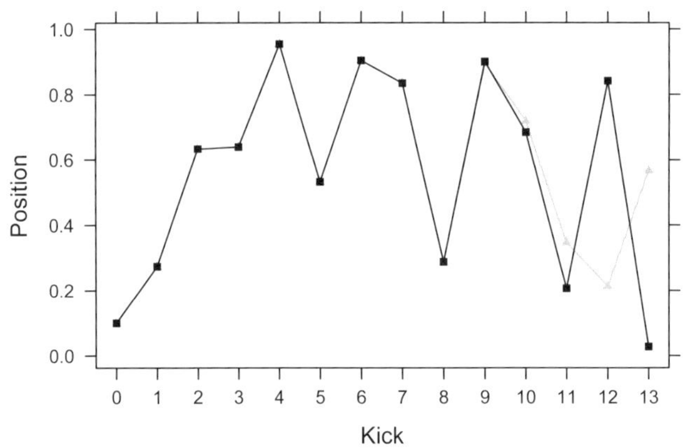

Fig. 1. Comparison of the relativistic (diamond) and Newtonian (square) positions

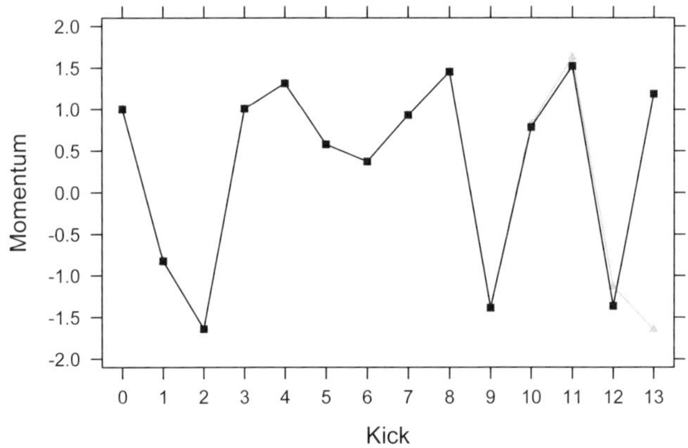

Fig. 2. Comparison of the relativistic (diamond) and Newtonian (square) momentums

This result and the previous results[5,6] raise the following question: when the two theories, Newtonian mechanics and special relativistic mechanics, yield completely different predictions for the trajectory of a *slow-moving* dynamical system, which of the two predictions agree with observation? In recent years, various ways[11–13] of testing the Lorentz invariance postulate of special relativity, which involve high-precision experiments[14,15] and astrophysical observations,[16,17] continue to show that Lorentz invariance is not violated. Therefore, because special relativistic mechanics is Lorentz invariant, the trajectory predicted by special relativistic mechanics is empirically correct for all speeds. This implies that, for a *slow-moving* dynamical system, once the trajectory predicted by Newtonian mechanics (Newtonian mechanics is not Lorentz invariant) is completely different from the empirically-correct trajectory predicted by special relativistic mechanics, Newtonian mechanics can no longer be used to study its dynamics. Special relativistic mechanics must be used instead.

References

1. A. Einstein, *Relativity: the special and the general theory* (Random House, New York, 1961, pp. 49–50).
2. H. C. Corben and P. Stehle, *Classical mechanics*, 2 edn. (Dover, New York, 1994, p. 287).
3. J. Ford and G. Mantica, *Am. J. Phys.* **60**, 1086 (1992).
4. W. D. McComb, *Dynamics and relativity* (Oxford University Press, Oxford, 1999).
5. B. L. Lan, *Chaos* **16**(033107) (2006).
6. B. L. Lan, Connecting Newtonian mechanics to special relativistic mechanics: Einstein's mistake., in *Proceedings of the 13th NPA Conference (April 3-7 2006, U. Tulsa)*, 3, 116(2006).
7. G. Schmidt and B. W. Wang, *Phys. Rev. A* **32**, 2994 (1985).
8. W. Wenzel, O. Biham and C. Jayaprakash, *Phys. Rev. A* **43**, 6550 (1991).
9. U. Feudel, C. Grebogi, B. Hunt and J. A. Yorke, *Phys. Rev. E* **54**, 71 (1996).
10. A. J. Lichtenberg and M. A. Lieberman, *Regular and stochastic motion* (Springer, New York, 1983).
11. M. Pospelov and M. Romalis, *Phys. Today* **57**, 40 (July 2004).
12. P. Ball, *Nature* **427**, 482 (2004).
13. A. Cho, *Science* **307**, 866 (2005).
14. P. Wolf, S. Bize, A. Clairon, A. N. Luiten, G. Santarelli and M. E. Tobar, *Phys. Rev. Lett.* **90**(060402) (2003).
15. H. Muller, S. Herrmann, C. Braxmaier, S. Schiller and A. Peters, *Phys. Rev. Lett.* **91**(020401) (2003).
16. V. A. Kostelecky and M. Mewes, *Phys. Rev. Lett.* **87**(251304) (2001).
17. T. Jacobson, S. Liberati and D. Mattingly, *Nature* **424**, 1019 (2003).

Visualisation of Atmospheric Pressure Plasma Electrical Parameters

V. J. Law, N. O'Connor and S. Daniels

National Center of Plasma Science and Technology
Dublin City University, Collins Avenue
Glasnevin, Dublin 9, Dublin, Ireland
Email: vic.law@dcu.ie

B. Twomey, J. Tynan, D. P. Dowling and G. Byrne

School of Electrical, Electronic and Mechanical Engineering
UCD, Belfield, Dublin 4, Ireland

Real-time principal component analysis of an atmospheric pressure plasma electrical (current, voltage and drive frequency) parameters is reported. Lissajous figure phase-space projection of the current and voltage waveforms and linear transformation of the electrical parameters on a Loading plot is also performed. Using LabVIEW graphical programming it is shown that the Lissajous figure captures a snap-shot of the plasma chemistry in the form of current waveform micro-scale events. The Loading plot provides a time-evolving view of the process that enables differentiation of plasma modes (bifurcation) and chemistry.

Keywords: Atmospheric pressure plasma; principal component analysis; phase-space; pc-space.

1. Introduction

Within the last few decades atmospheric pressure glow discharge (in the form of dielectric barrier discharge) has become established as a processing tool within the automotive, biomedical and microelectronics industries. The operating performance of these discharges is mainly carried out by monitoring their non sinusoidal periodic voltage and current waveforms.[1] Indeed the generative characteristics of the voltage and current waveforms are used to classify the mode of plasma production. It has also been establish that atmospheric discharges operated at drive frequencies below a few kHz and above 20 to 30kHz the plasma takes on a self-organised filamentary or column micro-discharges appearance.[2] This non-uniformity in appearance can be deleterious to some surface processes where pin-hole free and uniform processing is required. The occurrence of filaments, column micro-discharges (CMD) and other local disturbances modify the already distorted current and voltage waveform generated by the drive circuitry: the former disturbances created by the transformer circuitry and chamber topology (capacitance of the dielectric walls). A further complication associated with some low frequency transformer, and inductor-like, drive circuits is their tendency to be frequency pulled by the plasma load.[3]

The chaotic nature of these discharges (bifurcation, mode transition, system instability and visual appearance) makes them suitable for Multivariate Analysis and in particular Principal Component Analysis (PCA). However, to be of practical use in industrial processes the PCA needs to be in real-time with data refresh rates of the order of 1 second or less. The ability to capture the chaotic data from freely available electrical input parameters (current, voltage and frequency) and compress them into datasets that allows direct mapping of surface treatment outcomes is of great interest not only to the plasma processing industries, but also the plasma physics community.

2. Plasma and Measurement

This paper explores the PCA of the non-sinusoidal periodic voltage and current waveforms and the drive frequency of the Dow Corning® SE-1100 AP4 atmospheric pressure plasma system.[4] It comprises of 2 parallel-plate plasma chambers arranged in conjunction with a dedicated reel-to-reel web handling system. Each chamber measures 300 × 320mm with a 5mm electrode gap. A 10cm wide reel-to-reel PET (Poly (ethylene teraphthalate)) web passes through it. Applied powers of up to 2000W can be supplied to the electrodes using a Vetaphone power supply (frequency 16 to 23kHz). The power supply is self-matching. In this case a feedback circuit senses the load impedance (chamber plus plasma) and alters the supply conjugate impedance accordingly to maintain efficient power transfer to the discharge. This means that changes in the real and imaginary impedances in the discharge and system will cause a corresponding change in the resonant frequency of the system and thus the frequency of operation. An illustration of the plasma system and measurement set-up is shown in Figure 1.

The current and voltage are measured using a Person (1A:1V) 6585 current monitor and a North Star (1kV:1V) voltage divider, respectively. The signals are then sent to a Tektronix TDS3000B oscilloscope operating in the average ($n = 4$) acquisition mode. At this point the current and voltage

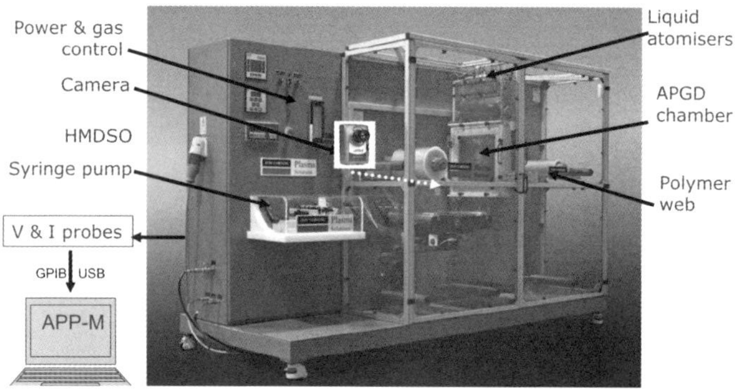

Fig. 1. The plasma system and measurement set-up

waveforms are passed via a GPIB-to-USB network to a computer where National Instruments LabVIEW 8.2 graphical programming language is used to display the raw data in the time-domain. This data is further manipulated to produce a Lissajous figure projection of the current and voltage waveforms to identify system instabilities and micro-scale events (MSE) in the current waveform. Principal Component Analysis[5,6] is used to compress the current, voltage and frequency data. This is performed by calculating the route-mean-square (rms) values of the current and voltage waveforms and adding to each the periodic frequency (f) in units of kHz, see equations (1) and (2). The rationale for this method of data compression is two fold. First, the rms value, rather than the peak value, is used as it is proportional to the signal energy. Second, frequency offset results in a score that always has a positive value.

$$\text{Current waveform} \rightarrow I_{\text{rms}} \rightarrow I_{\text{rms}} + f = \text{score} \tag{1}$$

$$\text{Voltage waveform} \rightarrow V_{\text{rms}} \rightarrow V_{\text{rms}} + f = \text{score} \tag{2}$$

A real-time loading plot is constructed with the X-axis score equal to the voltage$_{\text{rms}} \times f$ and the Y-axis score equal to current$_{\text{rms}} \times f$. The purpose of the PC-space and phase-space is to provide new coordinate systems where the surface treatment outcomes may be mapped directly. A screen shot of the LabVIEW front panel is shown in Figure 2: the Lissajous figure is at the bottom right and the loading plot is at the top right.

Finally digital images of the plasma are taken under varied gas helium and helium oxygen mixtures, and power conditions (400W to 2000W) for correlation with electrical data.

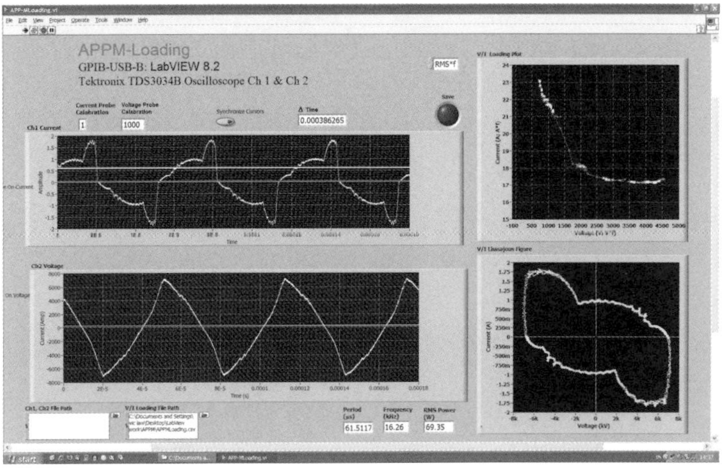

Fig. 2. LabVIEW front panel

3. Visualization

The atmospheric pressure plasma chemistry of helium and helium with (1%) diatomic oxygen dielectric barrier discharges have been extensively studied.[7–10] From this body of work it is known that the discharges support both Townsend-like glow discharge and column-micro-discharge. These modes of operation are characterised by their current and voltage non sinusoidal waveforms and the drive frequency. Furthermore for the helium discharge micro-scale events (MSE) within the current waveform are readily identified when using high resolution sampling oscilloscope which has been used here. Section 3.1 illustrates how these MSE are captured and displayed in the form of a Lissajous Figure, and Section 3.2 presents the PCA of the electrical data and displayed in the form of a Loading plot.

3.1. *Lissajous figure*

The current vs. voltage Lissajous figure is obtained directly by passing the current and voltage time-domain data (which share the same time-base) into the Y and X inputs of a general-purpose Cartesian mapping XY graph. Using this programming the current and voltage is sampled and displayed every second; hence a snap-shot is constructed of the process. In essence this approach is like current Parks vector patterns used in inductor motor fault analysis.[11]

Figure 3 provides four representative Lissajous figures of helium and helium with (1%) diatomic oxygen at discharge power levels of 1000W and 1400W, respectively. A comparison between the discharge chemistries shows that the MSE count is reduced in the oxygen rich discharge. The MSEs are observed to have amplitudes of ~ 0.1A in the two discharge periods and in the non discharge period. Their delay-time from the zero crossing-point can be estimated from the drive frequency. For example, if the drive frequency is 16.25kHz (60μs) one half revolution = 30μs. With this data the time-delay of the MSEs and the duration of the periods are estimated.

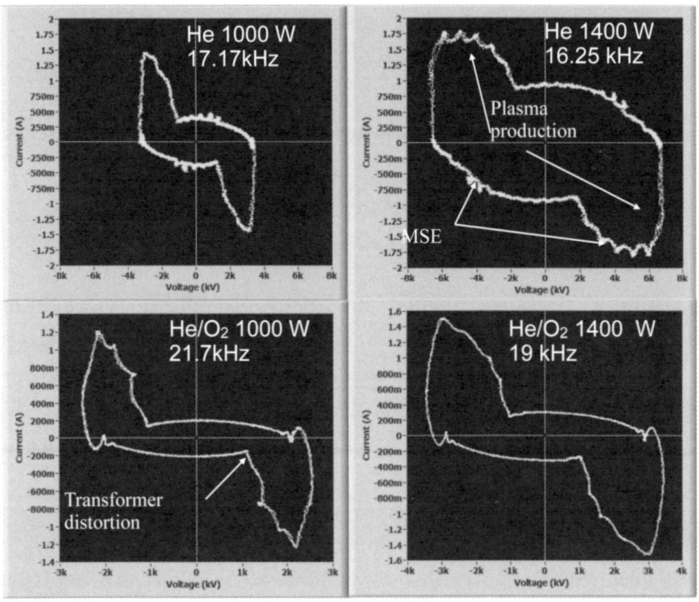

Fig. 3. Representative current vs. voltage Lissajous Figures for helium and helium with 1% oxygen discharge

When the phase-space information of the Lissajous figure is combined with the number of CMDs (obtained by digital photography, see Figure 4) the transition between discharge mode becomes self evident when plotted as function of applied power. Figures 5 and 6 illustrate the change in number of CMDs over the PET web film and the number of MSE per half cycle for helium and helium with (1%) diatomic oxygen as a function of applied power. Figure 5 shows that the CMDs increase in number in the 400W to 1000W range. Beyond this power level the discharge undergoes a mode transition, the CMDs merge in to a Townsend-like or pesudoglow appearance and the number of MSEs increase. Figure 6 shows how the number of MSE changes as function of applied power with the addition of 1% diatomic oxygen. This shows that the addition of oxygen quenches the number of MSEs. Possible reasons for this transition are discussed Section 4.

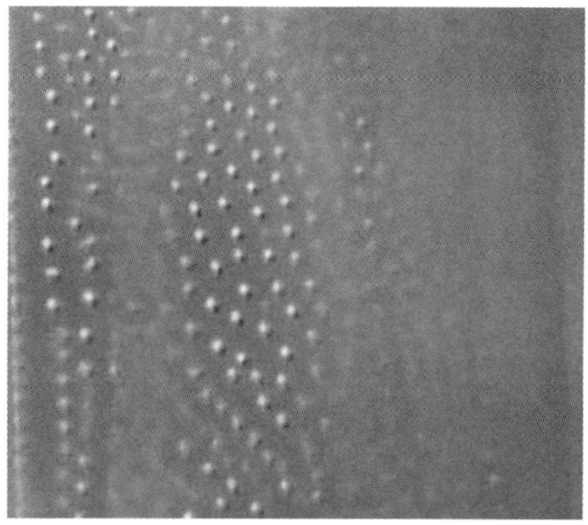

Fig. 4. Digital image of helium discharge at 400W on the PET web showing self-organised CMD. Each CMD has a typical diameter of ~ 2.5mm and spacing of 3mm

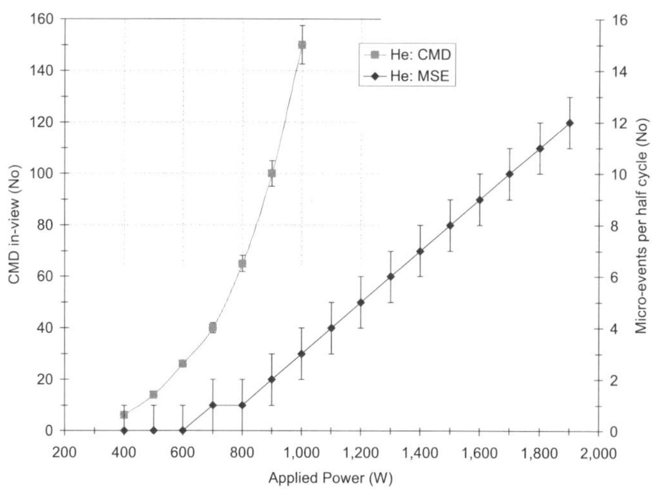

Fig. 5. Behaviour of CMDs and MSEs in a helium discharge as a function of power

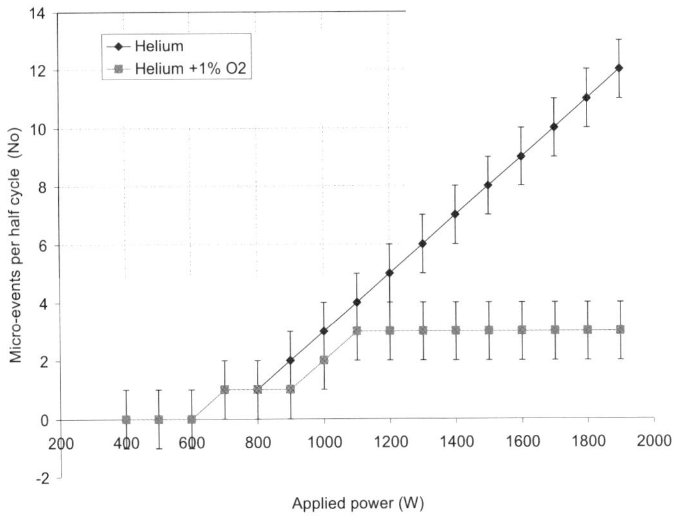

Fig. 6. MSE in helium and helium with 1% diatomic oxygen discharge

3.2. *Loading plots*

The LabVIEW Loading plot program is designed to calculate the drive frequency and the rms values from the current and voltage waveforms. Each rms value is then multiplied by the frequency (see equations (1) and (2)) and the resulting scores are passed on to the Y and X inputs of a Cartesian mapping XY graph, as illustrated in Figure 2. This process is repeated every time the data is sampled to build the Loading plot which now becomes a visual historical map of the process. The XY score data is also collected for off-line analysis. Using knowledge-domain information about the process, fault excursion limits and regions of process instability can be placed on the plot to high-light regions of process failure.[6]

Figure 7 provides an example of the power of the Loading plot. In this example the mean value of 20 sampled XY points at applied powers between 400 W and 2000 W for helium and helium with (1%) diatomic oxygen are plotted. There are three prominent features displayed within this plot. First with increasing applied power the two chemistries form linear trajectories with an abrupt inflection ($\sim$ 120 degrees) where system instabilities are observed. The instability region also marks the transition between the two modes of plasma operation. A further feature of note here is the helium with (1%) oxygen is seen to go through the mode transition at a high power. Second, there is a clear separation between the helium and helium with (1%) diatomic oxygen discharges. The existence of this separation has major advantages in identifying and controlling different plasma discharge chemistries. Thirdly, the CMDs and MSEs are grouped either side of the mode transition. At lower powers the CMDs are self-organized on the PET film, at high powers a Townsend-like or pesudoglow is formed. Taken together the plot provides a clear representation of the two plasma discharges modes and chemistries.

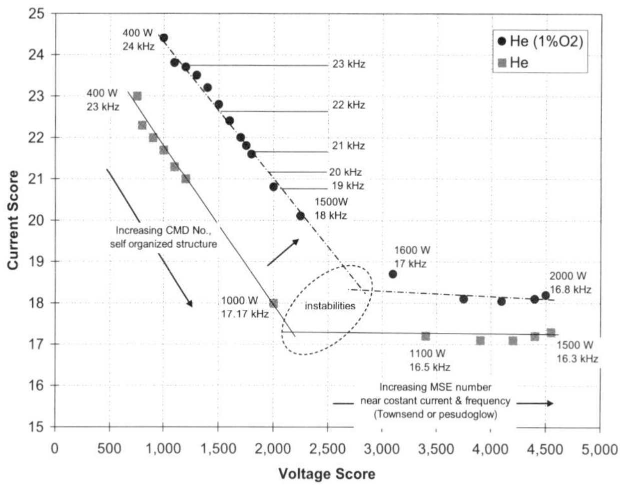

Fig. 7. Loading plot showing mean score value of helium and helium with 1% diatomic oxygen as a function of applied power. Overlaid on the plot is the direction of CMD and MSE growth, visual classification operating frequency

4. Discussion

Sections 2 and 3 have presented a method of real-time PCA of an atmospheric pressure discharge. The graphical results demonstrate the capture of the principal features of these plasmas (mode transitions, CMD and MSE). In addition to this there is the ability to differentiate between small (1%) percentages of diatomic oxygen in helium by the spatial separation of the data sets. However the results need to be tested and understood from a plasma physics view-point.[7–10]

The Loading plot in Figure 7 may be interpreted as follows. Consider first the helium discharge. At low powers and (voltages) the discharge is in the glow type mode with localised CMDs on the PET web. With increasing power and voltage the number of CMDs increases, with constant spacing between them. The spacing is determined by their mutual electrostatic repulsion, which is proportional to the current densities within the CMD, and the applied electric field. The number continues to increase until the discharge over the PET web surface becomes uniform, similar to that of the surrounding chamber volume. This reactance change pulls the frequency of the power supply to approximately 16.5kHz. Beyond this transition point the discharge is forced into a pseudoglow discharge where the discharge current and frequency remains nearly constant. In this mode the CMDs have coalesced, and an increasing number of MSEs are observed in the current waveform.

In the case of the 1% diatomic oxygen discharge there are many possible reaction pathways.[8] However, it has been shown that in order to establish a Townsend-like, or pesudoglow discharge, there must be a sufficient number of helium metastables (He^*) states remaining from the previous half cycle of the applied voltage. The metastables are created from interactions with high energy (fast) electrons and thus slow down the ionization process, halting the formation of streamer/filamentary discharges. The process is outlined in the balanced equation (3).

$$He^* + O_2 \rightarrow He + O_2^+ + e \tag{3}$$

This reaction drains the supply of metastables to facilitate the formation of a Townsend-like, or pesudoglow discharge where subsequent MSEs take place. The high electronegativity of diatomic oxygen also plays a role in this process. For these reasons, the CMD discharge is sustained to higher powers until a pseudoglow discharge forms. Here again the plasma and chamber reactance is similar to the helium discharge and processes frequency pulls the power supply to approximately 16.5kHz. The electronegativity of diatomic oxygen is responsible for the reduction in current density of the discharge with respect to that of pure helium, leading to closer spacing of the CMDs due to lesser mutual electrostatic repulsion. The overall effect is to delay the operational frequency transition in the current and voltage scores and hence displaces the oxygen rich discharge data up and to the right of the Loading plot, with respect to the same applied power for the helium discharge. Under these oxygen rich conditions, the number of MSE in the current waveform per half cycle are quenched with respect to the helium discharge. With this simple interpretation, the MSEs appear to be linked to the processes involved in the destruction of He^* species.

In conclusion, the Lissajous figure and the Loading plot representations of the complex electrical data can be interpreted in a qualitative way. The clarity of the visual information combined with ease of access to the information and manipulation using LabVIEW makes real-time PCA a powerful and practicable tool in capturing and monitoring the chaotic nature of the data obtained from atmospheric pressure plasma. Simple plasma physics analysis can also be applied to the data, giving further veracity to the application of real-time PCA.

Acknowledgement

Enterprise Ireland (grant: CFTD/7/IT/304) and National Instruments, UK & Ireland.

The authors would also like to thank the delegates of the CHAOS2008 conference, Chania, Crete, Greece, for their helpful and incisive comments: from which further non-parametric cluster analysis of the Loading plot data is expected to reveal detailed information on the bifurcations observed in this work.

References

1. F. Massines and G. Gouda, *Journal of Physics D: Applied Physics* **31**, 3411 (1998).
2. J. R. Roth (ed.), *Industrial Plasma Engineering* (IOP Publishing Ltd., London, 1995), London, pp. 453–461.
3. V. J. Law, *Vacuum* **82**(6), 630 (2008).
4. B. Twomey, M. Rahman, G. Byrne, A. Hynes, L. A. O'Hare, L. O'Neill and D. Dowling, *Plasma Processes and Polymers* **5**, 737 (2008).
5. A. T.-C. Koh, N. F. Thornhill and V. J. Law, *Electronic Letters* **35**(16), 1383 (1999).
6. V. J. Law and N. Macgearailt, *Meas. Sci. Technol.* **18**(3), 645 (2007.).
7. I. Radu, R. Bartnikas, G. Czeremuskin and M. Wertheimer, *IEEE Trans. Plasma Sci.* **31**(3), 411 (2003).
8. D. Lee, J. Myung, S. H. Hong and Y. Kim, *IEEE Trans. Plasma Sci.* **33**(2), 949 (2005).
9. X. Xu, O. Qiongrong, Z. Shaofeng, S. Xingsheng and M. Yuedong, *Plasma Sci. Tech.* **8**(3), 303 (2006).
10. D. Wang, Y. Wang and C. Liu, *Thin Solid Films* **506–507**, 384 (2006).
11. H. Nejjari and M. E. H. Benbouzid, *IEEE Trans. on Indust. Applications.* **36**(3), 730 (2000).

Chaos and Multifractals in the Solar System Plasma

W. M. Macek

Faculty of Mathematics and Natural Sciences
Cardinal Stefan Wyszyński University
Wóycickiego 1/3, 01-938 Warsaw, Poland
and
Space Research Centre, Polish Academy of Sciences
Bartycka 18 A, 00-716 Warsaw, Poland
Email: macek@cbk.waw.pl

We argue that dynamical behavior of space plasmas can often be approximately described by low-dimensional chaotic attractors in the inertial manifold, which is a subspace of a given system phase space. In fact, using nonlinear time-series analysis based on the method of topological embedding, we have identified a chaotic strange attractor in the solar wind data. In particular, we have shown that the multifractal spectrum of the solar wind attractor is consistent with that for the self-similar generalized weighted Cantor set with one probability measure parameter of the chaotic attractor and one or possibly two scaling parameters describing nonuniform compression in the phase space of the system. The values of the parameters fitted demonstrate small dissipation of the complex solar wind plasma and show that some parts of the attractor in phase space are visited much more frequently than other parts.

To quantify the multifractality of space plasma turbulence, we have recently considered the generalized two-scale weighted Cantor set also in the context of solar wind intermittent turbulence. We investigate the resulting multifractal spectrum of generalized dimensions depending on parameters of the new cascade model, especially for asymmetric scaling. In particular, we show that intermittent pulses are stronger for the model with two different scaling parameters; a much better agreement with the solar wind data is obtained, especially for the negative index of the generalized dimensions.

Therefore we argue that there is a need to use a two-scale cascade model. We hope that this generalized multifractal model will be a useful tool for analysis of intermittent turbulence in the Solar System plasma. We thus believe that fractal analysis of chaotic phenomena in the complex space environment could lead us to a deeper understanding of their nature, and maybe even to predict their seemingly unpredictable behavior.

Keywords: Nonlinear dynamics and chaos; multifractals; turbulence; solar wind plasma.

1. Importance of Chaos and Multifractality

The nature of the fluctuations in solar wind plasma parameters is still not sufficiently understood. The slow solar wind most likely originates from nonlinear processes in the solar corona. However, it appears that a certain kind of order does lie concealed within the irregular solar wind fluctuations, which can be described using methods of nonlinear time series analysis, based on fractal analysis and the theory of deterministic chaos. This involves the notions of fractal and multifractal sets, which could presumably be strange attractors in a certain state space of a given complex dynamical system. By employing the so-called false-nearest-neighbors method, we have argued that the deterministic component of solar wind plasma dynamics should be low-dimensional.[1]

In fact, the results we have obtained using the method of topological embedding indicate that the behavior of the solar wind can be approximately described by a low-dimensional chaotic attractor in the inertial manifold, which is a subspace of system phase space. A direct determination of a solar wind attractor from the data is known to be a difficult problem. This chaotic strange attractor has been identified in the solar wind data by Macek in [2] and further examined in [3]. In particular, we have calculated the correlation dimension of the reconstructed attractor in the solar wind[2] and have provided tests for this measure of *complexity* including statistical surrogate data tests.[4] Further, we have shown that the Kolmogorov entropy of the attractor is *positive* and finite, as it holds for a *chaotic* system.[3]

We have also considered the spectrum of generalized dimensions D_q as a function of a continuous index, $-\infty < q < \infty$, for the solar wind attractor, using a simple multifractal model with a measure of the self-similar weighted Cantor set with one parameter describing uniform compression and another parameter for the probability measure of the attractor of the system. The spectrum is found to be consistent with the data, at least for positive index q of the generalized dimensions D_q.[5–9] However, the full singularity spectrum is necessary to quantify the degree of multifractality. Notwithstanding of the well-known statistical problems with negative q (see [7]) we have recently succeeded in estimating the entire spectrum for solar wind attractor using a generalized weighted Cantor set with two different scales describing nonuniform compression.[10]

The question of multifractality is also of great importance because it allows us to investigate the nature of interplanetary hydromagnetic turbulence in the solar wind e.g. [11,12]. Starting from Richardson's version of turbulence, many authors try to recover the observed scaling exponents, using some simple and more advanced models of turbulence describing distribution of the energy flux between cascading eddies at various scales. In particular, the multifractal spectrum was investigated using Voyager (magnetic field) data in the outer heliosphere[13,14] and using Helios (plasma) data in the inner heliosphere.[15] The multifractal scaling has also been investigated using Ulysses observations[16,17] and with Advanced Composition Explorer (ACE) and WIND data.[18–20]

Recently, in order to further quantify the multifractality, we have considered the generalized weighted Cantor set also in the context of turbulence cascade.[21] Therefore we have argued that there is, in fact, need to use a two-scale cascade model. Here we investigate the resulting multifractal singularity spectrum depending on two scaling parameters and one probability measure parameter, demonstrating that a much better agreement has been obtained, especially for $q < 0$. We hope that this generalized new asymmetric multifractal model could shed light on the nature of turbulence and will be a useful tool for analysis of intermittent turbulence in various environments.

2. Two-Scale Cantor Set

A simple interesting example of mutifractals is the Cantor set with two scales $l_1 + l_2 \leq l$, as shown in Figure 1. At each step of construction we obtain 2^n closed narrow segments of various widths and probabilities. The resulting strange chaotic attractor for $n \to \infty$ is the weighted two-scale Cantor set. Even though one can find this generalized Cantor set in many classical textbooks,[22,23] it is still difficult to understand this strange attractor that exhibits multifractality in various complex real systems, also in case of intermittent turbulence.

According to a standard scenario, each of cascading eddies is breaking down into two new ones, but not necessarily equal and twice smaller. In particular, space filling turbulence could be recovered for $l_1 + l_2 = 1$ (see [24]). In the inertial region of the system of size L, $\eta \ll l \ll L = 1$ (normalized), the energy is not allowed to be dissipated directly until the Kolmogorov scale η is reached. However, in this range at each n-th step of the binomial multiplicative process, the flux of kinetic energy density ε transferred to smaller eddies (energy transfer rate) could be divided into non equal fractions p and $1 - p$ (see [21], Figure 1).

3. Solar Wind Data

We have already analyzed the Helios 2 data using plasma parameters measured *in situ* in the inner heliosphere[25] for testing of the solar wind at-

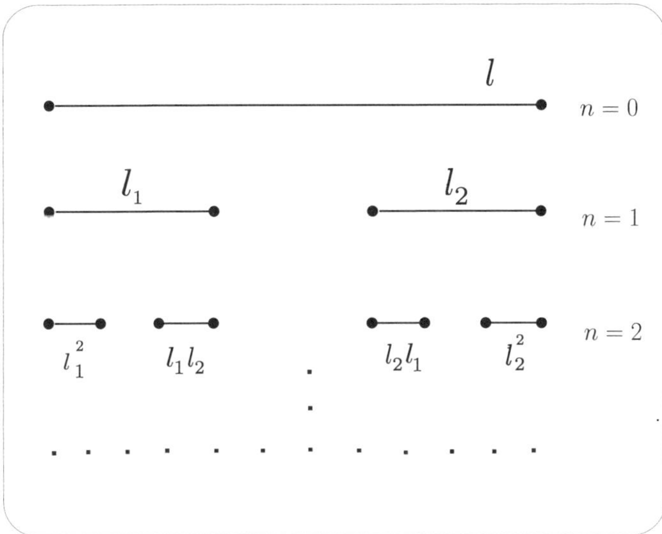

Fig. 1. Two-scale Cantor set.

tractor. The X-velocity (mainly radial) component of the plasma flow, v_x, has been already investigated by Macek[2,5,6] and Macek and Redaelli.[3] The Alfvénic fluctuations with longer (two-days) samples have been studied by Macek[7,10] and Macek et al.[8,9] To study the turbulence cascade, Macek and Szczepaniak[21] have selected four-day time intervals of v_x samples in 1976 (solar minimum) for both slow and fast solar wind streams measured at various distances from the Sun. In this paper we analyze time series of velocities of the solar wind measured by ACE in the ecliptic plane near the libration point $L1$, e.g., approximately at a distance of $R = 1$ AU from the Sun. Here we have selected even longer (five-day) time intervals of v_x samples, each of 6750 data points, interpolated with sampling time of 64 s, for both slow and fast solar wind streams during solar minimum (2006) and maximum (2001).

4. Methods of Data Analysis

4.1. *Generalized dimensions*

The generalized dimensions are important characteristics of *complex* dynamical systems.[26–29] Since these dimensions are related to frequencies with which typical orbits in phase space visit different regions of the attractors, they provide information about dynamics of the systems.[23] More precisely, one may distinguish a probability measure from its geometrical support, which may or may not have fractal geometry. Then, if the measure has different fractal dimensions on different parts of the support, the measure is multifractal.[30]

Let us consider the generalized weighted Cantor set where the probability of visiting one segment of size l_1 is p (say, $p \leq 1/2$), and for the remaining segment of size l_2 is $1-p$ (see Figure 1). For any q one obtains $D_q = \tau(q)/(q-1)$ by solving numerically the following transcendental equation[23]

$$\frac{p^q}{l_1^{\tau(q)}} + \frac{(1-p)^q}{l_2^{\tau(q)}} = 1 \tag{1}$$

4.2. *Turbulence scaling*

In the inertial range the transfer rate of the energy flux $\varepsilon(l)$ is widely estimated by

$$\varepsilon(l) \sim \frac{\left|u(x+l) - u(x)\right|^3}{l}, \tag{2}$$

where $u(x)$ and $u(x+l)$ are velocity components parallel to the longitudinal direction separated from a position x by a distance l. Therefore to each ith eddy of size l in the turbulence cascade ($i = 1, \ldots, N = 2^n$) we associate a probability measure defined by

$$p_i(l) = \frac{\varepsilon_i(l)}{\sum_{i=1}^{N} \varepsilon_i(l)} \tag{3}$$

This quantity can roughly be interpreted as a probability that the energy flux is transferred to an eddy of size $l = v_{\mathrm{sw}}t$. Here, for simplicity the third moment of structure function of velocity fluctuations in Eq. (2) is used for estimation of this measure.[15]

Similarly, we define a one parameter q family of (normalized) generalized pseudoprobability measures[31]

$$\mu_i(q,l) \equiv \frac{p_i^q(l)}{\sum_{i=1}^{N} p_i^q(l)} \tag{4}$$

Now, with an associated fractal dimension index $f_i(q,l) \equiv \log \mu_i(q,l)/\log l$ for a given q the multifractal singularity spectrum of dimensions is defined directly as the averages taken with respect to the measure $\mu(q,l)$ in Eq. (4) denoted from here on by $\langle \ldots \rangle$

$$f(q) \equiv \lim_{l \to 0} \sum_{i=1}^{N} \mu_i(q,l)\, f_i(q,l) = \lim_{l \to 0} \frac{\langle \log \mu_i(q,l) \rangle}{\log(l)} \tag{5}$$

and the corresponding average value of the singularity strength is given by[32]

$$\alpha(q) \equiv \lim_{l\to 0}\sum_{i=1}^{N}\mu_i(q,l)\ \alpha_i(l) = \lim_{l\to 0}\frac{\langle \log p_i(l)\rangle}{\log(l)} \tag{6}$$

One can easily verify that the multifractal singularity spectrum $f(\alpha)$ as a function of α satisfies the following Legendre transformation[29,32]

$$\alpha(q) = \frac{d\tau(q)}{dq}, \qquad f(\alpha) = q\alpha(q) - \tau(q). \tag{7}$$

5. Results and Discussion

5.1. *Dimensions for solar wind attractor and turbulence models*

To estimate the generalized dimensions for the solar wind attractor we should calculate for a given continuous index q and embedding dimension m the so-called generalized correlation sum $C_{q,m}(r)$ as a function of hyperspheres of radius r that cover the presumed attractor (see [10], Equation 1). This can roughly be interpreted as an average probability of finding q vectors in embedding space separated by a distance smaller than r. For large dimensions m and small distances r in the scaling region it can be argued that $C_{q,m}(r) \propto r^{\tau(q)}$, where $\tau(q)$ is an approximation of the ideal solution of Eq. (1).[27] Hence, the slopes of the natural logarithm of $C_{q,m}(r)$ versus $\ln r$ (normalized) provides

$$D_{q,m}(r) = \frac{1}{q-1}\,\frac{d\left[\ln C_{q,m}(r)\right]}{d(\ln r)} \tag{8}$$

If a plateau exists in a scaling region, $r_{\min} < r < r_{\max}$, which does not depend on m for some $m > m_0$, this plateau can be identified with the requested generalized dimension. Finally, the average slope for $6 \leq m \leq 10$ is taken as D_q.[2,7–10] We have verified that for $q > 0$ the slopes do not change substantially with the number of points used, providing that the dimension of the attractor is well below $2\log_{10} N \approx 9$, for $N = 26163$ (see, e.g., [33]). The results obtained using the moving average filter and singular-value decomposition linear filter for standard $q = 2$ are given in Figure 2 of [8], and are compared with $q = -2$ in Figures 2 (a) and (b) of [7], correspondingly, while those obtained for somewhat shorter samples ($N = 4514$) have been discussed in [2] and [3], using the nonlinear Schreiber filters. Next, the generalized dimensions D_q as a function of q (see [10],

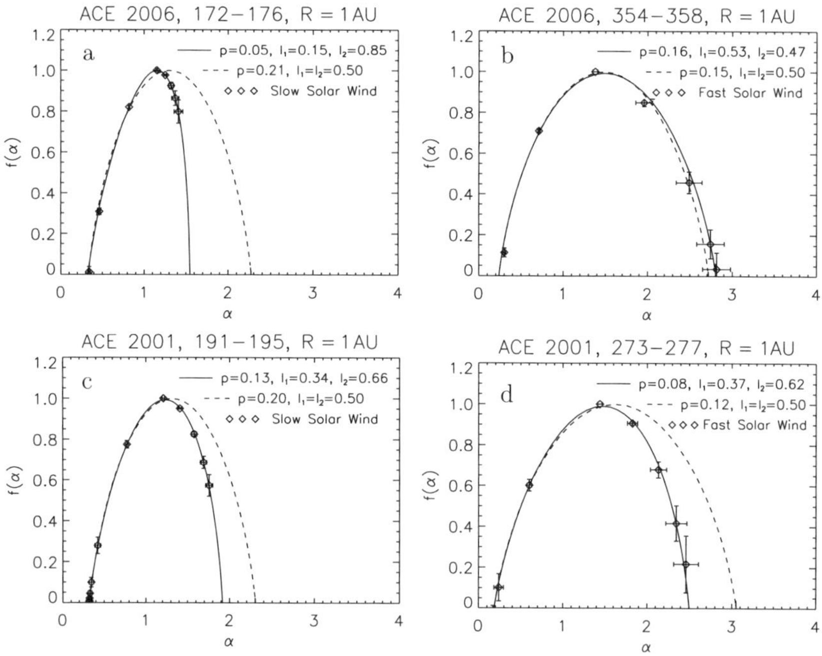

Fig. 2. The singularity spectrum $f(\alpha)$ as a function of α. The values obtained for one-dimensional turbulence are calculated for the usual one-scale (dashed lines) p-model and the generalized two-scale (continuous lines) model with parameters fitted to the multifractal measure $\mu(q,l)$ obtained using the v_x velocity components measured by ACE at 1 AU (diamonds) for the slow (a) and (c) and fast (b) and (d) solar wind during solar minimum (2006) and maximum (2001), correspondingly.

Equation 1) with the statistical errors of the average slopes obtained using weighted least squares fitting over the scaling range are shown in Figures 3 (a) and 4 (a) of [10] and compared with one-scale and two-scale Cantor set model (see [8], Figure 3).

5.2. *Multifractal spectrum for turbulence*

The results for the generalized dimensions D_q as a function of q, calculated from the data and compared with those obtained using Eq. (1) for solar wind turbulence for the slow (a) and (c) and fast (b) and (d) solar wind streams at distances of 0.3 AU and 0.97 AU, correspondingly, are presented in Figures 3 (a), (b), (c) and (d) of [21].

Here the results for the corresponding singularity spectra $f(\alpha)$ as a function of α are shown in Figures 2 (a), (b), (c) and (d). The values of

$f(\alpha)$ given in Eqs. (5) and (6), for one-dimensional turbulence, $d = 1$, are calculated using the radial velocity components $u = v_x$ (in time domain), cf. Figure 3 of [21]. It is well known that for $q < 0$ we have some basic statistical problems.[7,10] Nevertheless, in spite of large statistical errors in Figure 2, especially for $q < 0$, the multifractal character of the measure can still clearly be discerned. Therefore one can confirm that both the spectrum of dimensions and singularity spectrum still exhibit the multifractal structure of the solar wind in the inner heliosphere.[21]

For $q \geq 0$ these results agree with the usual one-scale p-model fitted to the singularity spectra as obtained analytically using $l_1 = l_2 = 0.5$ in Eq. (1) and the corresponding value of the parameter $p = 0.21$ and 0.20, 0.15 and 0.12 for the slow (a) and (c), and fast (b) and (d) solar wind streams at solar minimum and maximum, correspondingly, as shown by dashed lines. On the contrary, for $q < 0$ the p-model cannot describe the observational results.[15] Here we show that the experimental values are consistent also with the singularity spectrum obtained numerically from Eqs. (5-6) for the weighted two-scale Cantor set using an asymmetric scaling, i.e., using unequal scales $l_1 \neq l_2$, as is shown in Figures 2 (a), (b), (c), and (d) by continuous lines. We also confirm that the degree of multifractality and asymmetry of the solar wind in the inner heliosphere are different for slow and fast streams. One can say that in the slow streams the scaling is more asymmetric than those for the fast wind. It also seems that the degree of asymmetry for the slow wind is rather anticorrelated with the phase of the solar activity.

We see that the multifractal spectrum of the solar wind is only roughly consistent with that for the multifractal measure of the self-similar weighted symmetric one-scale weighted Cantor set only for $q \geq 0$. On the other hand, this spectrum is in a very good agreement with two-scale asymmetric weighted Cantor set schematically shown in Figure 1 for both positive and negative q. Obviously, taking two different scales for eddies in the cascade, one obtains a more general situation than in the usual p-model for fully developed turbulence,[34] especially for an asymmetric scaling, $l_1 \neq l_2$. Hence we hope that this generalized model will be a useful tool for analysis of intermittent turbulence in space plasmas.

6. Conclusions

In this way, we have supported our conjecture that trajectories describing the system in the inertial manifold of phase space asymptotically approach the attractor of low-dimension.[2] We have shown that the multifractal spectrum of the solar wind attractor is consistent with that for the multifractal measure of the generalized two-scale weighted Cantor set. The values of the parameters fitted for $l_1 + l_2 = 1$ and $p \sim 10^{-1}$, for the slow wind, demonstrate small dissipation of the complex solar wind plasma and show that some parts of the attractor in phase space are visited at least one order of magnitudes more frequently than other parts (see Figure 5 of [2]).

We have also studied the inhomogeneous rate of the transfer of the energy flux indicating multifractal and intermittent behavior of solar wind turbulence in the inner heliosphere. In particular, we have demonstrated that a much better agreement with the real data is obtained, especially for $q < 0$. Basically, the generalized dimensions for solar wind are consistent with the generalized p-model for both positive and negative q, but rather with different scaling parameters for sizes of eddies, while the usual p-model can only reproduce the spectrum for $q \geq 0$. In general, the proposed generalized two-scale weighted Cantor set model should also be valid for non space filling turbulence. Therefore we propose this cascade model describing intermittent energy transfer for analysis of turbulence in various environments.

Thus these results show multifractal structure of the solar wind in the inner heliosphere. Hence we suggest that there exists an inertial manifold for the solar wind, in which the system has *multifractal* structure, and where noise is certainly not dominant. The multifractal structure, convected by the wind, might probably be related to the complex topology shown by the magnetic field at the source regions of the solar wind.

Acknowledgements

This work has been supported by the Polish Ministry of Science and Higher Education (MNiSW) through Grant NN202412733.

References

1. W. M. Macek and M. Strumik, *Adv. Space Res.* **37**, 1544 (2006).
2. W. M. Macek, *Physica D* **122**, 254 (1998).
3. W. M. Macek and S. Redaelli, *Phys. Rev. E* **62**, 6496 (2000).
4. J. Theiler, S. Eubank, A. Longtin, B. Galdrikian and J. D. Farmer, *Physica D* **58**, 77 (1992).
5. W. M. Macek, Multifractality and chaos in the solar wind., in *Experimental Chaos*, eds. S. Boccaletti, B. J. Gluckman, J. Kurths, L. M. Pecora and M. L. Spano (American Institute of Physics, New York, 2002, vol. 622) pp. 74–79.
6. W. M. Macek, The multifractal spectrum for the solar wind flow, in *Solar Wind 10*, eds. M. Velli, R. Bruno and F. Malara (American Institute of Physics, New York, 2003, vol. 679) pp. 530–533.

7. W. M. Macek, *Space Sci. Rev.* **122**, 329 (2006).
8. W. M. Macek, R. Bruno and G. Consolini, *Phys. Rev. E* **72**(017202) (2005).
9. W. M. Macek, R. Bruno and G. Consolini, *Adv. Space Res.* **37**, 461 (2006).
10. W. M. Macek, *Nonlin. Processes Geophys.* **14**, 695 (2007).
11. E. Marsch and C.-Y. Tu, *Nonlin. Processes Geophys.* **4**, 101 (1997).
12. R. Bruno, V. Carbone, P. Veltri, E. Pietropaolo and B. Bavassano, *Planet. Space Sci.* **49**, 1201 (2001).
13. L. F. Burlaga, *Geophys. Res. Lett.* **18**, 69 (1991).
14. L. F. Burlaga, *J. Geophys. Res.* **106**, 15917 (2001).
15. E. Marsch, C.-Y. Tu and H. Rosenbauer, *Ann. Geophys.* **14**, 259 (1996).
16. T. S. Horbury, A. Balogh, R. J. Forsyth and E. J. Smith, *Adv. Space Res.* **19**, 847 (1997).
17. T. S. Horbury and A. Balogh, *J. Geophys. Res.* **106**, 15929 (2001).
18. B. Hnat, S. C. Chapman and G. Rowlands, *Phys. Rev. E* **67**(056404) (2003).
19. B. Hnat, S. C. Chapman, K. Kiyani, G. Rowlands and N. W. Watkins, *Geophys. Res. Lett.* **34**(L15108) (2007).
20. K. Kiyani, S. C. Chapman, B. Hnat and R. M. Nicol, *Phys. Res. Lett.* **98**(211101) (2007).
21. W. M. Macek and A. Szczepaniak, *Geophys. Res. Lett.* **35**(L02108) (2008).
22. K. Falconer, *Fractal Geometry. Mathematical Foundations and Applications.* (J. Wiley, New York, 1952).
23. E. Ott, *Chaos in Dynamical Systems.* (Cambridge University Press, Cambridge, 1993).
24. L. F. Burlaga, J. Perko and J. Pirraglia, *Astrophys. J.* **407**, 347 (1993).
25. R. Schwenn, Large-scale structure of the interplanetary medium, in *Physics of the Inner Heliosphere*, eds. R. Schwenn and E. Marsch (Springer-Verlag, Berlin, 1990) pp. 99–182.
26. P. Grassberger, *Phys. Lett. A* **97**, 227 (1983).
27. P. Grassberger and I. Procaccia, *Physica D* **9**, 189 (1983).
28. H. G. E. Hentschel and I. Procaccia, *Physica D* **8**, 435 (1983).
29. T. C. Halsey, M. H. Jensen, L. P. Kadanoff, I. Procaccia and B. I. Shraiman, *Phys. Rev. A* **33**, 1141 (1986).
30. B. B. Mandelbrot, *Pure and Applied Geophys.* **131**, 5 (1989).
31. A. Chhabra and R. V. Jensen, *Phys. Rev. Lett.* **62**, 1327 (1989).
32. M. H. Jensen, L. P. Kadanoff and I. Procaccia, *Phys. Rev. A* **36**, 1409 (1987).
33. J.-P. Eckmann and D. Ruelle, *Physica D* **56**, 185 (1992).
34. C. Meneveau and K. R. Sreenivasan, *Phys. Rev. Lett.* **59**, 1424 (1987).

On the Hyperchaotic Complex Lü System

G. M. Mahmoud* and M. E. Ahmed†

Department of Mathematics, Faculty of Science
Assiut University, Assiut 71516, Egypt
*Emails: *gmahmoud_56@yahoo.com, gmahmoud@aun.edu.eg, †m_ahmed@aun.edu.eg*

E. E. Mahmoud

Department of Mathematics, Faculty of Science, Sohag University, Egypt
Email: emad_eluan@yahoo.com

The aim of this paper is to introduce the new hyperchaotic complex Lü system. This system has complex nonlinear behavior which is studied and investigated in this work. Numerically the range of parameters values of the system at which hyperchaotic attractors exist is calculated. This new system has a whole circle of equilibria and three isolated fixed points, while the real counterpart has only three isolated ones. The stability analysis of the trivial fixed point is studied. Its dynamics is more rich in the sense that our system exhibits both chaotic and hyperchaotic attractors as well as periodic and quasi-periodic solutions and solutions that approach fixed points.

Keywords: Hyperchaotic; chaotic; attractors; periodic; quasi-periodic; Lyapunov exponents; complex.

1. Introduction

In the last two decades several researchers have focused their attention on the study of hyperchaotic systems with real variables in many important fields in applied nonlinear sciences, e.g., laser physics, secure communications, nonlinear circuits, synchronization, control, neural networks and active wave propagation [1−11]. Dynamical systems with at least two positive Lyapunov exponents are defined as hyperchaotic systems. These systems exhibit more complex nonlinear behavior.

In 1979, Rössler, the first one discovered hyperchaos in four−dimensinal hyperchaotic (4D) Rössler system [12], while in 1986, Matsumoto et al reported it in electronic circuits [4]. In recent years, other hyperchaotic systems are constructed and studies, for example, Lü, Chen, unified and

Lorenz [13–19]. All of these systems are 4D hyperchaotic systems.

In 2006, Chen et al [14] introduced the hyperchaotic Lü system by adding a state feedback controller as:

$$\dot{x} = a(y - x) + u, \dot{y} = cy - xz, \dot{z} = xy - bz, \dot{u} = xz + du \tag{1}$$

where a, b, c are constants of Lü system, d is a control parameter and x, y, z and u are real variables. This system has only three isolated fixed points.

In 2007, Mahmoud et al [20] introduced the complex Lü system as follows:

$$\dot{x} = \rho(y - x), \dot{y} = \nu y - xz, \dot{z} = 1/2\,(\bar{x}y + x\bar{y}) - \mu z, \tag{2}$$

where ρ, ν and μ are positive parameters, $x = u_1 + iu_2$, $y = u_3 + iu_4$ are complex variables, $i = \sqrt{-1}$, u_i and z are real variables, dots represent derivatives with respect to time and an overbar denotes complex conjugate variables. In laser physics problems and thermal convection of liquid flows, the electric field and atomic polarization amplitudes are complex quantities [21]. In secure communications doubling the number of variables or using complex variables (which means using higher dimensional chaotic systems) increase the content and security of the transmitted information [3]. For other nonlinear systems with complex variables, see for example, References [22-24] and references therein.

In this work we construct a new hyperchaotic complex Lü system by adding a state feedback controller to our system (2) as:

$$\begin{aligned} &\dot{x} = \rho(y - x) + w & &\dot{y} = \nu y - xz + w \\ &\dot{z} = 1/2\,(\bar{x}y + x\bar{y}) - \mu z & &\dot{w} = 1/2\,(\bar{x}y + x\bar{y}) - \sigma w \end{aligned} \tag{3}$$

where w is a real variable and σ is a control parameter. System (3) is a 6D hyperchaotic real system.

This paper is organized as follows: in the next section symmetry, invariance, fixed points of (3) and stability analysis of the trivial fixed point are discussed. In the third section the complex behavior of (3) is studied. Numerically the range of parameters values of the system at which hyperchaotic attractors exist is calculated based on the calculations of Lyapunov exponents.The signs of Lyapunov exponents provide a good classification of the dynamics of our system. The last section contains our concluding remarks.

2. Basic Properties of System (3)

In this section we study the basic properties of our new system (3). The real version of (3) which is a 6D hyperchaotic system reads:

$$\begin{array}{lll} \dot{u}_1 = \rho\,(u_3 - u_1) + u_6 & \dot{u}_2 = \rho\,(u_4 - u_2) & \dot{u}_3 = -u_1u_5 + \nu u_3 + u_6 \\ \dot{u}_4 = -u_2u_5 + \nu u_4 & \dot{u}_5 = u_1u_3 + u_2u_4 - \mu u_5 & \dot{u}_6 = u_1u_3 + u_2u_4 - \sigma u_6 \end{array} \tag{4}$$

The basic dynamical properties of system (4) are described in the following subsections.

2.1. *Symmetry and invariance*

In (4), we note that this system is invariant under the transformation

$$(u_1, u_2, u_3, u_4, u_5, u_6) \longrightarrow (u_1, -u_2, u_3, -u_4, u_5, u_6).$$

Therefore, if $(u_1, u_2, u_3, u_4, u_5, u_6)$ is a solution of system (4), then $(u_1, -u_2, u_3, -u_4, u_5, u_6)$ is also a solution of the same system.

2.2. *Dissipation*

The divergence of (4) is :

$$\nabla \cdot F = \sum_{i=1}^{6} \frac{\partial \dot{u}_i}{\partial u_i} = 2\nu - 2\rho - \mu - \sigma.$$

Therefore the system (4) is dissipative for the case

$$2\nu - 2\rho - \mu - \sigma < 0.$$

2.3. *Equilibria and their stability*

The fixed points of system (4) can be found by solving the following equations:

$$\begin{array}{rrr} \rho\,(u_3 - u_1) + u_6 = 0 & \rho\,(u_4 - u_2) = 0 & -u_1u_5 + \nu u_3 + u_6 = 0 \\ -u_2u_5 + \nu u_4 = 0 & u_1u_3 + u_2u_4 - \mu u_5 = 0 & u_1u_3 + u_2u_4 - \sigma u_6 = 0 \end{array} \tag{5}$$

Obviously, from second equation of (5) $u_4 = u_2$, by substituting in fourth equation of (5) we obtain $u_2\,(\nu - u_5) = 0$, which yields $u_2 = 0$ or $u_5 = \nu$.

When $u_5 = \nu$, we find $\rho = \nu$, $u_6 = \frac{\mu u_5}{\sigma} = \frac{\mu\nu}{\sigma}$ and system (3) has a whole *circle* of equilibria given by the expression:

$$(u_1 - \tfrac{\mu}{2\sigma})^2 + u_2^2 = \mu\nu + \tfrac{\mu^2}{4\sigma^2}, \tag{6}$$

in $(u_1,\, u_2)$ space. Equation (6) is a *circle* with center $(\frac{\mu}{2\sigma}, 0)$ and radius $r = \sqrt{\mu\nu + \frac{\mu^2}{4\sigma^2}}$.
Let $u_1 - \frac{\mu}{2\sigma} = r\cos\theta$ and $u_2 = u_4 = r\sin\theta$, where $\theta \in [0, 2\pi]$, we get the non-isolated nontrivial fixed points as:

$$E_\theta = (r\cos\theta + \tfrac{\mu}{2\sigma}, r\sin\theta, r\cos\theta - \tfrac{\mu}{2\sigma}, r\sin\theta, \nu, \tfrac{\mu\nu}{\sigma}), \text{ for } \theta \in [0, 2\pi]. \quad (7)$$

When $u_2 = 0$, system (3) has a trivial fixed point $(0, 0, 0, 0, 0)$, as well as two isolated nontrivial fixed points given by:

$$(s_1, 0, \tfrac{\sigma\rho s_1}{\sigma\rho + s_1}, 0, \tfrac{\sigma\rho s_1^2}{\mu(\sigma\rho + s_1)}, \tfrac{\rho s_1^2}{(\sigma\rho + s_1)}), (s_2, 0, \tfrac{\sigma\rho s_2}{\sigma\rho + s_2}, 0, \tfrac{\sigma\rho s_2^2}{\mu(\sigma\rho + s_2)}, \tfrac{\rho s_2^2}{(\sigma\rho + s_2)}), \quad (8)$$

where $s_1 = \frac{\mu}{2\sigma} + \frac{1}{2}\sqrt{(\frac{\mu}{\sigma})^2 + 4\mu\nu}$ and $s_2 = \frac{\mu}{2\sigma} - \frac{1}{2}\sqrt{(\frac{\mu}{\sigma})^2 + 4\mu\nu}$.
To study the stability of E_0 the Jacobian matrix of system (2) at E_0 is:

$$J_{E_0} = \begin{pmatrix} -\rho & 0 & \rho & 0 & 0 & 1 \\ 0 & -\rho & 0 & \rho & 0 & 0 \\ 0 & 0 & \nu & 0 & 0 & 1 \\ 0 & 0 & 0 & \nu & 0 & 0 \\ 0 & 0 & 0 & 0 & -\mu & 0 \\ 0 & 0 & 0 & 0 & 0 & -\sigma \end{pmatrix}.$$

The characteristic polynomial is:

$$(\lambda + \sigma)(\lambda + \mu)(\lambda - \nu)^2(\lambda + \rho)^2 = 0.$$

Then the eigenvalues are:

$$\lambda_1 = -\mu, \lambda_2 = -\sigma, \lambda_3 = \lambda_4 = \nu, \lambda_5 = \lambda_6 = -\rho.$$

The trivial fixed point $(0, 0, 0, 0, 0, 0)$ is a stable point if ν has negative values and μ, ρ, σ are positive. Otherwise it is an unstable fixed point.

3. Lyapunov Exponents and Attractors of (3)

In this section we calculate Lyapunov exponents and attractors of system (3).

3.1. *Lyapunov exponents*

System (4) in vector notation can be written as:

$$\dot{U}(t) = h(U(t); \eta), \quad (9)$$

where $U(t) = [u_1(t), u_2(t), u_3(t), u_4(t), u_5(t), u_6(t)]^t$ is the state space vector, $h = [h_1, h_2, h_3, h_4, h_5, h_6]^t$, η is a set of parameters and $[...]^t$ denoting transpose. The equations for small deviations δU from the trajectory $U(t)$ are:

$$\delta \dot{U}(t) = L_{ij}(U(t); \eta)\delta U, \qquad i, j = 1, 2, 3, 4, 5, 6 \tag{10}$$

where $L_{i,j} = \dfrac{\partial h_i}{\partial u_j}$ is the Jacobian matrix of the form:

$$L_{i,j} = \begin{pmatrix} -\rho & 0 & \rho & 0 & 0 & 1 \\ 0 & -\rho & 0 & \rho & 0 & 0 \\ -u_5 & 0 & \nu & 0 & -u_1 & 1 \\ 0 & -u_5 & 0 & \nu & -u_2 & 0 \\ u_3 & u_4 & u_1 & u_2 & -\mu & 0 \\ u_3 & u_4 & u_1 & u_2 & 0 & -\sigma \end{pmatrix}.$$

The Lyapunov exponents λ_i of the system is defined by:

$$\lambda_i = \lim_{t \longrightarrow \infty} \frac{1}{t} \log \frac{\|\delta u_i(t)\|}{\|\delta u_i(0)\|}. \tag{11}$$

To find λ_i, Eqs. (9) and (10) must be numerically solved simultaneously. Runge–Kutta method of order 4 is used to calculate λ_i.

For the choice $\rho = 42$, $\mu = 6$, $\nu = 25$ and $\sigma = 5$ and the initial conditions $t_0 = 0$, $u_1(0) = 1$, $u_2(0) = 2$, $u_3(0) = 3, u_4(0) = 4$, $u_5(0) = 5$ and $u_6(0) = 6$. we calculate the Lyapunov exponents which are : $\lambda_1 = 3.43$, $\lambda_2 = 0.55$, $\lambda_3 = 0$, $\lambda_4 = -7.22$, $\lambda_5 = -25.47, \lambda_6 = -36.23$.

This means that our system (4) for this choice of ρ, ν, μ and σ is a hyperchaotic system since two of the Lyapunov exponents λ_1 and λ_2 are positive and dissipative system because their sum is negative. The hyperchaotic attractors of (4) using the same choice of the parameters and initial conditions are plotted in Figure 1 in (u_1, u_2, u_3) and (u_4, u_5, u_6) spaces respectively.

3.2. *Attractors of the hyperchaotic complex Lü system*

System (4) is a 6D continuous dissipative dynamical system and assuming that its Lyapunov exponents are λ_i, $i = 1, 2, \ldots, 6$. The signs of Lyapunov exponents provide a good classification of the dynamics of our system (4). The attractors of (4) can be described as: (+, +, 0, −, −, −), a hyperchaotic attractor, (+, 0, −, −, −, −), a chaotic attractor, (0, −, −, −, −, −), a periodic attractor (limit cycle), (0, 0, −, −, −, −), a quasi–periodic and (−, −, −, −, −, −), a fixed point. From our calculations we found that λ_4, λ_5 and λ_6 are negative for all choices of the parameters values below.

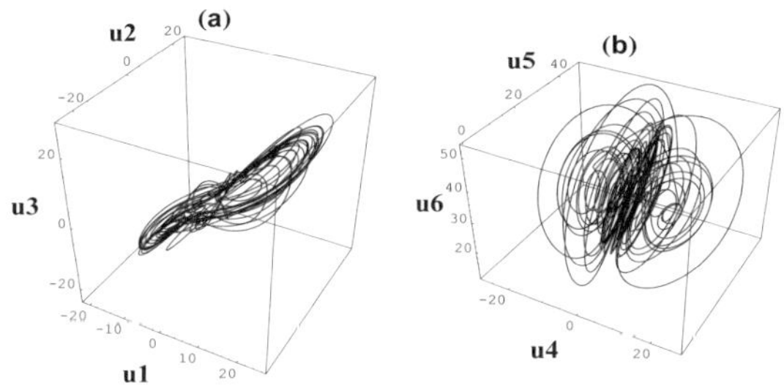

Fig. 1. A hyperchaotic attractor of (3) for $\rho = 42, \mu = 6, \nu = 25$ and $\sigma = 5$ with $t_0 = 0, u_1(0) = 1, u_2(0) = 2, u_3(0) = 3, u_4(0) = 4, u_5(0) = 5, u_6(0) = 6$. (a) in (u_1, u_2, u_3) space. (b) in (u_4, u_5, u_6) space.

3.2.1. *Fix* $\rho = 42$, $\mu = 6$, $\nu = 25$ *and vary* σ

In Figure 2a,b we plotted the corresponding Lyapunov exponents λ_i, $i = 1, 2, ..., 6$ of system (4) using the initial conditions of Figure 1. It is clear that from Fig. 2a, when $\sigma \in (0, 0.14]$ and (1.75, 16.4] the new system (4) has hyperchaotic attractors, while it has chaotic attractors for σ lies in the intervals (1.05, 1.6], (1.783, 1.75] and (16.4, 278). It has also periodic attractor for $\sigma \in$ (0.14, 1.05] and (1.6, 1.738]. In Figure 2a we consider the values of σ up to 20 and for the values >20, we have the case (+, 0, −, −, −, −) which means that our system has chaotic attractors. As is shown in Figure 2b the values of λ_4, λ_5 and λ_6 are negative.

3.2.2. *Fix* $\sigma = 10$, $\mu = 6$, $\nu = 25$ *and vary* ρ

From Figure 2c one can concludes that (4) has hyperchaotic attractors for $\rho \in$ [36.3, 37.3) and (40, 48.9], chaotic attractors for $\rho \in$ (34.8, 36.3), [37.3, 37.93], (38,40] and (48.9, 66.8], and quasi-periodic attractors for ρ lies in (27.1, 32]. The limit cycles of (4) are found for $\rho \in$ (17, 27.1] and (32, 34.8]. Solutions of our system (4) approach nontrivial fixed points for $\rho \in$ (66.8, 200).

3.2.3. *Fix* $\rho = 42$, $\sigma = 10$, $\nu = 25$ *and vary* μ

As we did before we plot only λ_1, λ_2 and λ_3 in Figure 2d and we see that (4) has hyperchaotic attractors for $\mu \in$ [2.05, 2.25], (2.75, 3.05], (3,45, 3.65], [4.15, 7.6], (7.8, 8.2], (9, 9.7] and (10.7, 13]. Solutions of (4) approach nontrivial fixed points for $\mu \in$ (33.2, 300). In between above values of μ our system has periodic and chaotic attractors as one sees from Figure 2d.

3.2.4. *Fix $\rho = 42$, $\mu = 6$, $\sigma = 10$ and vary ν*

As is shown in Figure 2e, the hyperchaotic attractors are exist only for $\nu \in [16, 26.5]$, $[27.1, 27.3]$ and $(28.6, 29.65]$, while the chaotic attractors for ν lies in the intervals $(26.5, 27.1]$, $(27.3, 28.6]$ and $(29.65, 31.5]$. The periodic solutions are exist for $\nu \in (31.5, 40]$ and for $\gamma < 16$, the solutions of (4) approach to nontrivial fixed points.

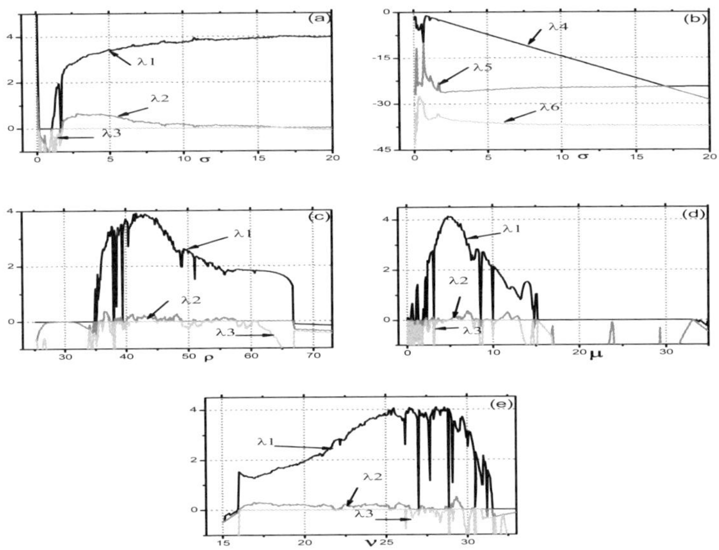

Fig. 2. Lyapunov exponents of (3) at the same initial conditions as in Figure 1. (a) λ_1, λ_2 and λ_3 versus σ (b) λ_4, λ_5 and λ_6 versus σ (c) λ_1, λ_2 and λ_3 versus ρ (d) λ_1, λ_2 and λ_3 versus μ (e) λ_1, λ_2 and λ_3 versus ν.

4. Conclusions

In this paper the hyperchaotic complex Lü system (3) is constructed by adding a state feedback controller. The values of Lyapunov exponents λ_i $(i = 4, 5, 6)$ are negative as shown in Fig. 2b and do not affect the classification of the behavior of our system. The new hyperchaotic system has both isolated and non-isolated fixed points. The projection of non-isolated fixed point in (u_1, u_2) space is a circle with center $(\frac{\mu}{2\sigma}, 0)$ which does not occur in chaotic real Lü system. The dynamic of system (3) is very complicated as shown in Figure 2, since it has, solutions approach to fixed points, periodic solutions, quasi-periodic solutions, chaotic behavior and hyperchaotic behavior. So the dynamics of our system is very rich. The values of the parameters at which (3) has hyperchaotic attractors are calculated. We hope that the results of these investigations increase the knowledge of higher dimensional systems as we have in the literature for 3D or 4D hyperchaotic systems.

References

1. R. Vicente, J. Daudén and R. Toral, IEEE. J. Quantum Electron **41**, 541 (2005).
2. S. Mascolo and G. Grassi, Proceedings of the 1998 IEEE Int. Conference on control Applications, Trieste, Italy 1–4 September 1016 (1998).
3. J. H. Peng, E. J. Ding, M. Ding and W. Yang, Phys. Rev. Lett **76**, 904 (1996).
4. T. Matsumoto, L. O. Chua and K. Kobayashi, IEEE Trans. CAS **33**, 1143 (1986).
5. A. Tamassevicius, A. Cenys, G. Mykolaitis, A. Namajunas and E. Lindberg, IEEE Electronics Lett **33**, 542 (1997).
6. T. kapitaniak, L. O. Chua and G. Q. Zhong, IEEE Trans. CAS **41**, 499 (1994).
7. C. Barbara and C. Silvano, Int. J. Circuits. Theor. Appl **30**, 625 (2002).
8. P. Arena, S. Baglio, L. Fortuna and G. Manganaro, Electron. Lett **31**, 250 (1995).
9. J. Y. Hsieh, C. C. Hwang, A. P. Li and W. J. Li, Int. J. Control **72**, 882 (1999) .
10. L. O. Chua, M. Hasler, G. S. Moschytz and J. Neirynck, IEEE Trans. CAS **42**, 559 (1995).
11. G. Grassi and S. Mascolo, Int. J. Bifurcation and chaos **9**, 705 (1999).
12. O. E. Rössler, Phys. Lett. A **71**, 155 (1979).
13. G. Wang, X. Zhang, Y. Zheng and Y. Li, physica A **371**, 272. (2006)
14. A. Chen, J. Lu, J. Lü and S. Yu, physica A **364**, 103 (2006).
15. M. Haeri and M. Dehghani, Phys. Lett. A **356**, 226 (2006).
16. M. T. Yassen, Nonlinear Analysis. To appear (2007).
17. X. Wu, J. Wang, Jun–an Lü and H. H. C. Iu, Chaos Solitons & Fractals **32**, 1485 (2007).
18. Q. Jia, Phys. Lett. A **366**, 217 (2007).
19. R. Barboza, Int. J. Bifurcation and Chaos **17**, 4285 (2007).
20. G. M. Mahmoud, T. Bountis and E. E. Mahmoud, Int. J. Bifurcation and Chaos **17**, 4295 (2007).
21. A. Rauh, L. Hannibal and N. Abraham, Physica D **99**, 45 (1996).
22. G. M. Mahmoud, S.A. Aly and M. A. Al-Kasfif, NonLinear Dynamics **51**, 171 (2008).
23. G. M. Mahmoud, M. A. Al-Kashif and A.A.Farghaly, J. of Physics A: Mathematical and Theoretical **41**, No.5 (2008).
24. G. M. Mahmoud, M. A. Al-Kashif and S. A. Aly, Int. J. Mod. Physc. C **18**, 253 (2007).

Chaotic Signal Analysis of Parkinson's Disease STN Brain Signals

P. A. Meehan* and P. A. Bellette

School of Engineering, University of Queensland, Australia
*Email: *meehan@uq.edu.au*

Techniques of signal processing from the field of non-linear time series analysis have been applied to microelectrode recordings from the subthalamic nucleus of patients suffering from Parkinson's disease. The recordings have been taken while the patients perform different language processing tasks from raw microelectrode recording data obtained from within the brain during surgery. Using classical chaotic signal analysis techniques, very complex behaviour is discerned but no robust trends for different tasks identified. However interesting behaviour is shown when the recordings are analysed using techniques from non-equilibrium statistical mechanics, indicative of different short term signal behaviour whilst performing the different language processing tasks. In particular, the evidence of discrimination between different linguistic tasks and left/right brain behaviour is revealed.

1. Introduction

The subthalamic nucleus (STN) is a small subcortical structure located within the basal ganglia of the brain which acts as an important relay for muscle control. Hyperactivity of the STN is a characteristic pattern of Parkinson's disease (PD) and therefore has been the target of recent successful procedures of deep brain stimulation (DBS) to relieve symptoms of PD. Although demonstrably very effective on some PD patients, a full understanding of how DBS affects brain functioning is yet to be obtained. Much research has therefore been performed recently to understand the success of DBS and therefore hone its performance and reliability.

Also the emergence of evidence for a role of the STN in language processing, through anatomical and functional neuroimaging studies, provides the impetus for a neurophysiological study investigating the direct link between STN activity and language processing. This may be achieved by live measurements of brain signals at the STN during controlled tasks using microelectrode recordings (MERs) and sophisticated investigative signal analysis to determine any relationships.[1] It is proposed that nonlinear analysis techniques, successfully used for characterizing other biological activity may provide further insight into STN signals and functioning.

In the present paper, characteristic measures of chaos and fractal dynamics from existing measured brain signal measurement, taken from PD patients have been evaluated so as to illuminate any relationships between these measures and linguistic task conditions.

2. Chaotic Measures and Biological Signals

The generally aperiodic, complex and non-linear behaviour of biological signals has motivated many researchers to consider the possibility of low dimensional chaos as a paradigm for modelling such systems (see for example [2]). To examine if a signal exhibits low dimensional chaotic behaviour requires the use of chaotic measures, the most common of which being the correlation dimension (CD) and maximum Lyapunov exponent (MLE). The CD measures the fractal nature of the phase space of the signal and the MLE tests for exponential divergence of neighbouring trajectories in phase space, indicating extreme sensitivity to initial conditions. Various algorithms exist for the estimation of these measures that have allowed them to become useful tools in the signal analysis of complex systems. In practical applications real signals are unlikely to ever strictly fulfil the theoretical requirements for using such measures of being completely noise free and non-stationary, however with careful application these measures have yielded many interesting results about the behaviour of many systems. In biological systems the apparent relationship between chaotic behaviour and proper biological functioning has been investigated, such as the existence of low dimensional chaos in heart rate variability measurements.[3]

The aforementioned real, complex phenomena are typically characterised by signals with non-stationarity and noise. Hence methods from statistical mechanics, focused on the complex dynamics of systems exposed to random fluctuations, may be more applicable in these cases, than methods based on low-dimensional chaos. One such method, that has yielded interesting results, is based upon the scaling behaviour in the signal through the Hurst Exponent. The analysis of the Hurst Exponent of measured time series has been applied to a variety of real, complex systems including financial,[4] physical,[5] and biological[6] processes. In a recent article by Darbin et al.[7] the Hurst exponent of spike interarrival times of STN activity of Rhesus Monkeys has been performed showing different forms of scaling behaviour, including persistent and non-persistent behaviour.

Another method along this line of enquiry has been performed by Yulmeteyev et al.[8] who have conducted studies on various physical and biological systems by investigating the so-called Non-Markovity and Relaxation Parameters arising from a discrete version of the Zwanzig-Mori chain of equations (see for example Zwanzig[9]). A qualitative scale for indication of the type of process may be defined as; $\epsilon_1 = 1$ indicates a Non-Markov Process (regular), $\epsilon_1 > 1$ indicates a Quasi-Markov Process (chaotic) and $\epsilon_1 >> 1$ indicates a Random Markov Process.[10] This same method of analysis has also been applied to Pakinson's Disease gait and finger tremor data.[10,11] This shows a low NMP for untreated PD patients, i.e. regular behaviour, and an increase, i.e. more chaotic, when under DBS or medication. The RP analysis has also been performed in[12] on magnetoencephalograms of people suffering from epilepsy.

3. Clinical Measurements

Recordings were available for 10 patients who had MERs taken from their STN prior to DBS implantation. They were taken on both hemispheres in 9 patients and on the left side only for one patient. The recordings are grouped into 3 sampling epochs, which are as follows:

(1) Prior to a semantic categorization task (Baseline measurement).
(2) Listening to two words from either the same or different semantic categories.
(3) Responding via pressing a button their categorization of the words as belonging to either same or different semantic categories.

MERs were acquired with a Tungsten microTargeting electrode (model mT-DWAR, FHC, Bowdoinham, ME) with a tip diameter of less than 50m and impedance of approximately 0.5 M (± 30%) at 1 kHz. MERs were filtered (500-5000 Hz) and recorded at a sampling rate of 24 kHz from LeadPoint (Medtronic Inc., Minneapolis, MN). The location of the STN was identified when a neurologist and neurosurgeon verified characteristic STN firing patterns. Once the optimal STN location was established, participants completed an auditory semantic categorisation task with the simultaneous acquisition of MERs. Prior to participation in the language task, patients were deemed to be sufficiently alert to perform the standard clinical assessments used during surgery for DBS.

4. Results – Analysis of Raw Micro-Electrode Recording (MER) Data from STN

The analysis was performed on the measured data in raw form as opposed to using any type of pre-filtering. The raw data effectively includes all "noise" contributions associated with the network of neurons surrounding the micro-electrode probe. It is noted that in a comparative investigation, the data was filtered to focus upon the spiking/firing of the local neurons to the probe, but these results are not detailed in the present paper. A detailed discussion of the raw recording results using each of the chaotic measures is provided in the following.

A sample of the time and frequency domain raw recordings in all three epochs can be seen in Figure 1.

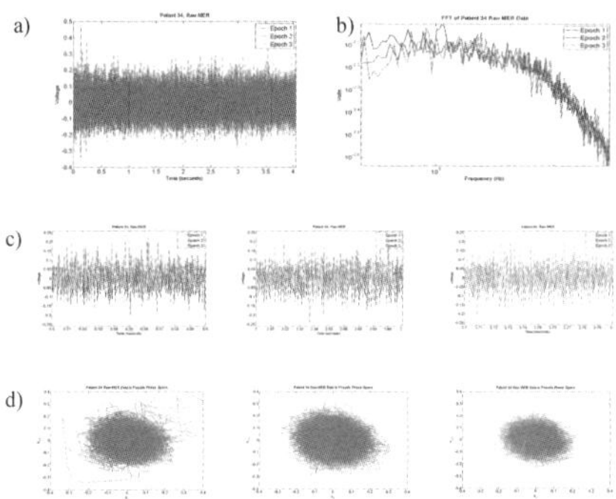

Fig. 1. Raw MER (a) Time domain (b) Frequency domain, (c) Zoomups of time domain and (d). Phase Spaces for each epoch. Epoch 1 RP = 3854±144, Epoch 2 RP = 3783±124 and Epoch 3 RP = 3687±28

Also included in this figure is a zoom onto small section of detail in the time history for all three sampling epochs as well as a delay embedding representation of the phase space, from which CD and MLE estimates were taken. The RP estimate for each of the sampling periods is also determined for comparison. Details of the aggregate results for all raw MER's are described below.

4.1. *Maximal Lyapunov Exponent (MLE)*

Unfortunately the trajectories diverge too quickly or are too saturated by noise at small time scales to be detected. This manifests as no linear scaling region for small trajectory separations. Successful determination of the

MLE would require the selection of data points that are very close together in phase space, which necessitates a higher sampling rate and/or more points to be included in the algorithm with a corresponding penalty in computation time.

4.2. *Correlation Dimension (CD)*

For the unfiltered data the CD estimation algorithm usually showed a small scaling region which gave CD estimates in the range of 12-20 for all the patients. A vector of 10000 points from the time series was used for the embedding. A number of estimates for the maximum discernable dimension exist (see for example Galka et al.[13]) and the region where scaling appears to occur is well above the limit predicted by such estimates, thus they cannot be considered as reliable estimates of the true correlation dimension. This is evidenced by the significant variation from sample to sample seen in the recorded estimates, which can be seen in Figure 2 where the CD results have been averaged together for all the tested words in each of the different sampling epochs for one patient's left side MER.

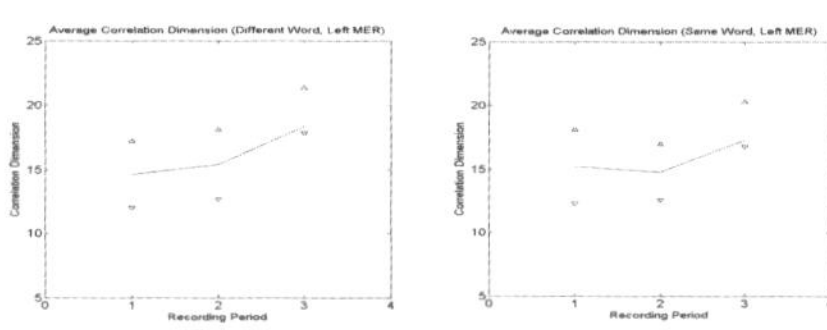

Fig. 2. Results for the average Correlation Dimension of Unfiltered MER data.

The mean value of the CD estimate in each of the sampling epochs over all the patients is provided in Table 1, where D and S refers to a test where words in different or the same semantic category are given and L and R refer to either the right or left side recording. The ± variation refers to one standard deviation.

Table 1. Mean Results for the average Correlation Dimension of Unfiltered MER data.

Epoch	1	2	3
DL	15.3±0.9	15.3±1.0	15.6±0.9
SL	15.5±0.7	15.3±1.1	15.6±1.2
DR	15.1±1.4	14.8±1.4	16.9±2.1
SR	15.4±2.3	15.5±1.4	16.5±1.4

The results indicate little or no correlation between the CD and semantic category and or epoch interval. Note again that the reliability of the CD estimation method is questionable for dimension estimates this high.

4.3. *Hurst Exponent (HE)*

Unfortunately the region where the scaling appears to occur, which yields an estimate of the HE, is at a very small time scale and hence cannot be reliably estimated from the data at the current sampling rate. A much higher sampling rate of at least 10x is required to derive a reliable estimate of the HE.

4.4. *Non-Markovity Parameter (NMP)*

Results for the unfiltered data NMP value, averaged over all word pairs for the different word categories are summarized in Table 2, where the relative change from the baseline value is shown. The baseline NMP is shown in brackets.

Table 2. Mean Results for the average NMP of Unfiltered MER data.

Epoch	1	2	3
DL	1 (2.9)	2.6±5.8	6.6±12.7
SL	1 (6.1)	1.9±3.1	2.3±3.7
DR	1 (2.8)	1.2±1.7	0.8±0.8
SR	1 (2.7)	3.3±6.9	5.3±10.3

Note that the NMP appears to be extraordinarily sensitive to the data, hence extreme fluctuations in the predicted value with a large standard deviation. No clearly discernable trends are immediately apparent from the NMP results.

4.5. *Relaxation Parameter (RP)*

The RP shows clear trends in nearly all of the patient data sets. An example of this is shown in Figure 3, which shows the averaged word pair results for one patient for both word pair categories and also for both hemisphere recordings. In this and subsequent such figures the star represents the mean parameter value and the triangle represents the one standard deviation error bar.

The results for all patients are shown in Table 3, where the units for the RP are $\sec^{-1}$. In Table 3, the average value of the size of the steps from the baseline value in epoch 1, either up or down have been grouped

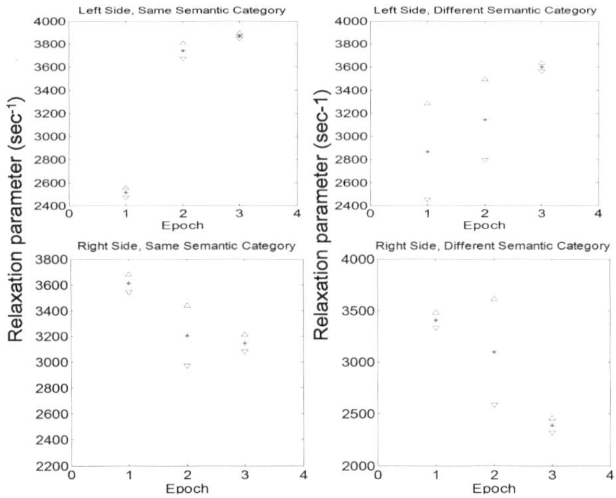

Fig. 3. Results of average relaxation parameter for all patient data sets.

Table 3. Mean Results for the average Relaxation Parameter of Unfiltered MER data.

Step Up				Step Down			
Epoch	1	2	3	Epoch	1	2	3
DL	0	$+200 \pm 197$	$+256 \pm 297$	DL	0	-78 ± 26	-111 ± 79
SL	0	$+418 \pm 422$	$+494 \pm 484$	SL	0	-56 ± 104	-85 ± 88
DR	0	$+218 \pm 64$	$+318 \pm 200$	DR	0	-304 ± 270	-453 ± 407
SR	0	$+250 \pm 35$	$+284 \pm 152$	SR	0	-187 ± 215	-252 ± 289

together. Note that patients 104, 103, 61, 54 and 38 all follow the same pattern, and 69 and 34 almost follow this pattern, of having a step up in the left side measurements and a step down in the right side measurements. The results were treated in this way to show that a bigger step up in RP usually occurred on the left side whilst a bigger step down occurred on the right. In most patients the left side usually showed an increase in RP while the right showed a decrease. This was reversed in two patients despite screening for handedness. The highlighting is to draw attention to the apparent symmetry on the diagonal also.

4.6. *Comparison of baseline deviation and measure sensitivity*

To determine the repeatability of these measurements, a comparison of the baseline deviation of the NMP and RP results were calculated and found to be NMP= 0.74 and RP= 0.036 respectively. These refer to the average ratio between the standard deviation and mean of the baseline measurement. Thus, a very repeatable measurement will have a ratio close to 0 and one that fluctuates significantly will tend towards infinity. In the present analysis we consider measures with values approaching 1 as being considered unreliable. Therefore the most repeatable baseline measurements are for the RP in the unfiltered case having the smallest fluctuations of approximately 4%.

5. Discussion

All measures provide evidence of highly complex, higher dimensional stochastic dynamics as expected from brain/neuronal activity. STN activity appears to be characterised by time scales much faster than previous investigations on mechanical motive motions of the body of PD patients (ie finger tremors), which is also expected. The statistical mechanics measures (NMP, HE, RP) confirms very high dimensional "chaotic-like" behaviour with mean-corrected "random-walk" behaviour. The signal behaviour is analogous that of a small ship navigating in very stormy seas based solely on a log/memory of prior compass headings.

For the raw MER data the algorithms for CD and MLE are unable to find reliable estimates of the chaotic measures. This would seem to be due to the highly complex (and high order) nature of the signal. It is difficult to accurately discern the scaling regions from statistical fluctuations for the number of sample points used. To accurately resolve dimension estimates this high would require the order of 108 data points,[14] which equates to a sample time (>1hr) too long to resolve any short term trends and approximately 108 times the computing time unless significant improvements to the algorithm are discovered.

The reason for the very large fluctuation of the NMP is currently not clear. At present it is thought that it may be due to the scale over which it is calculated. A bounded chaotic process over a long period of time may appear to be regular since it is confined within limits. Hence a suitable choice of time scale which represents the change in signal may achieve better results for the NMP.

The most promising and interesting trends are shown in the RP data for the raw MER. Firstly there is a consistent rise in RP usually on the left brain hemisphere during the different tasks and a corresponding decrease in the right side. It should be noted that the patients were screened for right handedness but not for right/left brain dominance. There also appears to be differences in the magnitude of the step size for different and same word semantic categorization tasks for the different hemispheres. An increase in RP indicates an increase in the short term fluctuations of the signal and thus may be showing an increase in neuron activity around the microelectrode and vice versa for a decrease in RP. This measure was also the most consistent for baseline fluctuations and demonstrated good sensitivity to changes in the signal. The RP can be mathematically interpreted as the initial rate of decorrelation of the signal (or rate of loss of memory) at high frequencies and physically interpreted as the diffusion coefficient for anomalous diffusion processes such as brownian motion. The effectiveness of the RP over the NMP in the present case may indicate that the important brain activity is occurring at high rather than low frequencies.

6. Conclusions

Non-linear signal analysis techniques have been applied to MERs from the STN of PD patients while undergoing different language processing tasks. The MER were not filtered to isolate the response of the closest neuron to the microelectrode ie they include any networked neuronal behaviour. Of the non-linear measures the CD and MLE were unsuccessful in obtaining useful signal information, most likely due to the complexity of the unfiltered data and noise and incompatibility with the filtering technique. The RP for the raw data seemed to be the most consistent and sensitive measure of signal changes, showing different behaviour during different language processing tasks. The most interesting trends, included an increase in STN activity in one hemisphere and a corresponding decrease on the opposite side during the different processing task phases. It also showed correlation of different magnitude changes depending on whether the linguistic task/word group belonged to the same or different semantic categories. The following main conclusions were drawn:

(1) The statistical mechanics based measures were more successful at characterising the behaviour of STN signals which exhibited very complex, high dimensional, combined chaotic yet stochastic behaviour.
(2) The Relaxation Parameter reveals evidence of discrimination between different linguistic tasks and left/right brain behaviour. This may indicate that the aggregate behaviour of the network of neurons in the STN is important although analysis on filtered data is desirable for further investigation.
(3) Of the statistical mechanics parameters investigated, the Relaxation Parameter appeared to be the most repeatable/reliable mea-

sures for STN activity, most likely because the behaviour is characterised by smaller time scales than that suited to the non-Markovity parameter. A more rigorous statistical analysis of the results may confirm and reveal unrecognized trends.

(4) Higher sampling rates may resolve performance of the Hurst exponent with the raw data to provide validation and /or more clarity to the RP results.

Acknowledgements

The authors are greatly indebted to Profs Helen Chenery & Andrew Bradley of the Faculties of Electrical Engineering and Health Sciences, University of Queensland and Prof Peter Silburn of St. Andrew's War Memorial and The Wesley Hospitals, Australia for their motivation, guidance, interdisciplinary expertise and funding.

References

1. A. Angwin, H. Chenery, D. Copland, B. Murdoch and P. Silburn, *Cognitive Brain Research* **25**(1), 78 (2004).
2. P. B. Pascolo, A. Marini, R. Carniel and F. Barazza, *Chaos, Solitons and Fractals* **24**, 1343 (2004).
3. F. Beckers, B. Verheyden and A. Aubert, *American Journal of Physiology-Heart and Circulatory Physiology* **290**(6), H2560 (2006).
4. K. Bassler, G. Gunaratne and J. McCauley, *Physica A-Statistical Mechanics and its Applications* **369**(2), 343 (2006).
5. A. Ponomarev, A. Zavyalov, V. Smirnov and D. Lockner, *Tectonophysics* **277**(1-3), 57 (1997).
6. A. Knezevic and M. Martinis, *International Journal of Bifurcation and Chaos* **16**(7), 2103 (2006).
7. O. Darbin, J. Soares and T. Wichmann, *Brain Research* **1118**, p. 84:93 (2006).
8. R. Yulmetyev, P. Hanggi and F. Gafarov, *Physical Review E* **62**(5), 6178 (2000).
9. R. Zwanzig, *Non-equilibrium statistical mechanics* (Oxford University Press, New York, 2001).
10. R. Yulmetyev, S. Demin, O. Y. Panischev, P. Hanggi, S. Timashev and G. Vstovsky, *Physica A* , p. 655.
11. R. Yulmetyeva, S. Demin, N. Emelyanova, F. Gafarov and P. Hanggi, *Physica A* **319**, p. 432 446 (2003).
12. R. Yulmetyeva, P. Hanggi, D. Yulmetyeva, E. K. S. Shimojo, K. Watanabe and J. Bhattacharya, *Physica A* **383**, p. 443454 (2007).
13. A. Galka, T. MaaB and G. Pfister, *Physica D* **121**, 237 (1998).
14. H. Kantz and T. Schreiber, *Chaos* **5**(1), 143 (1995).

Analysing Time Series using Symbolic Representations

R. Monetti, W. Bunk and F. Jamitzky

Max-Planck-Institut für extraterrestrische Physik
Giessenbachstr. 1, 85748 Garching, Germany

We present a method to characterize synchronization in time series based on symbolic representations. A symbol is linked to a sequence of numbers through the rank-order of its values. A representation of a time series is obtained after mapping all sequences into symbols. We propose a transcription scheme between symbolic representations to study the dynamics of coupled systems. This scheme allows us to use elements of group theory and to derive information measures to assess the degree of synchronization. We apply our method to a prototype non-linear system which displays a rich coupled dynamics.

Keywords: Synchronization; symbolic representations; transcriptions; order classes; complexity; information measures.

1. Introduction

Synchronization phenomena are widespread in Nature. They are observed not only in coupled periodic systems but also in coupled systems with complex dynamics. Examples of synchronization appear in different fields of science like electronics (e.g. coupled circuits), physiology (e.g. between cardiac and respiratory systems or EEG signals),[1,2] ecological systems[3] or in non-linear optics (e.g. coupled laser systems). In such systems, a number of different synchronization states may arise as complete synchronization,[4] phase[5,6] and lag synchronization,[7] generalized synchronization,[8,9] etc (for a review about synchronization in chaotic systems see Boccaletti et al.[10]).

We present a methodology to assess synchronization in coupled systems using information measures based on symbolic representations of time series.

2. Method

Let x be a time series and $q = (x_0, \ldots, x_{p-1})$ be a sequence of length p extracted from x. The symbol Q associated to q is defined as the rank-ordered

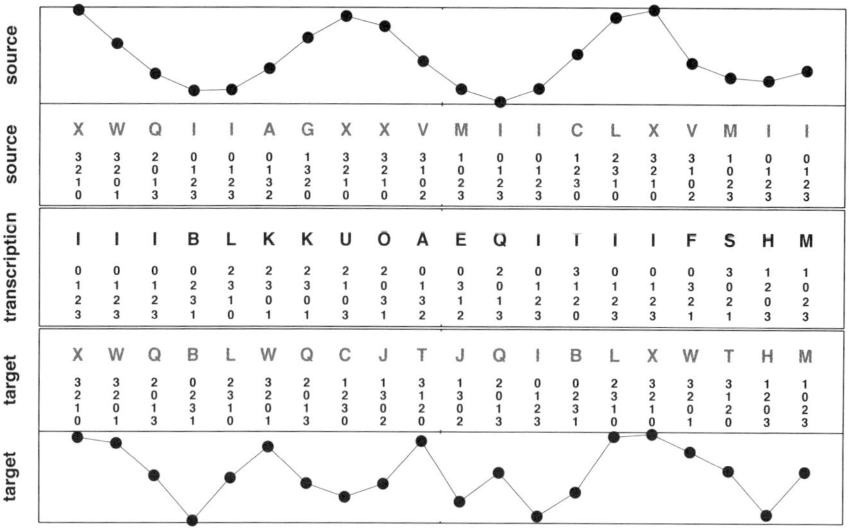

Fig. 1. Example of symbolic representations of time series for sequence length $p = 4$. Symbols in light grey correspond to the representations and the dark grey symbols indicate the transcriptions that have to be applied to the upper symbols (source) to obtain the lower ones (target). Note that this operation is not commutative.

indices of the components of q. For instance, for $q = (1.6, 1.3, 1.4, 1.5)$, the symbol associated to q is $Q = (3, 0, 1, 2)$. This symbolic representation was first introduced by Bandt et al.[11] in the context of complexity analysis of time series and motivated some studies of the characterization of similarities in time series.[12,13] It should be mentioned that the occurrence of identical values in q has not been considered. When the sequence contains equal values, one may add a small random perturbation to avoid this case.

Figure 1 shows symbolic representations of two time series (light grey symbols) for $p = 4$. Given two symbols A_1 and A_2 there always exists a symbol T, in the following called transcription, such that $T[A_1] = A_2$. The action of symbol T is defined as follows. Let $A_1 = (j_0, j_1, \ldots, j_{p-1})$ and $T = (k_0, k_1, \ldots, k_{p-1})$. Then,

$$T[A_1] = (j_{k_0}, j_{k_1}, \ldots, j_{k_{p-1}}). \tag{1}$$

It should be noted that with the operation T, the set of symbols form a finite non-Abelian group of order $p!$ known as the symmetric group S_p. Dark grey symbols in Fig. 1 indicate the transcriptions between the symbolic representations of the time series. The group S_p can be partitioned into non-overlapping classes $\mathcal{C}_i$ ($S_p = \bigcup \mathcal{C}_i$) satisfying a power relation, namely

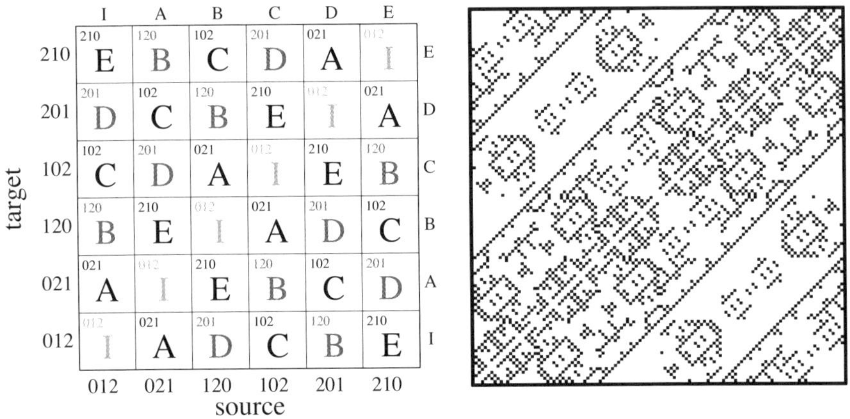

Fig. 2. Left: Transcription matrix for sequence length $p = 3$. The black (dark grey) symbols belong to order 2 (3) class, respectively. The identity symbol (light grey) is a singular one symbol class satisfying $T^n = I$, $\forall n$. Source symbols are displayed at the bottom and target symbols on the left of the transcription matrix. Right: For $p = 5$ the transcription matrix has $(5!)^2$ elements. This matrix shows the positions of the elements belonging to order 3 class, i.e. the structure generated by order 3 transcriptions for $p = 5$.

if $T \in \mathcal{C}_n$ then $T^n = I$, where $I = (0, 1, \ldots, p-1)$ is the identity symbol and T^n is the composition $T[T^{n-1}]$ with $n \geq 1$ and $T^0 \equiv I$. Figure 2 (left panel) shows the transcription matrix for $p = 3$, where the three existing order classes, i.e. $T = I$ (light grey symbol), $T^2 = I$ (black symbols), and $T^3 = I$ (dark grey symbols) are shown. It is worth discussing the action of transcriptions for different order classes. The identity transcription leaves symbols unchanged thus it is the simplest transcription. For $p = 3$, consider for instance transcription $A = (0, 2, 1)$ which belongs to order 2 class and apply it to $E = (2, 1, 0)$ (see Fig. 2), thus $A[E] = D = (2, 0, 1)$. Then, the action of A is identical to one transposition, i.e. the interchange of 0 and 1 in symbol E. However, if we consider transcription $B = (1, 2, 0)$ that belongs to order 3 class and apply it to $E = (2, 1, 0)$ the result is $C = (1, 0, 2)$. We have to perform either two transpositions or one cyclic permutation on E to obtain C. Note that for $p = 3$ all order 2 transcriptions cause a one transposition change while all order 3 transcriptions lead to two transpositions change. Thus, we interpret order 3 transcriptions as "more complex" than order 2 transcriptions. For longer sequences, we can still identify order classes in term of the action of their component symbols although the description becomes more difficult. From this point of view, the order of a class offers a rough estimation of the "complexity" of the transcription.

The order classes satisfy an important property of invariance. Let A and B two symbols connected by the transcription T, i.e. $T[A] = B$ and suppose that $T^N = I$, i.e. $T \in \mathcal{C}_N$. Let Y an arbitrary transcription such that $Y[A] = C$ and $Y[B] = D$. There always exists a transcription T' such that $T'[C] = D$. One can prove that $T = Y^{-1}T'Y$ thus T' belongs to order N class as well. This property of invariance also implies that T and T^{-1} belong to the same order class. However, an order class is not a group since it does not satisfy closure. Order classes are conjugacy classes since order class elements satisfy $T = Y^{-1}T'Y$. Figure 2 (right panel) shows the structure generated by the set $\mathcal{C}_3$ for $p = 5$. The symmetry displayed by this structure is a general property found in all order classes.

The action of a transcription is equivalent to applying permutations. It is known that any permutation can be written as a product of disjoint cyclic permutations (DCP). Using this fact, one can prove that the order of any transcription is the least common multiple (LCM) of the lengths of the DCP factors. Since the sum of the lengths of the DCP equals the sequence length p, the succession of order classes is never interrupted up to order p. For $p \geq 7$, gaps of missing order classes always appear. For $p = 7$, order 8, 9, and 11 classes are missing since there is no possible splitting of a sequence of length 7 in DCP which satisfy the LCM condition. However, order 10 and 12 classes are present since one can have a combination of DCP of lengths 2 and 5 for order 10 and 3 and 4 for order 12.

We now focus on the probability density of transcriptions. Consider a source and a target symbolic representations generated by the actual coupled dynamics of the time series. Given a sequence of length p, the set of all feasible symbols $\mathcal{S}^1 = \{X_i\}$ and $\mathcal{S}^2 = \{X_j\}$ conform the state spaces for the source and the target representations, respectively. The probability density of transcriptions $P_T(p)$ can be written as follows

$$P_{T_k}(p) = \sum_{\Omega=\{(i,j):\, T_k[X_i]=X_j\}} P^C(X_i, X_j), \tag{2}$$

where $X_i \in \mathcal{S}^1$, $X_j \in \mathcal{S}^2$, and $P^C(X_i, X_j)$ is the joint probability density. Let $P^{(1)}(X_i)$ and $P^{(2)}(X_j)$ be the marginal probability densities of the symbols X_i and X_j in state spaces $\mathcal{S}^1$ and $\mathcal{S}^2$, respectively. The matrix $M_{i,j} = P^{(1)}(X_i)P^{(2)}(X_j)$ is the probability density matrix of transcriptions for two independent processes. In this case, the probability density of transcriptions $P_T^{\text{ind}}(p)$ can be evaluated as follows

$$P_{T_k}^{\text{ind}}(p) = \sum_{\Omega=\{(i,j):\, T_k[X_i]=X_j\}} M_{i,j}, \tag{3}$$

where $X_i \in \mathcal{S}^1$ and $X_j \in \mathcal{S}^2$. The aim is to find an information measure to assess how much P_T deviates from P_T^{ind}. A natural choice to quantify the contrast between probability densities is the Kullback-Leibler (KL) entropy defined as follows

$$E_{KL}(P, P^{\text{ind}}) = \sum_i P_{T_i}(p) \log(P_{T_i}(p)/P_{T_i}^{\text{ind}}(p)). \tag{4}$$

Since the E_{KL} is not a symmetric quantity, we use the following symmetrized form[14]

$$S_{KL}(p) = \frac{E_{KL}(P, P^{\text{ind}}) E_{KL}(P^{\text{ind}}, P)}{E_{KL}(P, P^{\text{ind}}) + E_{KL}(P^{\text{ind}}, P)}. \tag{5}$$

We demonstrated above that order classes are also conjugacy classes and that T and T^{-1} belong to the same order class. Thus, $S_{KL}(p)$ for transcriptions inside a class is a suitable invariant measure under the interchange of source and target time series. This property of invariance also allows us to calculate the Kullback-Leibler entropy $S_{KL}^{\mathcal{C}}(p)$ using the probability density of order classes $P_{\mathcal{C}}$ (see Fig. 4). In this case, equations analog to Eqs. (2 - 5) can easily be derived.

3. Applications

We apply the method to a bi-directionally coupled Roessler-Roessler system[7] defined by the following set of equations

$$\begin{aligned} \dot{x}_{1,2} &= -w_{1,2} y_{1,2} - z_{1,2} + k(x_{2,1} - x_{1,2}), \\ \dot{y}_{1,2} &= w_{1,2} x_{1,2} + 0.165 y_{1,2}, \\ \dot{z}_{1,2} &= 0.2 + z_{1,2}(x_{1,2} - 10) \end{aligned} \tag{6}$$

where $w_1 = 0.99$ and $w_2 = 0.95$ are the mismatch parameters. All time series were generated using a fourth-order Runge-Kutta method with an increment $\delta t = 0.001$ and the following initial conditions: $x_1(0) = -0.4$, $y_1(0) = 0.6$, $z_1(0) = 5.8$, $x_2(0) = 0.8$, $y_2(0) = -2$, and $z_1(0) = -4$. Results were saved at intervals $\Delta t = 0.01$. This chaotic system exhibits a rich synchronization behavior which ranges from phase ($k \approx 0.036$) to lag ($k \approx 0.14$) and finally complete synchronization as the coupling parameter k is increased.[7] The results presented here were obtained using the x-components of the Roessler subsystems. Before transforming the time series into symbolic representations, they were sampled using a sampling time $\tau = 150\Delta t$ and time series of length $L = 2^{19}$ ($\sim$ 775 orbits) were considered. This sampling time fulfills the condition of minimum mutual

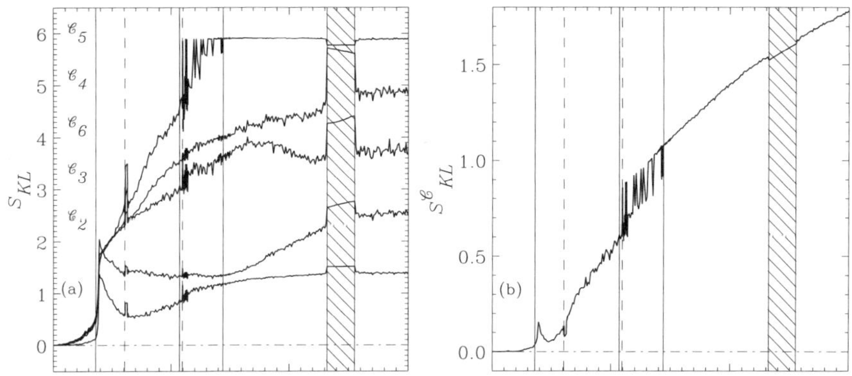

Fig. 3. (a) Kullback-Leibler entropy S_{KL} obtained using the probability density of transcriptions for all available order classes for $p = 6$. Curves for the different order classes are indicated in the plot from top to bottom. (b) $S^{\mathcal{C}}_{KL}$ obtained using the probability density of order classes for $p = 6$. Vertical full lines from left to right indicate transitions to phase-, intermittent lag-, and lag-synchronization, respectively. Vertical dashed lines and hatched areas indicate periodic windows. The values of the coupling constant for transitions (vertical full lines) and the first periodic window were taken from Rosenblum et al.[7]

information of the delay coordinates $(x_1(t), x_1(t+\tau))$ for the uncoupled system $(k = 0)$.[15]

Figures 3(a) show S_{KL} for transcriptions in all feasible order classes for $p = 6$. Figure 3(b) shows $S^{\mathcal{C}}_{KL}$ obtained using the probability density of order classes $P_{\mathcal{C}}$ for $p = 6$. For small values of the coupling constant k, the time series behave independently since the Roessler subsystems are uncoupled. For $k \in [0, 0.036]$, S_{KL} indicates that the actual dynamics hardly deviates from that of the independent processes. S_{KL} sharply increases at $k \sim 0.036$ indicating the onset of phase synchronization. At $k \approx 0.061$ all curves display a peak which corresponds to the presence of a period 3 window.[7] Some curves also indicate the presence of a period 6 window at $k \approx 0.11$. To our best knowledge, the presence of this periodic window has not been reported before probably due to the extremely narrow range of k values ($k \in [0.1094, 0.1096]$) where it takes place. Curves also display a step within the coupling range $k \in [0.232, 0.256]$ which indicates the presence of period 5 windows. Figure 3(a) shows that S_{KL} saturates for order 5 class. Saturation occurs when an order class vanishes, i.e. no transcription belonging to this order class is generated by the coupled dynamics. When this occurs, the KL entropy is not defined thus Eq. 5 can not be used. However, for independent processes the probability density of transcriptions in this particular order class P_T^{ind} is non-vanishing. In these cases, we found that a reasonable choice is to set S_{KL} to the Shannon information entropy $S_{KL} = -\sum_i P_{T_i}^{\text{ind}} \log P_{T_i}^{\text{ind}}$ for transcriptions in the vanishing class.

Figure 3 also unveils another interesting feature of this coupled chaotic system. For $k \approx [0.11, 0.14]$, S_{KL} displays fluctuations which are particularly strong in Fig. 3(b) and for order 5 class, and sharply decrease for $k > 0.14$. This result provides evidence of the existence of a typical behavior known as intermittent-lag-synchronization,[7,16] characterized by synchronization periods interrupted by bursts of non-synchronized behavior. The intermittent bursts of activity induce the large fluctuations displayed by S_{KL} in this range of coupling values. The absence of these fluctuations for higher coupling values indicate that bursts of non-synchronized behavior are no longer present thus lag-synchronization completely develops. In particular, S_{KL} for $\mathcal{C}_5$, which is the most sensitive measures to this intermittent behavior, saturates at $k \sim 0.14$. This value of the coupling constant is in agreement with the one reported in[7] ($k = 0.14$) for the onset of lag-synchronization. Figure 3(b) shows that $S_{KL}^{\mathcal{C}}$ for the probability density of order classes also reveals features above discussed and describes the overall behavior of the coupled system.

Figure 4 shows plots of the probability density $P_{\mathcal{C}_i}$ of the order classes for $p = 6$. Note that Fig 3(b) shows the contrast between the probability densities shown in Fig. 4 and the ones for independent processes. Figure 3(b) indicates that for $k = 0.005$ the contrast is vanishing ($S_{KL}^{\mathcal{C}} \sim 0$) thus the Roessler subsystems behave independently. Then, the probability density $P_{\mathcal{C}}$ for $k = 0.005$ is similar to that generated by two independent processes. Note that even for two random independent processes, the probability density of order classes is not uniform since the cardinality of order classes is different. In the vicinity of the transition to phase synchronization, $P_{\mathcal{C}}$ deviates from that of the independent processes (see Fig. 3(b)) and higher-order classes dominate the coupled dynamics (see Figs. 4 for $k = 0.039$). This trend is reversed when increasing k and already at $k = 0.062$ $\mathcal{C}_2$ is the most relevant class. Figure 4 shows that $\mathcal{C}_2$ dominates up to large values of k where finally $\mathcal{C}_I$ prevails.

As discussed above, the order of a transcription roughly estimates its "complexity". Thus, the probability density of order classes indicates how "complex" the relationship between the time series is. Notice that the probability densities of higher-order classes decrease when increasing k and $\mathcal{C}_5$ vanishes. In fact, simpler synchronization states like intermittent lag and lag synchronization ($k > 0.11$) are described by lower order classes ($\mathcal{C}_2$ and

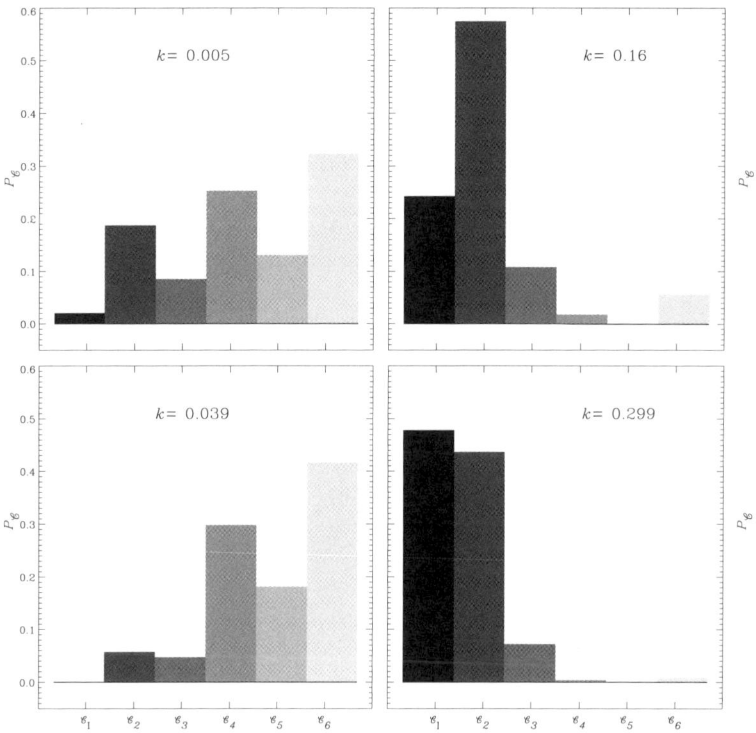

Fig. 4. Probability density $P_{\mathcal{C}}$ of the existing order classes for different values of the coupling constant k for $p = 6$. Note that class $\mathcal{C}_I$ comprises only one transcription ($I = (0, 1, 2, 3, 4, 5)$).

$\mathcal{C}_I$). Clearly, the simplest synchronization state, namely complete synchronization, will only be described by $\mathcal{C}_I$. However, for more complex synchronization states like phase synchronization ($k > 0.036$), higher-order transcriptions play an important role.

4. Conclusions

We presented a method for studying similarities in time series by means of transcription properties that characterize the transition between symbolic representations of the respective time series. In exploiting some basic properties of these transcriptions, namely the power relation, we were able to define order classes of transcriptions with respect to the "complexity" of the underlying transition dynamics from one to the other time series. Our results in the context of studying coupled Roessler-systems show that the defined order classes provide complementary information about the coupled dynamics. In particular we found by means of the Kullback-Leibler contrast to the independent process that the density distributions of order classes can be interpreted as a measure of the "complexity" of the interdependency of the coupled systems. Thereby, more complex synchronization states (e.g. phase synchronization) are associated with a predominance of higher order classes, while simpler synchronization states (e.g. lag synchronization) come along with low-order classes. Our approach that combines elements of group theory with information theory provides a new frame to study synchronization phenomena in time series by their transition dynamics. Since our methodology is easily applied to multi-dimensional time series, we expect it to be useful for the analysis of a wide range of real world coupled systems. It is particularly helpful for those systems, where the detection of synchronization phenomena are of diagnostic interest e.g. in the analysis of physiological signals like electroencephalograms or polysomnographies.

Acknowledgments

We would like to thank Thomas Aschenbrenner for valuable comments.

References

1. C. Schafer, M. Rosenblum, J. Kurths and H. Abel, *Nature* **392**, p. 239 (1998).
2. P. Tass *et al.*, *Phys. Rev. Lett.* **81**, p. 3291 (1998).
3. B. Blasius, A. Huppert and L. Stone, *Nature* **399**, p. 354 (1999).
4. L. Pecora and T. Carroll, *Phys. Rev. Lett.* **64**, p. 821 (1990).
5. M. Rosenblum, A. Pikovsky and J. Kurths, *Phys. Rev. Lett.* **76**, p. 1804 (1996).
6. E. Rosa, E. Ott and M. Hess, *Phys. Rev. Lett.* **80**, p. 1642 (1998).
7. M. Rosenblum, A. Pikovsky and J. Kurths, *Phys. Rev. Lett.* **78**, p. 4193 (1997).
8. N. Rulkov, M. Sushchik, L. Tsimring and H. Abarbanel, *Phys. Rev.E* **51**, p. 980 (1995).
9. L. Kocarer and U. Parlitz, *Phys. Rev. Lett.* **79**, p. 1816 (1996).
10. S. Boccaletti, J. Kurths, G. Osipov, D. Valladares and C. Zhou, *Phys. Rep.* **336**, p. 1 (2002).
11. C. Bandt and B. Pompe, *Phys. Rev. Lett.* **88**, p. 174102 (2002).
12. Z. Liu, *Europhys. Lett.* **68**, p. 19 (2004).
13. K. Keller and K. Wittfeld, *Int. J. Bifurc. Chaos* **14**, p. 693 (2004).
14. D. Johnson and S. Sinanović, *Available online at http://citeseer.ist.psu.edu/johnson01symmetrizing.html* (2001).
15. A. Fraser and H. Swinney, *Phys. Rev. A* **33**, p. 1134 (1986).
16. S. Boccaletti and D. Valladares, *Phys. Rev. E* **62**, p. 7497 (2000).

Detection of Directionality of Information Transfer in Nonlinear Dynamical Systems

A. Papana* and D. Kugiumtzis†

Department of Mathematical, Physical and Computational Sciences
Aristotle University of Thessaloniki
Thessaloniki, Greece
*Emails: *agpapana@gen.auth.gr, †dkugiu@gen.auth.gr*

For the quantification of strength and identification of direction of coupling between two sub-systems of a complex dynamical system, observed from bivariate time series, a number of measures have been proposed that can be grouped in measures of phase synchronization, state space and information. We review all these measures and, in particular, for the information measures we examine different estimates of the probability distributions. We propose also a modification of the transfer entropy measure to span larger time windows and thus be more appropriate for flows. Simulations on systems of different types and for varying coupling strengths showed that information measures, and the modified transfer entropy measure in particular, detect best the coupling strength and direction. This is also found when applying the measures to pairs of EEG channels in order to detect the propagation of pre-epileptic brain activity.

Keywords: Information flow; directionality; coupling; time series; chaos.

1. Introduction

The interaction or coupling between variables or sub-systems of a complex dynamical system is a developing area of nonlinear dynamics and time series analysis.[1–3] The recently developed measures of interaction go beyond the standard cross-correlation and exploit nonlinear properties of dynamical systems. Further, indexes have been derived from these interaction measures in order to detect the direction of interaction. Interaction measures can be split in three classes: the state-space based measures that use the inter-distances of reconstructed points from the time series,[4–6] the measures making use of the idea of synchronization and quantify the concordance of signal quantities such as phases or events,[7,8] and the information based measures that quantify information flow through probability distribution functions.[9,10] All measures involve specific parameters and their performance depends critically on them.

The information measures make no assumptions on the system dynamics as opposed to phase or event synchronization measures that assume strong oscillatory behavior or distinct event occurrences, respectively, and the state space methods that require local dynamics being preserved in neighborhoods of reconstructed points. For the probability density functions used in information measures, different estimates have been proposed. In a recent work,[11] we have compared the histogram-based,[12] kernels,[13] and k-nearest neighbors[14] estimators for the mutual information on scalar time series and found through simulations that for nonlinear systems they are consistent only when noise is added to the data, meaning that the estimates converge with the time series length. Moreover, the estimation and convergence depends on the method specific parameter, especially for the histogram-based methods, where both equidistant and equiprobable binning have been used.

Here, we extend the simulation study to measures of the information flow that involve joint and conditional probability density estimation from bivariate time series. Information flow measures included in this investigation are the transfer entropy,[9] the coarse-grained transinformation rate[15] and the conditional mutual information.[16] The information measures are also compared to state space and synchronization directionality measures.

The structure of the paper is as follows. In Section 2, the directional coupling measures are discussed. In Section 3, the evaluation procedure and the simulation systems are presented and in Section 4, the results are discussed. In Section 5, we present an application of the measures to pre-epileptic EEG signals and in Section 6, the main conclusions are drawn.

2. Directional Coupling Measures

Let us assume a bivariate time series $\{x_t, y_t\}_{t=1}^N$, where x_t and y_t are observed variables from systems that may have unidirectional or bidirectional interaction. For simplicity, we refer to the variables with X and Y, respectively. We are basically concerned here with interaction of nonlinear dynamical systems and therefore we study measures that have the power to detect this type of interaction, skipping correlation and coherence measures used in linear multivariate analysis.[17,18] We also do not consider methods of mutual nonlinear prediction.[19]

2.1. *State space measures*

This class of interaction measures involves the state space reconstruction from each time series separately using standard delay embedding, i.e. the reconstructed points from each time series are $\mathbf{x}_t = [x_t, x_{t-\tau}, \ldots, x_{t-(m-1)\tau}]'$ and $\mathbf{y}_t = [y_t, y_{t-\tau}, \ldots, y_{t-(m-1)\tau}]'$ for the same embedding dimension m and delay τ. The measures operate on the neighboring reconstructed points.

The *nonlinear interdependence* measure (NI) assumes the time indices $r_{t,j}$ and $s_{t,j}$, $j = 1, ..., k$ of the k-nearest neighbors of $\mathbf{x}_t$ and $\mathbf{y}_t$, respectively.[4] The mean squared Euclidean distance of the k-nearest neighbors to $\mathbf{y}_t$ is $R_t^{(k)}(Y) = \frac{1}{k}\sum_{j=1}^{k} \|\mathbf{y}_t - \mathbf{y}_{s_{t,j}}\|^2$, and the X-conditioned mean squared Euclidean distance of $\mathbf{y}_t$ is $R_t^{(k)}(Y|X) = \frac{1}{k}\sum_{j=1}^{k} \|\mathbf{y}_t - \mathbf{y}_{r_{t,j}}\|^2$. Then the NI measure is defined as

$$S(Y|X) = \frac{1}{N'}\sum_{t=1}^{N'} \frac{R_t^{(k)}(Y)}{R_t^{(k)}(Y|X)}, \tag{1}$$

where $N' = N - (m-1)\tau$. Values of $S(Y|X)$ close to zero suggest independence of Y and X, while significant positive values of $S(Y|X)$ suggest dependence of Y on X. The measure is non-symmetric, so if $S(Y|X) > S(X|Y)$ then Y depends more on X than vice versa. It has been reported that NI is robust against noise and capable of detecting weak inter-dependence.[4,20,21]

Recalling that neighboring points correspond to nearby trajectories or equivalently recurrences of a single (chaotic) trajectory at the neighborhood of reference, NI bears a great deal of similarity to the measure of *mean conditional probability of recurrence* (MCR).[6] According to the visualizing method of recurrence plot,[22] the recurrence matrixes of X and Y are respectively $R_{i,j}^X = \Theta(\varepsilon_x - \|\mathbf{x}_i - \mathbf{x}_j\|)$ and $R_{i,j}^Y = \Theta(\varepsilon_y - \|\mathbf{y}_i - \mathbf{y}_j\|)$, where Θ is the Heaviside function and ε_x, ε_y are distances (radii when the distance metric is Euclidean). In addition, the joint recurrence matrix of (X, Y) is defined as $J_{i,j}^{X,Y} = \Theta(\varepsilon_x - \|\mathbf{x}_i - \mathbf{x}_j\|)\Theta(\varepsilon_y - \|\mathbf{y}_i - \mathbf{y}_j\|)$. The concept of recurrence has been used to quantify a weaker form of synchronization, whereas MCR is an extension of it that detects the direction of the coupling. Then MCR of Y given X is

$$M_{CR}(Y|X) = \frac{1}{N'}\sum_{i=1}^{N'} \frac{\sum_{j=1}^{N'} J_{i,j}^{X,Y}}{\sum_{j=1}^{N'} R_{i,j}^X}. \tag{2}$$

In loose terms, MCR estimates the probability of recurrence on Y when there is synchronous recurrence on X. The independence and direction of dependence is defined as for NI, so if $M_{CR}(Y|X) > M_{CR}(X|Y)$ then X drives Y and if $M_{CR}(Y|X) \simeq M_{CR}(X|Y)$ then the coupling is symmetric (same strength of coupling in both directions). The main difference between NI and MCR is that MCR uses counts of neighboring points according to a distance threshold, whereas NI uses the distances for a fixed number of neighboring points.

2.2. *Synchronization measures*

Reverting from state space dynamics to phase dynamics, the measure of *directionality index* (DI) quantifies the degree the phase dynamics of one oscillator is influenced by the phase dynamics of another oscillator.[8] To form the DI measure it is assumed that the two time series indeed exhibit oscillating behavior, so that phases $\phi_X(t)$, $t = 1, \ldots, N$, can be extracted from $\{X_t\}_{t=1}^N$, typically using the Hilbert transform (the same for Y, also for the following expressions). It is also assumed that the phase increments $\Delta_X(t) = \phi_X(t+\tau) - \phi_X(t)$ are generated by an unknown two-dimensional map $\Delta_X(t) = F_X[\phi_X(t) - \phi_Y(t)]$ fitted by a finite Fourier series $\sum_{m,l} A_{m,l} e^{im\phi_X + il\phi_Y}$. The fitted functions for X and Y are used to quantify the cross dependencies of phase dynamics of the two systems given as

$$c_X^2 = \int_0^{2\pi}\int_0^{2\pi} \frac{\partial F_X}{\partial \phi_X} \mathrm{d}\phi_X \mathrm{d}\phi_Y, \qquad c_Y^2 = \int_0^{2\pi}\int_0^{2\pi} \frac{\partial F_Y}{\partial \phi_Y} \mathrm{d}\phi_X \mathrm{d}\phi_Y. \tag{3}$$

The directionality index is defined as:

$$d_{X,Y} = \frac{c_Y - c_X}{c_X + c_Y}, \tag{4}$$

where $d_{X,Y}$ close to -1 or 1 suggests unidirectional coupling driven by X or Y, respectively, while $d_{X,Y} \simeq 0$ suggests symmetric bidirectional coupling or no coupling. The main disadvantage of this method is that it is not always possible to extract phases from scalar time series.

Event synchronization is the relative timing of certain "events" in the time series (like spikes, local minima or maxima), i.e. quasi-simultaneous appearances of these events in the two time series.[23] Let $c^\tau(X|Y)$ denote the number of times an event appears in X shortly after it appears in Y (for details see [23]), allowing a time lag τ between two synchronous events. The strength of the coupling, termed as *event synchronization* (ES), is expressed by the normalized total of synchronized events

$$Q_\tau = \frac{c^\tau(Y|X) + c^\tau(X|Y)}{\sqrt{n_x n_y}}, \tag{5}$$

and the coupling direction, termed as *event delay* (EvD), by the normalized difference

$$q_\tau = \frac{c^\tau(Y|X) - c^\tau(X|Y)}{\sqrt{n_x n_y}}, \tag{6}$$

where n_x and n_y are the total number of occurrences of events in X and Y. The measures are normalized so that $0 < Q_\tau \leq 1$ and $-1 \leq q_\tau \leq 1$. For $Q_\tau = 1$ the events of the signals are fully synchronized while $Q_\tau = 0$ suggests no synchronization. When q_τ is close to 1 an event in X is likely to precede an event in Y and thus X drives Y, and respectively when q_τ is close to -1 suggests that Y drives X. As also mentioned in [23], τ is not fixed but adapted to the time interval between events at each step (it is half of the minimum of times from the current event to the preceding and to the succeeding event for both X and Y).

2.3. *Information measures*

The information measures make use of the Shannon entropy for the uncertainty or information in X, $H(X) = -\sum_x p_X(x) \ln p_X(x)$, where $p_X(x)$ is the probability function defined on a proper binning of X. Further, the conditional entropy is $H(X|Y) = -\sum_{x,y} p_{X,Y}(x,y) \ln \frac{p_{X,Y}(x,y)}{p_Y(y)}$ and the mutual information is $I(X,Y) = -\sum_{x,y} p_{X,Y}(x,y) \ln \frac{p_{X,Y}(x,y)}{p_X(x)p_Y(y)}$, where $p_{X,Y}(x,y)$ is the joint probability of (X,Y) (assuming proper binning of Y as well). In the presence of a third variable Z, the conditional mutual information $I(X,Y|Z)$ of the variables X, Y given the variable Z is

$$I(X,Y|Z) = H(X|Z) + H(Y|Z) - H(X,Y|Z). \tag{7}$$

For a scalar time series, mutual information is defined in terms of a delay, $I(X,Y) = I(x_t, x_{t-\tau}) = I(\tau)$ and for bivariate time series a natural extension is the cross mutual information $I(x_t, y_{t-\tau})$. If instead of $y_{t-\tau}$ the difference $\Delta_\tau y_t = y_{t+\tau} - y_t$ is considered, then the (cross) mutual information conditioned on y_t is $I(x_t, \Delta_\tau y_t|y_t)$ and can be found from (7). The average of $I(x_t, \Delta_\tau y_t|y_t)$ for delays up to a maximum delay $\tau_{\max}$ is the measure of *mean conditional mutual information* (MCMI)[16]

$$i_{X\to Y} = \frac{1}{\tau_{\max}} \sum_{\tau=1}^{\tau_{\max}} I(x_t, \Delta_\tau y_t|y_t) \tag{8}$$

that quantifies the information transferred from X to Y at a later time conditioning on the current state of Y, i.e. the level of driving of X to Y. $i_{Y\to X}$

is defined similarly. A different expression for the conditional mutual information is used to define the measure of *Coarse-grained transinformation rate* (CGTR).[10,24] Averaging on time increments τ as for MCMI, CGTR measures the average rate of the net amount of information transferred from X to Y and is defined as:

$$Ci_{X\to Y} = \frac{1}{\tau_{\max}} \sum_{\tau=1}^{\tau_{\max}} I(x_t, y_{t+\tau}|y_\tau) - \frac{1}{2\tau_{\max}} \sum_{\tau=-\tau_{\max},\tau\neq 0}^{\tau_{\max}} I(x_t, y_{t+\tau}). \quad (9)$$

If $Ci_{X\to Y} \simeq 0$ then there is no information flow, while if $Ci_{X\to Y} > 0$ then X influences Y. If $Ci_{X\to Y} > Ci_{Y\to X}$ then X influences Y more than vice versa.

The states of X and Y in MCMI and CGTR are scalars x_t and y_t, respectively, but can also be given by vectors $\mathbf{x}_t^{m_x} = [x_t, x_{t-1}, \ldots, x_{t-(m_x-1)}]'$ and $\mathbf{y}_t^{m_y} = [y_t, y_{t-1}, \ldots, y_{t-(m_y-1)}]'$. The latter is used in the measure of *transfer entropy* (TE)[9]

$$T_{X\to Y} = \sum p(y_{t+1}, \mathbf{y}_t^{m_y}, \mathbf{x}_t^{m_x}) \ln \frac{p(y_{t+1}|\mathbf{y}_t^{m_y}, \mathbf{x}_t^{m_x})}{p(y_{t+1}|\mathbf{y}_t^{m_y})}. \quad (10)$$

TE quantifies the information flow from X to Y by the amount of information explained in Y at one step ahead from the state of X, accounting for the concurrent state of Y. The concept of transfer entropy extends the Shannon entropy for transition probabilities and quantifies how the conditioning on X change the transition probabilities of Y. The inclusion of vectors in the probabilities does not allow for reliable estimation of TE for small data sets, although there are several methods for efficient coarse graining. It has been shown that with proper conditioning, transfer entropy $T_{X\to Y}$ is the exact equivalent to the conditional mutual information $I(x_t, y_{t+\tau}|y_t)$.[2,3] In the computation of TE, only one time step (lag) ahead is considered, as opposed to MCMI where all lags up to some $\tau_{\max}$ contribute to the estimation.

TE involves the time window lengths m_x and m_y (or embedding dimensions when using delay one). While m_x is selected using standard embedding criteria, m_y has commonly been set either equal to m_x or most frequently to 1. For time series from flows, we propose the states to be defined through separate delay embedding, i.e. $\mathbf{x}_t^{m_x} = [x_t, x_{t-\tau_x}, \ldots, x_{t-(m_x-1)\tau_x}]'$ and $\mathbf{y}_t^{m_y} = [y_t, y_{t-\tau_y}, \ldots, y_{t-(m_y-1)\tau_y}]'$ for proper selection of the delay parameters. Moreover, there is little information flow to observe for one step ahead when the data are densely sampled and in line with MCMI we use a delay (or better a step ahead) τ instead of 1.

We consider two estimators for the probability functions in all information measures, the one estimator uses equidistant binning (ED)[12] and the other estimator uses correlation sums.[25]

3. Setup

The evaluation of the measures is assessed by means of Monte Carlo simulation on different types of systems. Directionality coupling measures are computed on coupled systems for increasing coupling strengths in order to evaluate the ability of the measures to detect the degree and direction of coupling. We use 100 realizations from the different simulation systems (linear and nonlinear/ chaotic systems, unidirectional and bidirectional), with time series lengths $n = 1024, 2048, 4096$. Analytically, the simulation systems are:

- Two $AR(1)$ systems with unidirectional coupling:

$$\begin{aligned} x_{t+1} &= 0.5x_t + e_t^x \\ y_{t+1} &= 0.5y_t + cx_t + e_t^y \end{aligned}$$

  with coupling strength $c = 0, 0.1, 0.2, 0.3, 0.4, 0.5$ and e_t^x, e_t^y normal iid with variance 1.
- Two unidirectionally coupled Henon maps:

$$\begin{aligned} x_{t+1} &= 1.4 - x_t^2 + 0.3x_{t-1} \\ y_{t+1} &= 1.4 - cx_t y_t + (1-c)y_t^2 + 0.3y_{t-1} \end{aligned}$$

  with coupling strength $c = 0, 0.1, 0.2, 0.3, 0.4, 0.5$.
- A Rössler system (x_1, y_1, z_1) driving a Lorenz system (x_2, y_2, z_2):

$$\begin{aligned} \dot{x}_1 &= -6(y_1 + z_1) & \dot{x}_2 &= 10(x_2 + y_2) \\ \dot{y}_1 &= -6(x_1 + 0.2y_1) & \dot{y}_2 &= 28x_2 - y_2 - x_2 z_2 + c y_1^2 \\ \dot{z}_1 &= -6(0.2 + z_1(y_1 - 5.7)) & \dot{z}_2 &= x_2 y_2 - \frac{8}{3} z_2 \end{aligned}$$

  with coupling strength $c = 0, 0.5, 1, 1.5, 2$.
- Two bidirectionally coupled Henon maps:

$$\begin{aligned} x_{t+1} &= 1.4 - x_t^2 + 0.3x_{t-1} + c_2(x_t^2 - y_t^2) \\ y_{t+1} &= 1.4 - y_t^2 + 0.3y_{t-1} + c_1(y_t^2 - x_t^2) \end{aligned}$$

  with coupling strengths (c1, c2) = (0.05,0.05), (0.1,0.05), (0.1,0.1), (0.15,0.05), (0.2,0.05).

The range of the coupling strengths is bounded from zero (independent systems) up to the level of almost complete synchronization. All the time series are first normalized to have mean 0 and standard deviation 1. The values for measure parameters are set as follows: number of bins $b = 8$ for measures requiring coarse graining, radius for finding neighbors $r = 0.15$, number of neighbors $k = 10$, Theiler window $W = 10$ (to exclude time correlated points). For all systems apart from the unidirectionally coupled Rössler–Lorenz system we set: embedding dimension $m = 2$, time lag $\tau = 1$, maximum lag for the average of time increments in information measures $\tau_{\text{max}} = 10$, and for transfer entropy $\tau = 1, m_x = m_y = 1, \tau_x = \tau_y = 1$. For the unidirectionally coupled Rössler–Lorenz system we set: $m = 5, \tau = 5, \tau_{\text{max}} = 20$, and for transfer entropy: $\tau = 5, m_x = m_y = 2, \tau_x = \tau_y = 5$.

4. Results

For systems that are unidirectionally coupled, directionality measures should increase with coupling strength at the direction of coupling and be stable (and close to zero) at the opposite direction and when there is no coupling ($c = 0$). We start with the results from the simulations of unidirectionally coupled $AR(1)$ system. As expected, state space measures and DI do not discriminate between the linear stochastic systems, even for large time series lengths, due to the lack of nonlinear dynamics (in local state space) and phase dynamics, respectively. On the other hand, EvD regarding the local maxima of the time series, seems to discriminate the directionality at a level that increases with the coupling strength, as shown in Fig. 1a. Both estimators of MCMI (with correlation sum and equidistant binning) detect the directionality of the coupling correctly (see Fig. 1b and c), and even for small time series lengths, while CGTR does not seem to detect the directionality of the coupling. Both estimators of TE (from binning and correlation sum) detect the direction of information flow correctly (see Fig. 1d and e).

For the unidirectionally coupled Henon map we observe that MCR detects the increase of coupling in the correct direction, but it wrongly indicates the same but at a lesser degree in the other direction (see Fig. 2a). On the other hand, and as Fig. 2b shows, NI is at the order of 10^{-3} and therefore no information flow is detected. DI detects correctly the strength and the directionality of the coupling but with a large variance, even for large time series lengths, and does not allow for a clear indication of directionality for all cases (see Fig. 2c). EvD also does not detect any information flow. On the other hand, all information measures detect correctly the strength

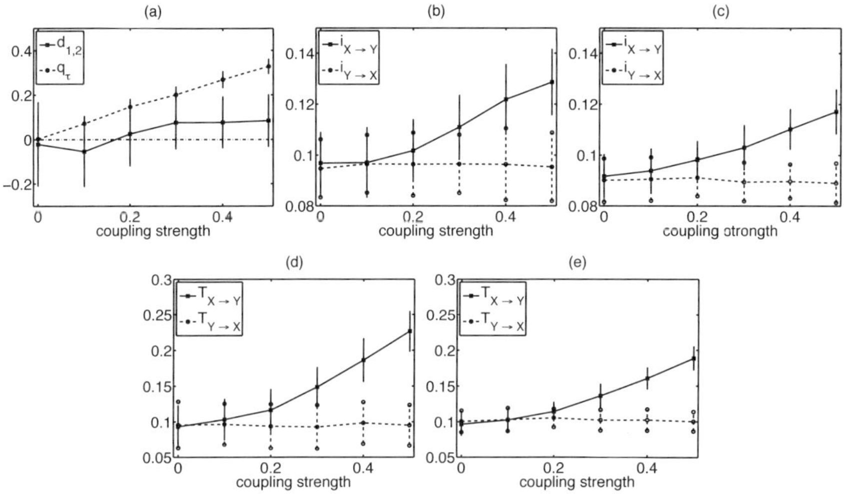

Fig. 1. Average and standard deviation (shown as error bars) of the directionality measures for 100 realizations of unidirectionally coupled $AR(1)$ system. (a) DI and EvD as shown in the legend, for $n = 4096$. (b) MCMI computed by correlation sums, $n = 1024$. (c) As in (b) but for equidistant partition. (d) TE computed by correlation sum, $n = 1024$. (e) As in (d) but for equidistant partition.

and the directionality of the coupling even for small time series lengths (results for CGTR and TE are shown in Fig. 2d-f).

For the unidirectionally coupled Rössler–Lorenz system, we note that the detection of the directionality of the coupling is expected to be more difficult as the two flows are of different type. The simulations showed that EvD, NI, DI and MCMI could not detect the direction and strength of the coupling, whereas MCR and CGTR could both detect it, even for small time series lengths. Using the TE measure with the standard parameters ($\tau = 1, m_x = 2$, $m_y = 1$ and $\tau_x = \tau_y = 1$) we get only slight difference of the TE on the two directions (somehow larger difference is obtained using the histogram based estimator as shown in Fig. 3a, d). However, estimation of TE using proper reconstruction of state space, detects the coupling only in the correct direction and again better for the histogram based estimate (see Fig. 3b and e). The detection improves with the increase of time series length as shown in Fig. 3c and f. The strong dependence on data size is due to the estimation of probability densities regarding the reconstructed points rather than the samples.

For the bidirectionally coupled Henon system the synchronization measures (for phase and event) fail to detect the direction of stronger coupling,

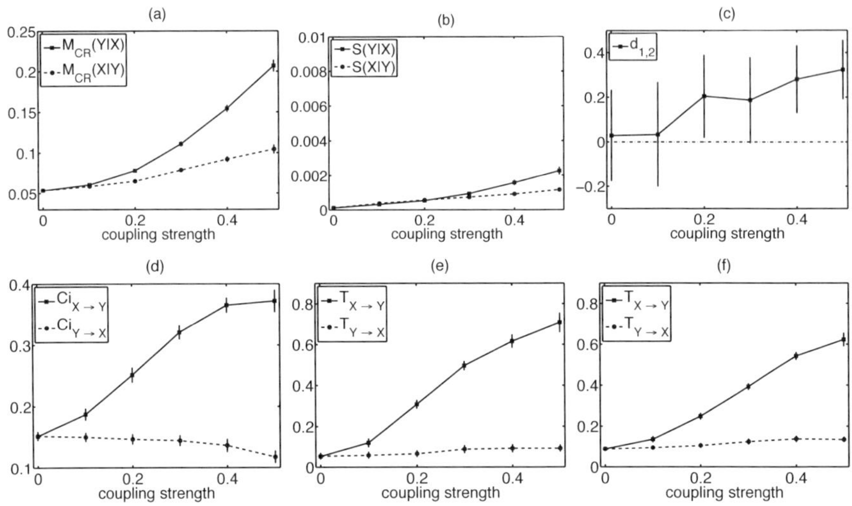

Fig. 2. Average and standard deviation (shown as error bars) of the directionality measures for 100 realizations of unidirectionally coupled Henon system. (a) MCR, for $n = 4096$, (b) NI, for $n = 4096$ and (c) DI, for $n = 4096$. (d) CGTR, for $n = 1024$. (e) TE computed by correlation sums, for $n = 1024$. (f) As in (e) but for equidistant partition.

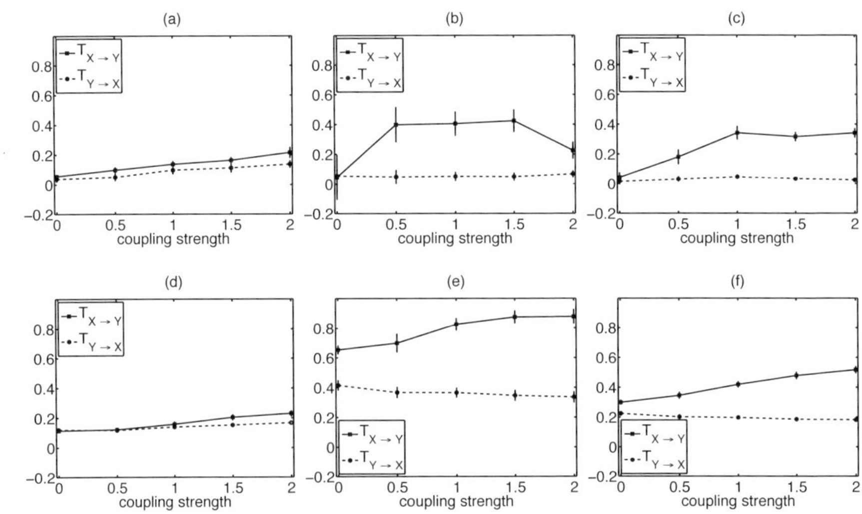

Fig. 3. Average and standard deviation (shown as error bars) of the directionality measures for 100 realizations of unidirectionally coupled Rössler–Lorenz system. (a) TE computed by correlation sums, for $\tau = 1, m_x = 2, m_y = 1, n = 1024$. (b) As in (a) but for $\tau = 5, m_x = m_y = 2, \tau_x = \tau_y = 5, n = 1024$. (c) As in (b) for $n = 2048$. (d), (e), (f) As in (a), (b), (c) respectively, but for equidistant partition.

as there are no distinct phases or events in the time series. Regarding the state space measures, MCR correctly estimates the driver-response interaction only when it is present as shown in Table 1, whereas NI gives always small values independently of the coupling setting. All information measures perform properly (see Table 1). They all give about the same value for both directions when the coupling is the same in both directions. When there is stronger coupling in one direction, TE turns out to give the larger difference for the two directions. For example, for $c_1 = 0.15$ and $c_2 = 0.1$, $T_{X\to Y}$ increases by about 136% from $T_{Y\to X}$ for the estimator using correlation sums and 70% for the estimator using histograms, whereas the increase for MCMI is 74% (correlation sums estimator) and 45% (estimator using histograms) and for CGTR is also 45%. Note that these results are for $n = 1024$, and for larger time series the difference in the presence of stronger coupling in one direction is much more clear.

Table 1. Average of the directionality measures for 100 realizations of bidirectionally coupled Henon map for $n = 1024$.

	Coupling strengths				
Measures	(0.05,0.05)	(0.1,0.05)	(0.1,0.1)	(0.15,0.1)	(0.2,0.05)
$M_{CR}(X\|Y)$	0.0874	0.1018	0.1655	0.1363	0.1873
$M_{CR}(Y\|X)$	0.0875	0.0940	0.1649	0.1133	0.1453
$S(X\|Y)$	0.0022	0.0032	0.0075	0.0045	0.0074
$S(Y\|X)$	0.0022	0.0032	0.0075	0.0051	0.0101
$i_{X\to Y}$(Cor. Sum)	0.1519	0.2273	0.3876	0.3159	0.3609
$i_{Y\to X}$(Cor. Sum)	0.1473	0.1555	0.3870	0.1820	0.1847
$i_{X\to Y}$(hist. based)	0.1398	0.1769	0.2267	0.2172	0.2551
$i_{Y\to X}$(hist. based)	0.1386	0.1387	0.2269	0.1503	0.1519
$Ci_{X\to Y}$	0.1749	0.2036	0.1119	0.2336	0.2629
$Ci_{Y\to X}$	0.1734	0.1663	0.1124	0.1609	0.1391
$T_{X\to Y}$(Cor. Sum)	0.1007	0.1818	0.2787	0.3121	0.3984
$T_{Y\to X}$(Cor. Sum)	0.0990	0.1044	0.2804	0.1323	0.1318
$T_{X\to Y}$(hist. based)	0.1021	0.1415	0.1660	0.2075	0.3021
$T_{Y\to X}$(hist. based)	0.1024	0.0999	0.1658	0.1224	0.1300

5. Application

As an application, we assess the examined information flow measures in detecting the propagation of the epileptic activity that may be used as a predictor of an impending seizure. The problem of direct and indirect transfer of information among EEG channels is not considered here (for this see e.g. [26,27]). Specifically, we investigate whether changes in directionality of coupling and information flow occur in brain areas of an epileptic patient at two different periods prior to seizure onset. The data are extra-cranial multichannel recordings of electroencephalographs (EEG) from a patient with back left temporal (LT) lobe epilepsy. We compare the measures on recordings from 60 to 50min before seizure onset (early preictal state) and on recordings from 10min before and up to seizure onset (late preictal state). We use two electrodes from the epileptic focus area (LT1 and LT2) and investigate the information flow to other areas close or farther from it: occipital channels (OC1, OC2), middle channel (MI) and right temporal (RT) channel.

For each of the two periods and for each channel we derive 20 consecutive non-overlapping time series of 30sec each and compute all directionality measures for pairs of LT1 and LT2 to all other selected channels. The parameters of the directionality measures are: $m = 10$, $\tau = 5$, $k = 10$, $W = 10$, $r = 0.15$, $m_x = m_y = 2$, $\tau_x = \tau_y = 5$, $b = 8$, $\tau_{\max} = 20$. We observed a slight decrease in information flow from focus to other areas (mostly from LT to OC and MI) moving from early to late preictal state, but this could be observed only with some measures, as shown in Fig. 4. Although transfer entropy seemed to perform better than other measures in the simulation study, in this application it did not perform any better than the other measures.

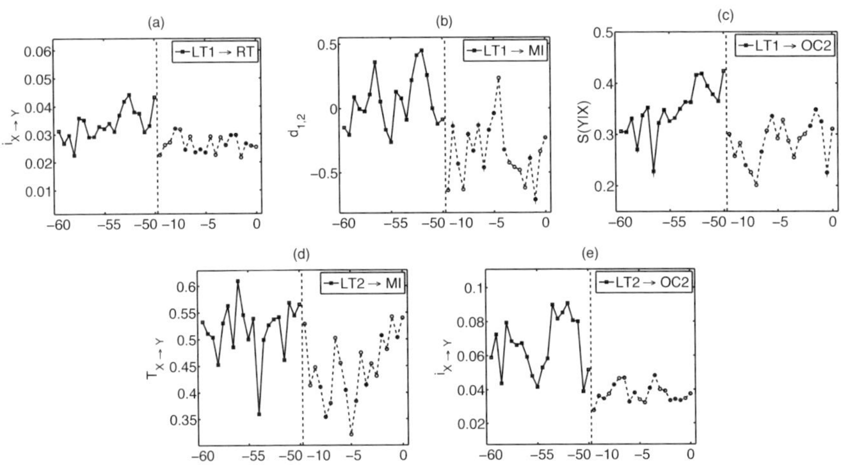

Fig. 4. Estimated values from the two preictal states of (a) MCMI (histogram based) from channel LT1 to RT, (b) DI from LT1 to MI, (c) NI from LT1 to OC2, (d) TE (histogram based) from LT2 to MI, and (e) MCMI (histogram based) from channel LT2 to OC2. The states are indicated by the time in sec, with reference to time 0 at seizure onset and they are separated by a vertical dashed line.

6. Conclusions

From the simulation study we educed that information measures and specifically transfer entropy were able to detect the directionality in coupled systems independently of the type of the system (linear or non-linear, map or flow). Moreover, information measures seem to perform better for short time series lengths than state space and synchronization measures do. The proposed generalized form of TE introducing reconstructed points and larger time increment seems to perform well, however the selection of the embedding parameters is crucial. In the simulations, we used standard values for measure parameters, as given in the literature, and further investigation with optimized parameters is needed to completely assess the usefulness of the measures. In the EEG application we found that only a few directionality measures could give a slight change in the information flow from the epileptic focus to other brain areas when comparing the period about an hour prior to seizure to the period just before seizure onset.

Acknowledgments

This research project is implemented within the framework of the "Reinforcement Programme of Human Research Manpower" (PENED) and is co-financed at 90% jointly by E.U.-European Social Fund (75%) and the Greek Ministry of Development-GSRT (25%) and at 10% by Rikshospitalet, Norway. We thank Pål G. Larsson at Rikshospitalet for providing the EEG data.

References

1. E. Pereda, R. Quiroga and J. Bhattacharya, *Progress in Neurophysiology E* **77**(1-2), 1 (2005).
2. K. Hlavackova-Schindler, M. Palus, M. Vejmelka and J. Bhattacharya, *Physics Reports* **441**(1), 1 (2007).
3. M. Palus and M. Vejmelka, *Physical Review E* **75**, p. 056211 (2007).
4. J. Arnhold, P. Grassberger, K. Lehnertz and C. Elger, *Physica D* **134**(4), 419 (1999).
5. S. Schiff, P. So and T. Chang, *Physical Review E* **54**(6), 6708 (2000).
6. M. Romano, M. Thiel, J. Kurths and C. Grebogi, *Physical Review E* **76**, p. 036211 (2007).
7. R. Quiroga, J. Arnhold and P. Grassberger, *Physical Review E* **61**(5), 5142 (2000).
8. M. Rosenblum and A. Pikovsky, *Physical Review E* **64**(4), p. 045202 (2001).
9. T. Schreiber, *Physical Review Letters* **85**(2), 461 (2000).
10. M. Palus, V. Korarek, Z. Hrncir and K. Sterbova, *Physical Review E* **63**(4), p. 046211 (2001).

11. A. Papana and D. Kugiumtzis, *Nonlinear Phenomena in Complex Systems* **11**(2), 225 (2008).
12. G. Darbellay and I. Vajda, *IEEE Transactions on Information Theory* **45**(4), 1315 (1999).
13. Y. Moon, B. Rajagopalan and U. Lall, *Physical Review E* **52**(3), 2318 (1995).
14. A. Kraskov, H. Stögbauer and P. Grassberger, *Physical Review E* **69**(6), p. 066138 (2004).
15. M. Palus, V. Korarek, T. Prochazka, Z. Hrncir and K. Sterbova, *IEEE Engineering in Medicine and Biology* **20**(5), 65 (2001).
16. M. Vejmelka and M. Palus, *Physical Review E* **77**, p. 026214 (2008).
17. M. Kaminski, *Philosophical Transaction of the Royal Society B, Biological Sciences* **360**(1457), p. 947952 (2005).
18. M. Winterhalder, B. Schelter, W. Hesse, K. Schwab, L. Leistritz, D. Klan, R. Bauer, J. Timmer and H. Witte, *Signal Processing* **85**(11), 2137 (2005).
19. L. Faes, A. Porta and G. Nollo, *Physical Review E* **78**(2), p. 026201 (2008).
20. R. Quiroga, A. Kraskov, T. Kreuz and P. Grassberger, *Physical Review E* **65**(4), p. 041903 (2002).
21. D. Krug, H. Osterhage, C. Elger and K. Lehnertz, *Physical Review E* **76**, p. 041916 (2007).
22. J.-P. Eckmann, S. Kamphorst and D. Ruelle, *Europhysics Letter* **4**(9), 973 (1987).
23. R. Quiroga, T. Kreuz and P. Grassberger, *Physical Review E* **66**(4), p. 041904 (2002).
24. M. Palus, *Physica D* **93**(1-2), 64 (1996).
25. S. Manzan and C. Diks, *Studies in Nonlinear Dynamics and Econometrics* **6**(2), p. 1005 (2002).
26. R. Kus, M. Kaminski and K. Blinowska, *IEEE Transactions on Biomedical Engineering* **51**(9), p. 1501 (2004).
27. M. Winterhalder, B. Schelter, W. Hesse, K. Schwab, L. Leistritz, J. Timmer and H. Witte, *Biomedizinische Technik* **51**, p. 281287 (2006).

Preparing Materials with a Desired Refraction Coefficient and Applications

A. G. Ramm
Mathematics Department, Kansas State University
Manhattan, KS 66506-2602, USA
Email: ramm@math.ksu.edu

A recipe is given for creating material with a desired refraction coefficient by embedding many small particles in a given material. To implement practically this recipe, some technological problems are to be solved. These problems are formulated. Applications of the materials with a desired refraction coefficient are mentioned. One of the possible applications is preparing materials with desired wave-focusing properties. These are materials which have a desired radiation pattern at a fixed frequency and at a fixed incident direction. The other application is preparing materials with negative refraction.

Keywords: Metamaterials; nanotechnology; refraction coefficient; negative refraction, wave focusing, wave scattering by small particles.

1. Introduction

In a series of papers, cited in the references, and in the monograph [1] the author has developed wave scattering theory by many small bodies of arbitrary shapes and, on this basis, proposed a method for creating materials with some desired properties. The goal of this paper is to explain this method and to formulate two nanotechnological engineering problems that should be solved in order that the proposed method can be implemented practically. Some of the possible applications are mentioned. These applications include creating materials with desired wave-focusing properties and materials with negative refraction.

Our presentation in this paper follows [2], but is slightly expanded.

Let us formulate the statement of the problem. Let $D \subset \mathbb{R}^3$ be a bounded domain filled in by a material with a known refraction coefficient. The scalar wave scattering problem consists of finding the solution to the

Helmholtz equation:

$$L_0 u := [\nabla^2 + k^2 n_0^2(x)]u = 0 \quad \text{in} \quad \mathbb{R}^3, \quad \operatorname{Im} n_0^2(x) \geq 0, \tag{1}$$

$$u = u_0 + v, \quad u_0 := e^{ik\alpha\cdot x}, \tag{2}$$

$$v = A_0(\beta, \alpha)\frac{e^{ikr}}{r} + o\left(r^{-1}\right), \quad r := |x| \to \infty, \quad \beta := \frac{x}{r}. \tag{3}$$

The function $n_0(x)$ is assumed bounded, piecewise-continuous,

$$n(x) = 1 \quad \text{in} \quad D' := \mathbb{R}^3 \backslash D, \quad \operatorname{Im} n_0^2(x) \geq 0. \tag{4}$$

The function $A(\beta, \alpha)$ is called the scattering amplitude. The wavenumber $k = \frac{2\pi}{\lambda}$, where λ is the wavelength in D', $\alpha \in S^2$ is the direction of the incident plane wave u_0, S^2 is the unit sphere in $\mathbb{R}^3$, $\beta \in S^2$ is the direction of the scattered wave. The solution to problem (1)-(3) is called the scattering solution. It is well-known ([3]) that this solution exists and is unique under our assumptions, namely $n_0^2(x)$ is bounded $\sup_{x\in\mathbb{R}^3} |n_0^2(x)| \leq n_0 = const, \operatorname{Im} n_0^2(x) \geq 0$, $n_0^2(x) = 1$ in D', so that $A_0(\beta, \alpha)$ is determined uniquely if $n_0^2(x)$ is given. We assume $k > 0$ fixed and do not show the dependence of $A_0(\beta, \alpha)$ on k. The operator L_0 at a fixed $k > 0$ can be considered as a Schroedinger operator $L_0 = \nabla^2 + k^2 - q_0(x)$, where $q_0(x) := k^2 n_0^2(x) - k^2$.

Problem (P): *We want to construct a material in D with a desired refraction coefficient, that is, with a desired function $n^2(x), \operatorname{Im} n^2(x) \geq 0$.*

Why is this problem of practical interest? We give two reasons.

First, creating a refraction coefficient such that the corresponding material has *negative refraction* is of practical interest.

One says that a material has negative refraction if the group velocity in this material is directed opposite to the phase velocity.

Secondly, creating such a refraction coefficient that the corresponding scattering amplitude $A(\beta) = A(\beta, \alpha)$ at a fixed $\alpha \in S^2$ (and a fixed $k > 0$) approximates with any desired accuracy $\epsilon > 0$ an arbitrary given function $f(\beta) \in L^2(S^2)$ is of practical interest.

This problem we call the problem of creating material with a desired wave-focusing property.

In Section 2 we explain how to create such a material. In Section 3 we discuss the technological aspects of our recipe for creating material with a desired refraction coefficient. Creating material with negative refraction is discussed in Section 4.

2. Creating Material with a Desired Refraction Coefficient

The basic idea of our method is to embed in D many small particles D_m, $1 \leq m \leq M$, with a prescribed boundary impedance ζ_m, defined in formulas (7) and (9) below. The number of the embedded particles in an arbitrary open subdomain $\Delta \subset D$ is given asymptotically as

$$\mathcal{N}(\Delta) = \frac{1}{\phi(a)} \int_\Delta N(x) dx [1 + o(1)], \qquad a \to 0. \tag{5}$$

Here $\phi > 0$ if $a > 0$, $\phi(0) = 0$, ϕ is strictly monotonically growing, $N(x) \geq 0$ is a piecewise-continuous function which we can choose as we want. The smallness of the embedded particles means that $ka \ll 1$, where $a = \frac{1}{2} \max_m \operatorname{diam} D_m$, where D_m is the m-th small particle. Let $M = M(a)$ be the total number of the embedded particles. For simplicity we assume that D_m is a ball, centered at a point $x_m \in D$, of radius a. The distance $d := \min_{m \neq j} \operatorname{dist}(D_m, D_j)$ between neighboring particles is much larger than a, $d \gg a$, but *d is not assumed much larger than λ*, so in our theory there can be many small particles on the wavelength in D. This means that the distribution of the embedded particles *is not necessarily "diluted"*.

The scattering problem in the region with embedded particles is formulated as follows:

$$L_0 U = 0 \quad \text{in} \quad \mathbb{R}^3 \setminus \bigcup_{m=1}^{M} D_m, \tag{6}$$

$$U_N = \zeta_m U \quad \text{on} \quad S_m, \quad 1 \leq m \leq M, \tag{7}$$

$$U(x) = u(x) + V, \quad \frac{\partial V}{\partial r} - ikV = o\left(r^{-1}\right), \quad r \to \infty. \tag{8}$$

Here N is the unit normal to S_m, pointing out of D_m, S_m is the surface of D_m, M is the total number of the embedded particles D_m, ζ_m is the boundary impedance, $\operatorname{Im}\zeta_m \leq 0$, u is the scattering solution in the absence of the embedded particles, i.e., the solution to problem (1)-(3).

Let us assume that

$$\zeta_m = \frac{h(x_m)}{a^\kappa}, \quad -1 < \kappa < 2, \quad \kappa = \text{const}, \tag{9}$$

where $h(x)$ is a piecewise-continuous function in D,

$$\operatorname{Im} h(x) \leq 0. \tag{10}$$

The function $N(x) \geq 0$ in (4) and $h(x), \operatorname{Im} h \leq 0$, we can choose as we wish. These functions will define the desired refraction coefficient of the new material which is obtained in the limit $a \to 0$.

From formula (9) it follows that the impedance ζ_m tends to infinity as $a \to 0$. If $\zeta_m = \infty$, then the particle D_m is acoustically soft, $U = 0$ on S_m. The rate of growth of the impedance ζ_m as $a \to 0$ is very important for our theory: it allows one to prove the existence of the limit, as $a \to 0$, of the effective field in the medium. Asymptotic behavior of the solution to the exterior boundary-value problem for the Helmholtz equation as the impedance tends to infinity was studied in the recent paper [4], which was originated from the theory described in this paper.

We assume that

$$M = M(a) = O(a^{\kappa-2}), \qquad d = O(a^{\frac{2-\kappa}{3}}). \tag{11}$$

The function $\phi(a)$ in (5) is

$$\phi(a) = a^{2-\kappa}. \tag{12}$$

The solution to problem (6)-(8) depends on a, $U = U(x,a)$, but we do not show this dependence for brevity. We also do not show the dependence of U on α and k, because they are fixed. Our standing assumptions are (4), (5), (9)–(12). They are not repeated in the statements of Theorems.

Denote by $g(x,y)$ the Green's function of the operator L_0 in the absence of the embedded particles:

$$L_0 g(x,y) = -\delta(x-y) \qquad \text{in} \quad \mathbb{R}^3,$$

and g satisfies the radiation condition.

Theorem 2.1. *Problem* (6)-(8) *has a solution, this solution is unique, and it is of the form:*

$$U(x) = u(x) + \sum_{m=1}^{M} \int_{S_m} g(x,s)\sigma_m(s)ds, \tag{13}$$

where the function σ_m are uniquely determined by the data, i.e., by $u_0(x,\alpha)$, ζ_m, D_m, $1 \le m \le M$, and $n_0^2(x)$.

Theorem 2.2. *The solution U to problem* (6)–(8) *has the following asymptotic as $a \to 0$*

$$U(x) = u(x) + \sum_{m=1}^{M(a)} g(x,x_m)Q_m + o(1), \quad a \to 0, \quad \inf_m |x - x_m| \ge d, \tag{14}$$

where

$$Q_m := \int_{S_m} \sigma_m(s)ds = -4\pi h(x_m)u_e(x_m)a^{2-\kappa}[1+o(1)], \quad a \to 0, \tag{15}$$

$$\sigma_m(s) = -h(x_m)u_e(x_m)a^{-\kappa}[1 + o(1)], \quad a \to 0, \tag{16}$$

and

$$u_e(x_m) := u_e^{(m)}(x_m), \quad u_e^{(m)}(x) := u(x) + \sum_{m' \neq m} \int_{S_{m'}} g(x,s)\sigma_{m'}(s)ds. \tag{17}$$

Theorem 2.3. *Assume that f is an arbitrary bounded piecewise continuous function in D, and the points x_m are distributed in D so that (5) holds. Then*

$$\lim_{a \to 0} \phi(a) \sum_{m=1}^{M(a)} f(x_m) = \int_D f(x)N(x)dx. \tag{18}$$

Theorem 2.4. *Under the assumptions (4), (5), (9)–(12) the following limit exists:*

$$\lim_{a \to 0} u_e^{(m)}(x) = u_e(x), \quad \forall x \in D. \tag{19}$$

This limit does not depend on m and is the unique solution to the equation

$$u_e(x) = u(x) + \int_D g(x,y)p(y)u_e(y)dy, \tag{20}$$

where

$$p(x) = 4\pi N(x)h(x). \tag{21}$$

Theorem 2.5. *The limiting material in D, which is obtained as $a \to 0$, is described by the equation*

$$Lu_e = 0 \quad in \quad D, \quad L := \nabla^2 + k^2 n^2(x) := \nabla^2 + k^2 - q(x), \tag{22}$$

$$n^2(x) := n_0^2(x) - k^{-2}p(x), \quad p(x) := 4\pi N(x)h(x), \quad q(x) := q_0(x) + p(x), \tag{23}$$

$$u_e = u(x) + v_e(x), \tag{24}$$

where $v_e(x) := \int_D g(x,y)p(y)u_e(y)dy$.

The scattering amplitude, corresponding to $n^2(x)$, (or, which is equivalent to the potential $q(x)$) is:

$$A(\beta,\alpha) = A_0(\beta,\alpha) + \frac{1}{4\pi}\int_D u(y,-\beta)p(y)u_e(y)dy, \tag{25}$$

where $u(x,-\beta)$ is the solution to problem (1)–(3) with $u_0 = e^{-ik\beta\cdot x}$, that is, the scattering solution in the absence of the embedded particles when the incident direction of the plane wave u_0 equals to $-\beta$.

Remark 2.1. Formula (25) is based on the Ramm's lemma (see [3, p. 46]) which says

$$g(x,y) = \frac{e^{ik|y|}}{4\pi|y|}u(x,-\beta) + O(\frac{1}{|y|^2}), \quad |y| \to \infty, \quad \beta := \frac{y}{|y|}, \tag{26}$$

where $u(x,-\beta)$ is the scattering solution with the incident direction $-\beta \in S^2$ and the remainder $O(\frac{1}{|y|^2})$ is uniform with respect to x running through any fixed bounded domain.

We can now formulate *a recipe for creating material with a desired refraction coefficient:*

Recipe: *To obtain material with a desired refraction coefficient $n^2(x)$, given a bounded domain D, filled with a material with the coefficient $n_0^2(x)$, one embeds into D small balls D_m, centered at x_m and of radius a, distributed according to* (5), *with $\phi(a)$ defined in* (12), *so that conditions* (9)-(11) *hold. Then one finds $h(x)$ and $N(x)$ from the equation*

$$k^2[n_0^2(x) - n^2(x)] := p(x) = 4\pi h(x)N(x). \tag{27}$$

Equation (27) *for $N(x) \geq 0$ and $h(x), \operatorname{Im} h \leq 0$, where $p(x)$ is given, has infinitely many solutions. For example, one can choose $N(x) > 0$ arbitrary, and then find uniquely h_1 and h_2 by the formulas:*

$$h_1 := \operatorname{Re} h = \frac{p_1}{4\pi N(x)}, \quad h_2 := \operatorname{Im} h = \frac{p_2}{4\pi N(x)} \tag{28}$$

where $p = p_1 + ip_2$, $p_1 = \operatorname{Re} p$, $p_2 = \operatorname{Im} p$.

By Theorem 2.5 the material with the embedded small particles has the desired coefficient $n^2(x)$ with the error that goes to zero as $a \to 0$.

We do not give proofs of Theorem 2.1–2.5. These proofs can be found in the cited papers. Our goal is to formulate clearly the recipe for creating materials with a desired $n^2(x)$, so that engineers and physicists can try to implement this recipe practically.

The non-uniqueness of the choice of $h(x)$ and $N(x)$ from the equation (27): $p(x) = 4\pi h(x)N(x)$ can be used practically: one may wish to choose $N(x)$ small in order to decrease the total number of the particles to be embedded, for example.

3. A Discussion of the Recipe

There are two technological problems that should be solved in order that the recipe be implemented practically.

Problem 1: *How does one embed many small particles in a given material so that the desired distribution law (5),(11)-(12) is satisfied?*

Problem 2: *How does one prepare a small particle with the desired boundary impedance $\zeta_m = \frac{h(x_m)}{a^\kappa}$?*

How can one prepare a small particle with the desired dispersion of the boundary impedance, that is, the desired $\omega-$ dependence of $h = h(x,\omega)$?

The motivation for the last question one finds in Section 3.1: this question is related to preparing materials with negative refraction.

The first problem, possibly, can be solved by stereolitography.

The second problem one should be able to solve because the limiting cases $\zeta_m = 0$ (hard particles) and $\zeta_m = \infty$ (soft particles) can be prepared, so that any intermediate value of ζ_m one should be able to prepare as well. The author formulates the above technological problems in the hope that engineers get interested and solve them practically.

3.1. *Negative refraction*

Material with negative refraction is, by definition, a material in which group velocity is directed opposite to the phase velocity, (see [5] and references therein). Group velocity is defined by the formula $\mathbf{v}_g = \nabla_{\mathbf{k}}\omega(\mathbf{k})$. Phase velocity $\mathbf{v}_p$ is directed along the wave vector $\mathbf{k}^0 = \frac{\mathbf{k}}{|\mathbf{k}|}$. In an isotropic material $\omega = \omega(|\mathbf{k}|)$, and $\omega = \frac{c|\mathbf{k}|}{n(x,\omega)}$, so $\omega n(x,\omega) = c|\mathbf{k}|$. Differentiating this equation yields

$$\nabla_{\mathbf{k}}\omega\big[n(x,\omega) + \omega\frac{\partial n}{\partial\omega}\big] = c\mathbf{k}^0. \tag{29}$$

Thus,

$$\mathbf{v}_g\big[n + \omega\frac{\partial n}{\partial\omega}\big] = c\mathbf{k}^0. \tag{30}$$

Wave speed in the material with refraction coefficient $n(x,\omega)$ is $|\mathbf{v}_p| = \frac{c}{n(x,\omega)}$, where c is the wave speed in vacuum (in D').

For $\mathbf{v}_g$ to be directed opposite to $\mathbf{k}^0$, that is, opposite to $\mathbf{v}_p$, it is necessary and sufficient that

$$n + \omega\frac{\partial n}{\partial\omega} < 0. \tag{31}$$

If the new material has $n(x,\omega)$ satisfying (31), then the new material has negative refraction.

One can create material with $n(x,\omega)$ satisfying (31) by choosing $h = h(x,\omega)$ properly. Namely, (27) implies

$$n^2(x,\omega) = n_0^2(x) - 4\pi k^{-2}N(x)h(x,\omega). \tag{32}$$

Assuming $n_0^2 > 4\pi k^{-2}N(x)h(x,\omega)$, $\text{Im}\, h = 0$, one has (31) satisfied if

$$\sqrt{n_0^2 - 4\pi k^{-2}N(x)h(x,\omega)} < \frac{2\pi k^{-2}N(x)\frac{\partial h(x,\omega)}{\partial\omega}}{\sqrt{n_0^2 - 4\pi k^{-2}N(x)h(x,\omega)}},$$

or

$$t(x) := \frac{n_0^2(x)}{2\pi k^{-2}N(x)} < \frac{\partial h}{\partial\omega} + 2h. \tag{33}$$

This inequality has many solutions, e.g., the function $h(x,\omega) = e^{-2\omega}h_0 + \frac{1}{2}e^{2\omega}t(x)$ solves (33).

Let us formulate the technological problem solving of which allows one to implement practically our method for creating materials with negative refraction.

Problem 3: *How does one prepare a small particle with the impedance* $\zeta_m = \frac{h(x_m,\omega)}{a^\kappa}$, *where* $h(x,\omega)$ *satisfies* (33)*?*

3.2. *Wave-focusing property*

Let us formulate the problem of preparing a material with desired wave-focusing property. The refraction coefficient is related to the potential by the formula $q(x) = k^2 - k^2n^2(x)$. We assume in this Section that $k > 0$ is fixed, and $\alpha \in S^2$, the incident direction (that is, the direction α of the incident plane wave $e^{ik\alpha\cdot x}$) is also fixed.

Problem: Given an arbitrary fixed $f(\beta) \in L^2(S^2)$, *an arbitrary small fixed* $\epsilon > 0$, *an arbitrary fixed* $k > 0$, *and an arbitrary fixed* $\alpha \in S^2$, *can one find* $q \in L^2(D)$, $q = 0$ *in* $D' = \mathbb{R}^3\backslash D$, *such that*

$$\|A_q(\beta) - f(\beta)\|_{L^2(S^2)} < \epsilon, \tag{34}$$

where $A_q(\beta) := A_q(\beta,\alpha,k)$ *is the scattering amplitude, corresponding to* q, *at fixed* α *and* k, *and* $q := k^2 - k^2n^2(x)$*?*

The answer is *yes*, and an algorithm for finding such a q, and, therefore, such an $n^2(x)$, is given in [6]. See also [7,8].

There are many q which solve the above problem. It is possible to choose from the set of these potentials the one which satisfies some additional properties, for example, one can choose q to be arbitrarily smooth in D, one can choose q with non-negative imaginary part, etc.

The desired radiation pattern $f(\beta)$ can, for example, be equal to 1 in a given solid angle and equal to 0 outside this solid angle. In this case the scattered field is scattered predominantly into the desired solid angle.

References

1. A. G. Ramm, *Wave scattering by small bodies of arbitrary shapes* (World Sci. Publishers, Singapore, 2005).
2. A. G. Ramm, *Nonlinear Analysis: Theory, Methods and Applications* (to appear).
3. A. G. Ramm, *Scattering by obstacles* (D.Reidel, Dordrecht, 1986).
4. H.-D. Alber and A. G. Ramm, *J. Math. Anal. Appl.* **349**(1), 156 (2009).
5. V. M. Agranovich and Y. Gartstein, *Physics-Uspekhi* **176**(10), 1051 (2006).
6. A. G. Ramm, *Nonlinear Analysis: Theory, Methods and Applications* **69**(4), 1478 (2008).
7. A. G. Ramm, *Latin-American Journ. of Solids and Structures* **5**, 119 (2008).
8. A. G. Ramm, *Modern Phys. Lett. B* **22**(23), 2217 (2008).
9. A. G. Ramm, *Inverse problems* (Springer, New York, 2005).
10. A. G. Ramm, *Phys. Lett. A* **312**(3-4), 256 (2003).
11. A. G. Ramm, *Phys. Lett. A* **360**(6), 735 (2007).
12. A. G. Ramm, *Phys. Lett. A* **370**(5-6) (2007).
13. A. G. Ramm, *Europ. Phys. Lett.* **80**, p. 44001 (2007).
14. A. G. Ramm, *J. Math. Phys.* **48**(10), p. 103511 (2007).
15. A. G. Ramm, *Phys. Lett. A* **367**, 156 (2007).
16. A. G. Ramm, *Phys. Lett. A* **368**(1-2), 164 (2007).
17. A. G. Ramm, *Communic. in Nonlinear Sci. and Numer. Simulation* **12**(7), 1115 (2007).
18. A. G. Ramm, *Jour. Stat. Phys.* **127**(5), 915 (2007).
19. A. G. Ramm, *Phys. Lett. A* **372**(13), 2319 (2008).
20. A. G. Ramm, *Phys. Lett. A* **372**(17), 3064 (2008).
21. A. G. Ramm, *Phys. Lett. A* **372**(23), 4298 (2008).
22. A. G. Ramm, *J. Phys A* **41**(212001) (2008).
23. A. G. Ramm, *Phys. Lett. A* **372**, 6518 (2008).

Surrogates with Random Fourier Phases

C. Räth and R. Monetti

Max-Planck Institut für extraterrestrische Physik
Giessenbachstr. 1, 85748 Garching, Germany

The method of surrogates is widely used in the field of nonlinear data analysis for testing for weak nonlinearities. The two most commonly used algorithms for generating surrogates are the amplitude adjusted Fourier transform (AAFT) and the iterated amplitude adjusted Fourier transfom (IAAFT) algorithm. Both the AAFT and IAAFT algorithms conserve the amplitude distribution in real space and reproduce the power spectrum (PS) of the original data set very accurately.

The basic assumption in both algorithms is that higher-order correlations can be wiped out using a Fourier phase randomization procedure. In both cases, however, the randomness of the Fourier phases is only imposed before the (first) Fourier back transformation. Until now, it has not been studied how the subsequent remapping and iteration steps may affect the randomness of the phases.

Using the Lorenz system as an example, we show that both algorithms may create surrogate realizations containing Fourier phase correlations. We present two new iterative surrogate data generating methods being able to control the randomization of Fourier phases at every iteration step. The resulting surrogate realizations which are truly linear by construction display all properties needed for surrogate data.

Keywords: Surrogates; AAFT; IAAFT; IPAFT; non-linear prediction error; Lorenz system.

1. Introduction

The detection of non-linear behavior in data sets is a general problem which arises in diverse disciplines.[1–9] Tests for non-linearity which compare a data set to the null hypothesis of a Gaussian linear process are relevant either to constrain models or support theories of a particular system. In this context, the surrogate data tests are model independent tests for non-linearity[10–13] which can be applied with any non-linear statistics that characterizes a data set. Ideal surrogate data should possess not only the same power spectrum and amplitude distribution in real space as the original data set but also be free of higher-order correlations. The Amplitude Adjusted Fourier Transform (AAFT) and the Iterative Amplitude Adjusted Fourier Transform (IAAFT) algorithms[10,12,13] are the most popular algorithms to generate an ensemble of surrogate realizations. AAFT and IAAFT algorithms conserve the amplitude distribution in real space and reproduce the power spectrum (PS) of the original data set quite accurately. The basic assumption in both algorithms is that higher-order correlations can be wiped out using a Fourier phase randomization procedure since the PS is invariant under Fourier phase permutations. Previous studies[10,14,15] focused on the ability of surrogate generating algorithms to accurately reproduce the PS of the original data. However, it has been shown that for an ARMA process an ensemble of surrogates generated by the AAFT algorithm displays an anomalous small variability of the PS when compared to an ensemble of identical ARMA processes.[16] This poses the question of the definition of the null hypothesis for the surrogate test which will certainly depend upon the system under study. More important than exactly reproducing the PS of the original data is to create surrogate data free of higher-order correlations which by definition are linked to the properties of the Fourier phases. However, how phase correlations are related to higher-order correlations is an important yet unresolved open problem. AAFT and IAAFT algorithms impose constraints on the Fourier amplitudes and the amplitude distribution in real space without controlling Fourier phases. As shown in the next section, these constraints may generate Fourier phase correlations. In the rest of this contribution we propose and discuss novel surrogate generating algorithms in which the randomness of the phases is explicitly controlled.

2. Phase Correlations in Surrogates

In this section we introduce the model systems which we use in our study. Furthermore we demonstrate, how phase correlations, thus non-linearities, can be induced in surrogate data sets.

2.1. *Model systems*

As a linear system we consider the first order ARMA process described by

$$x_{n+1} = 0.7x_n + \xi_n \,, \tag{1}$$

with ξ_n being Gaussian noise.

Furthermore, we use the z-component of the Lorenz system[17]

$$\dot{x} = a(y - x) \tag{2}$$

$$\dot{y} = x(b - z) - y \tag{3}$$

$$\dot{z} = xy - cz \tag{4}$$

in the chaotic regime, i.e. $a = 10$, $b = 28$ and $c = 8/3$, as an example for a time series derived from a deterministic chaotic system, which explicitly contains non-linear correlations. The original data were sampled at $\tau' = 0.001$ and a rank ordered remapping onto a Gaussian distribution was performed before applying the surrogate algorithms. The original time series resulting from the rank ordered remapping onto a Gaussian distribution is called z_L throughout the text. The length of the times series is $M = 2^{15}$.

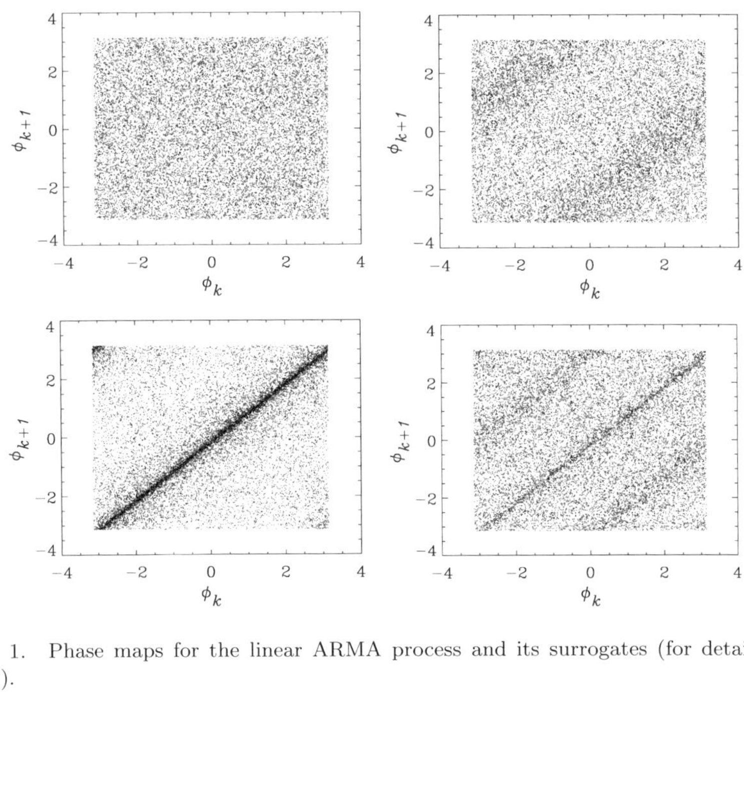

Fig. 1. Phase maps for the linear ARMA process and its surrogates (for details see text).

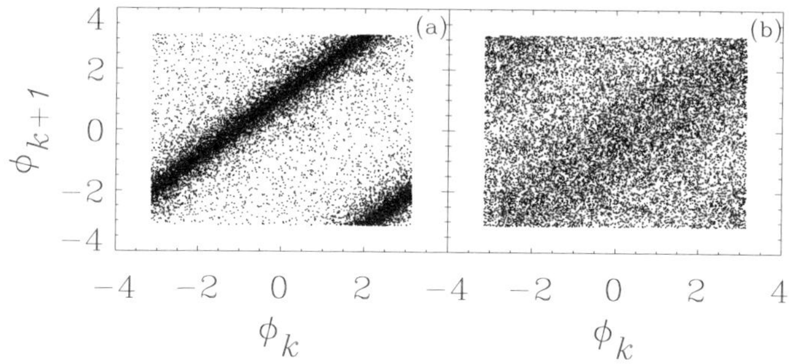

Fig. 2. a) Fourier phase map of an IAAFT surrogate realization of z_L for $\Delta = 1$. b) same as (a) but using an AAFT surrogate realization.

2.2. *Phase maps*

A Fourier phase map is a two-dimensional set of points $G = \{\phi_k, \phi_{k+\Delta}\}$ where ϕ_k is the phase of the k-th mode of the Fourier transform and Δ a phase delay. All possible modes k up the Nyquist frequency are considered. Using this representation, phase correlations can show up in the maps as high/low density regions.

2.3. *Inducing phase correlations*

First, we consider the ARMA process described above. The amplitude distribution in real space for the ARMA process is Gaussian. Fig. 1 upper left panel shows a Fourier phase map for a relization of this ARMA process. As expected, this phase map shows no signatures of phase correlations. Suppose that we replace two data point by two spikes thus the distribution is almost Gaussian. Fig. 1 upper right panel shows the Fourier phase map for the spiky time series. It is clear that the presence of spikes has induced Fourier phase correlations. Now, we generate an IAAFT surrogate realization of the spiky times series. Fig. 1 lower left panel shows the Fourier phase map for this IAAFT surrogate where once again the Fourier phases are strongly coupled. The first guess is that this strong coupling in the phases is only due to the spikes still present in the IAAFT realization and that this will vanish after removing the spikes. However, as shown in Fig. 1 lower right panel, phases correlations still remain after removing the spikes. Then, this residual phase correlations have been induced during the iterative process leading to an IAAFT surrogate which displays signatures of non-linearity. We conclude that Gaussian remapping must always be performed since this step warrants that we obtain random uniformly distributed Fourier phases at the first iteration step of IAAFT.

Second, we calulate AAFT and IAAFT surrogate realisations of the (Gaussian remapped) z-component of the Lorenz system z_L and represent the phase information using phase maps (Fig. 2). It becomes immediately obvious that both surrogate generating algorithms can induce phase correlations.

3. The IPAFT Algorithm

We introduce two alternative iterative methods to generate surrogate realizations where the randomization of higher-order correlations is controlled at every iteration step by imposing uncorrelated uniformly distributed random Fourier phases. For data having a Gaussian amplitude distribution in real space, the resulting surrogate realization will have a similar distribution in real space, a well-reproduced auto-correlation function, and uncorrelated uniformly distributed Fourier phases.

Algorithm A consists of the following steps. Consider a time series y_n, $n = 1, \ldots, M$. (i) Perform a rank-ordered remapping of the original data onto a Gaussian distribution $x_n = G(y_n)$. Remapping a time series onto a Gaussian distribution conserves dynamic non-linearities[13] and reduces the amount of whitening of the PS induced by the iterations. Evaluate the Fourier transform of x_n. In addition, generate a random shuffling realization of x_n and calculate its Fourier transform. Since the amplitude distribution in real space is Gaussian, we obtain uncorrelated uniformly distributed Fourier phases in the interval $[-\pi, \pi]$. If we avoid the remapping step onto a Gaussian distribution then the shuffling step cannot ensure uncorrelated uniformly distributed Fourier phases as shown above.[4,5] (ii) Combine the Fourier amplitudes of x_n with the random Fourier phases and perform an inverse Fourier transformation. Let us call the resulting time series x'_n. (iii) Sort x_n as x'_n ($z_n = S_{x'_n}(x_n)$) and evaluate the Fourier transform of z_n. Steps (i), (ii), and (iii) describe the AAFT algorithm, i.e. z_n is an AAFT surrogate realization. Instead of replacing the Fourier amplitudes of z_n by the original Fourier amplitudes, as is the case for IAAFT, we replace the Fourier phases by the random phases obtained in step (i). Using the new Fourier amplitudes and the random phases we repeat steps (ii) and (iii). Iterations stop after convergence of the probability distribution in real space. In the last iteration, step (iii) is avoided thus we obtain a surrogate realization containing uncorrelated uniformly distributed Fourier phases. Step (iii) may actually reintroduce uncontrolled Fourier phase correlations as shown in Fig. 2 (b). Imposing random Fourier phases at every iteration step will prevent the algorithm to develop phase correlations during evolution and so it maintains the randomization of higher-order correlations. However, the original PS is imposed only at the first iteration[s] and evolves freely through subsequent iterations. This freedom given to Fourier amplitudes introduces, however, only small deviations to the PS after the first iteration step, which is the one that brings the initial flat PS to the desired one. Convergence was verified in several tests even in higher dimensions. It is expected since a Gaussian linear process is characterized by both a Gaussian amplitude distribution and uncorrelated uniformly distributed Fourier phases.

Using the ARMA process, we quantified the spectral variability of an ensemble of phase-controlled surrogates via $\alpha = \frac{1}{N}\sum_{k=1}^{N} \frac{(P_k - \langle P_k \rangle)^2}{\langle P_k \rangle^2}$, where P_k is the power of the k-mode of a surrogate realization and $\langle P_k \rangle$ is the mean power.[16] The spectral variability is compared with the actual variability of an ensemble of ARMA processes using $\langle \alpha \rangle_S$ and $\sigma_S(\alpha)$. Our results (see Table below) show that in spite of allowing for freely evolving Fourier amplitudes, the spectral variability of phase-controlled surrogates is almost two orders of magnitude smaller than that of the ensemble of ARMA processes. Algorithm B tackles this problem by allowing a variance of P_k over the set of surrogates given by $\sigma_k^2 = P_k^2$ [see 18]. It is identical to Algorithm A except for introducing the variance in step (ii) only at the first iteration.

Table 1. Spectral variability. Results were obtained using 100 realizations. A and B stand for algorithms A and B, respectively.

Ensemble	$\langle \alpha \rangle_S$	$\sigma_S(\alpha)$
ARMA process	1.04	0.03
Phase-controlled A	0.016	0.006
Phase-controlled B	0.67	0.02

4. Results

4.1. *Convergence*

We consider the convergence of the amplitude distribution in real space as the stopping criterion for the algorithms. Convergence is assessed via the relative deviation of the amplitude distribution defined as $\Delta I^{(i)} = \sum_{k=0}^{M-1}(r_k^{(i)} - r_k)^2 / \sum_{k=0}^{M-1} r_k^2$, where $r_k^{(i)}$ is the rank-ordered time series re-

Table 2. Mean value and standard deviation of the relative deviation $\Delta I^{(\infty)}$ for all surrogate classes. Results were obtained using 50 surrogate realizations.

Ensemble	$\langle \Delta I^{(\infty)} \rangle_S$	$\sigma_S(\Delta I^{(\infty)})$
AAFT	0.0054	0.0028
IAAFT	5.9x10^{-7}	4.4x10^{-7}
Phase-controlled A	1.7x10^{-8}	1x10^{-9}
Phase-controlled B	1.7x10^{-8}	1.1x10^{-9}

sulting after step (ii) at iteration i, and r_k is the original rank-ordered time series. By skipping the last rank-ordered remapping step in AAFT and IAAFT algorithms, it is possible to compare the performance of all algorithms to reproduce the amplitude distribution in real space. The Table above shows the results for $\Delta I^{(\infty)}$, i.e. the value of $\Delta I^{(i)}$ after convergence. This table indicates that phase-controlled surrogates reproduce the amplitude distribution in real space one order (six orders) of magnitude better than IAAFT (AAFT), respectively.

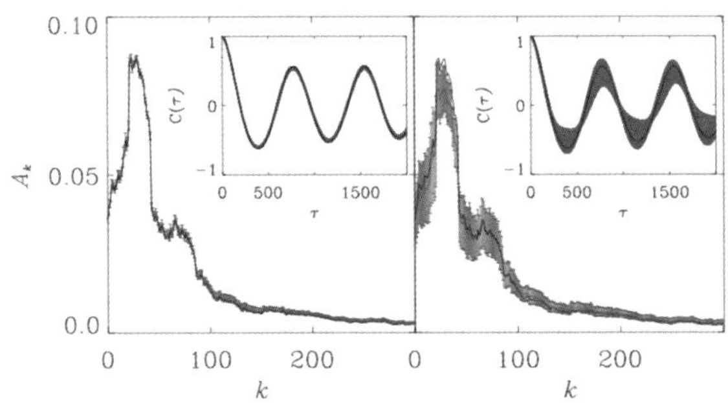

Fig. 3. (left) Smoothed Fourier amplitude A_k versus the wave number k for z_L. The inset shows the autocorrelation function $C(\tau)$ versus the time lag τ for z_L. The shaded region shows 1σ error bars obtained using 50 phase-controlled surrogate realizations generated by algorithm A. (right) Same as the left figure but using 50 phase-controlled surrogate realizations generated by algorithm B.

Fig. 3 shows A_k versus k and $C(\tau)$ versus τ for the z component of the Lorenz system z_L for algorithms A and B after convergence. These results indicate that both algorithms lead to well-behaved surrogate realizations. A comparison of Figs. 3 (left) and 3 (right) shows that introducing a variance in the PS leads to a widening of the 1σ error bars.

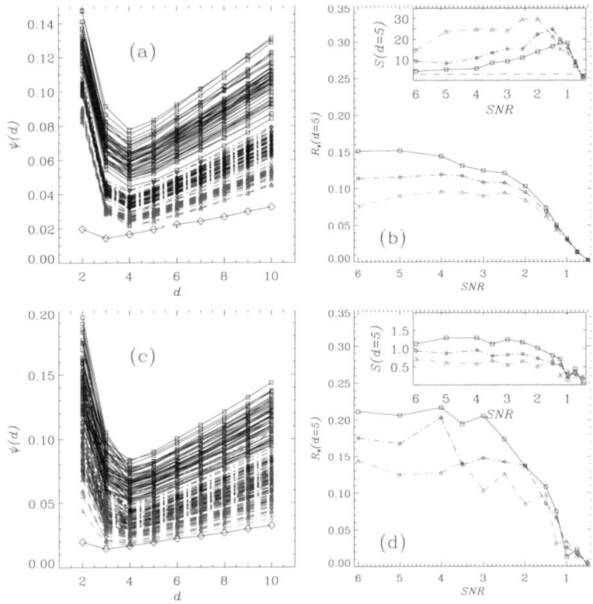

Fig. 4. a) ψ versus the embedding dimension for z_L (black curve), an ensemble of IAAFT (gray, dashed) , AAFT (black, dash-dotted), and phase-controlled (A) surrogate realizations. b) The relative deviation $R_\psi(d = 5)$ versus SNR for the three classes. $R_\psi(d = 5) = |\psi_O - \langle\psi\rangle_S|/\langle\psi\rangle_S$ where ψ_O is the NLPE value for the original data and $\langle\psi\rangle_S$ is the mean value of ψ as derived from surrogate realizations, for embedding dimension $d = 5$. The inset shows the σ-normalized deviation versus SNR for the three classes. $S(d) = |\psi_O(d) - \langle\psi\rangle_S|/\sigma_S(\psi)$, where $\sigma_S(\psi)$ is the standard deviation of ψ as derived from surrogate realizations. c) and d) The same as a) and b) but using phase-controlled (B) surrogate realizations, respectively.

4.2. *Applications to the Lorenz system*

We performed a comparative study of the four different surrogate generating algorithms, namely AAFT, IAAFT, phase-controlled A, and phase-controlled B surrogates. To this purpose, we applied the non-linear prediction error (NLPE) method[12,19] to test for non-linearity. To calculate NLPE, the time series is embedded in a d-dimensional space using the method of delay coordinates: $\vec{x}_n = (x_{n-(d-1)\tau}, x_{n-(d-2)\tau}, \ldots, x_n)$, where τ is the delay time. Then, we define the NLPE as

$$\psi(d, \tau, T, N) = \frac{1}{(M - T - (d-1)\tau)} \left(\sum_{n=(d-1)\tau}^{M-1-T} \left[\vec{x}_{n+T} - F(\vec{x}_n)\right]^2 \right)^{1/2} \tag{5}$$

where F is a locally constant predictor, M is the length of the time series, and T is the lead time. The predictor F is calculated by averaging over future values of the N ($N = d + 1$) nearest neighbors in the delay coordinate representation. We have studied the behavior of ψ as a function of the lead time T. We found that for $T > 5$ ψ remains rather constant, thus a value of $T = 10$ was used for this test. Using ψ, we evaluated the σ-normalized deviation S and the relative deviation R (see caption of Fig. 4). In all cases, τ was chosen in order to satisfy the criterion of zero autocorrelation.[20] Fig. 4(a) shows ψ versus the embedding dimension d for IAAFT, AAFT and phase-controlled A surrogates. NLPE distinguishes between the three groups of surrogates and indicates that both AAFT and IAAFT surrogate realizations are more predictable than phase-controlled A surrogates, i.e. $\|\psi_O - \langle\psi\rangle_S\|$ is larger for phase-controlled A surrogate realizations for all embedding dimensions. It is remarkable that the three groups do not overlap. Fig. 4(a) also shows that the largest variability is observed for the set of phase-controlled A surrogate realizations and it decreases for AAFT and IAAFT surrogate realizations, respectively.

The σ-normalized deviation, which is a function of the surrogate ensemble variability and $\|\psi_O - \langle\psi\rangle_S\|$, takes the values $S(5) = 5.1$, 6.9, and 5.8 for IAAFT, AAFT, and phase controlled A surrogates, respectively.

We also analyzed the behavior of all surrogate classes in the presence of noise. We took z_L and superimposed additive Gaussian white noise with varying standard deviation σ_n. The signal to noise ratio is defined as $SNR = \sigma_s/\sigma_n$ where σ_s is the standard deviation of the original time series. In the presence of noise, the behavior of ψ versus d resembles that of Fig. 4(a) but differences among the three surrogate classes are now smaller. The inset plot of Fig. 4(b) shows $S(5)$ versus SNR for IAAFT, AAFT and phase-controlled A surrogates. For $SNR \leq 0.5$, the three groups lead to non-significant results as expected. For $SNR > 0.5$, $S(5)$ grows rapidly and displays exceptional high values for all surrogate classes for relatively low SNR values. In fact, maxima values of $S(5) \sim 20, 25$, and 30 are observed at $SNR \sim 1, 1.5$, and 2 for phase-controlled A, AAFT, and IAAFT surrogates, respectively. All surrogate classes display a counterintuitive decreasing behavior when further increasing SNR. Even for the highest SNR value here studied ($SNR = 6$), IAAFT and AAFT display $S(5)$ values higher than the value obtained in the absence of noise. However, for phase-controlled A surrogates this feature is only observed for $SNR < 3.5$. Two different effects are responsible for this trend, namely the increase of $|\psi_O - \langle\psi\rangle_S|$ and the decrease of $\sigma_S(\psi)$ with decreasing SNR. However, this trend is more pronounced for IAAFT and AAFT surrogates than due to the smaller variability observed for the standard surrogate data which already becomes clear in Fig. 4 (a). $\sigma_S(\psi)$ is approximately one order of magnitude (5 times) smaller for IAAFT (AAFT) surrogates than for phase-controlled A surrogates. For $SNR \sim 0.5$, $|\psi_O - \langle\psi\rangle_S|$ sharply drops to zero thus yielding the expected non-significant results for all surrogate classes. Fig. 4(b) shows the relative deviation $R_\psi(5)$ versus SNR. Note that the order of the curves has been mirrored with respect to their order in the inset plot. Since $S = \frac{R_\psi \langle\psi\rangle_S}{\sigma_S(\psi)}$, then $\sigma_S(\psi)$ is responsible for this reordering. $R_\psi(5)$ not only shows the expected trend for increasing SNR but it also takes larger values when evaluated in the absence of noise ($R_\psi(5) = 0.33, 0.51$, and 0.70 for IAAFT, AAFT, and phase-controlled A surrogates respectively). Thus, it provides a suitable relative measure of significance for different SNR. This behavior is also observed for other embedding dimensions.

Fig. 4(c) shows ψ versus d for IAAFT, AAFT and phase-controlled B surrogates. In order to allow for a fair comparison among surrogate classes, we modified both the IAAFT and AAFT algorithms by introducing a variability in the PS as done in algorithm B. Phase-controlled B surrogates are still distinguished as a single group by the NLPE and display the largest variability. In the absence of additive noise phase-controlled B surrogates lead to the most significant result $S(5) = 5.9$ followed by AAFT ($S(5) = 3.6$), while IAAFT is non-significant ($S(5) = 2.1$). The inset plot of Fig. 4(d) shows that in the presence of noise the σ-normalized deviation has dropped well below the 3σ threshold for all surrogate classes. However, it displays the expected increasing trend with SNR values thus providing an equivalent description as the relative deviation $R_\psi(5)$.

To investigate the differences between surrogate classes, we considered all the 50 AAFT and 50 IAAFT surrogate realizations of z_L for $SNR = 4$ and generated 20 phase-controlled A surrogates for each of them. Red dots in Fig. 5 show the values of $\psi(5)$ for AAFT and IAAFT surrogate realizations and black dots show the values of $\psi(5)$ for the respective phase-controlled A surrogates. We observe that for both AAFT and IAAFT surrogates, phase-controlled A surrogates lead to larger $\psi(5)$ values in all cases. It is worth noting that in all $2 \cdot 50 = 100$ tests, none of the phase-controlled A surrogate realizations showed lower (more predictable) values of $\psi(5)$ than the respective AAFT (IAAFT) surrogate. These differences in $\psi(5)$ between standard surrogates and phase-controlled A surrogates generated from them become even larger for higher SNR values. Since algorithm A accurately reproduces the PS of the original data and phase correlations

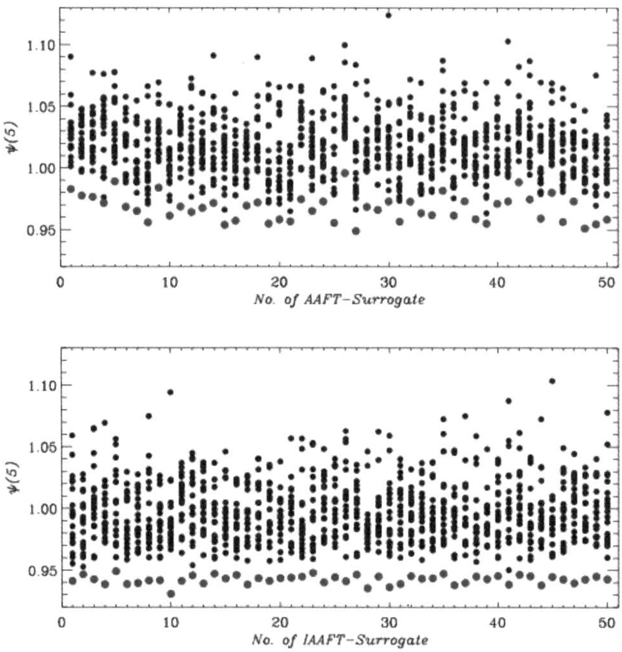

Fig. 5. $\psi(5)$ values for 50 AAFT (above) and 50 IAAFT (below) surrogate realizations of z_L for $SNR = 4$ (gray dots) and for 20 phase-controlled (A) surrogates generated from each of them (black dots).

in standard surrogates become stronger for higher values of SNR, we attribute the differences unveiled by the NLPE to the presence of Fourier phase correlations.

5. Conclusions

We demostrated that the AAFT and IAAFT algorithm may generate surrogate realizations containing Fourier phase correlations. Motivated by this finding, we presented two new surrogate generating algorithm being able to control the randomization of Fourier phases at every iteration step. Using the NLPE method we found clear differences among the surrogate classes that we attribute to the randomization level of the Fourier phases of the surrogate realizations. These differences appear due to the constraints imposed on the surrogate generating procedure. As shown above, freely evolving Fourier amplitudes still lead to ensembles of surrogates having a small variability of the PS. Algorithm B deal with this problem by allowing for a variability in the PS. The choice of the algorithm will depend upon the definition of the null hypothesis for every specific problem.

In view of our results, phase-controlled surrogate realizations are data sets which best reproduce the amplitude distribution in real space and the PS of the original data while explicitly fulfilling the constraint that only linear correlations are contained.

References

1. P. Coles and L. Chiang, *Nature* **406**, p. 376 (2000).
2. L. Chiang, P. Naselsky, O. Verkhodanov and M. Way, *ApJ* **590**, p. L65 (2003).
3. L. Chiang, P. Naselsky and P. Coles, *ApJ* **602**, p. L1 (2004).
4. T. Matsubara, *ApJ* **591**, p. L79 (2003).
5. C. Hikage, T. Matsubara and Y. Suto, *ApJ* **600**, p. 553 (2004).
6. C. Räth and P. Schuecker, *Mon. Not. R. Astron. Soc.* **344**, 115 (2003).
7. P. Bak, R. Yoshino, N. Asakura and T. Nakano, *Phys. Rev. Lett.* **83**, p. 1339 (1999).
8. K. Lehnertz *et al.*, *J. Clin. Neurophys.* **18**, p. 209 (2001).
9. B. Sivakumar, *Journal of Hydrology* **227**, p. 1 (2000).
10. T. Schreiber and A. Schmitz, *Phys. Rev. Lett.* **77**(4), 635 (1996).
11. T. Schreiber, *Phys. Rev. Lett.* **80**, p. 2105 (1998).
12. T. Schreiber and A. Schmitz, *Physica D* **142**, p. 346 (2000).
13. J. Theiler, S. Eubank, A. Longtin, B. Galdrikian and J. D. Farmer, *Physica D* **58**, 77 (1992).
14. C. Räth, W. Bunk, M. B. Huber, G. Morfill, J. Retzlaff and P. Schuecker, *Mon. Not. R. Astron. Soc.* **337**, 413 (2002).
15. D. Kugiumtzis, *Phys. Rev. E* **60**, p. 2808 (1999).
16. K. T. Dolan and M. L. Spano, *Phys. Rev. E* **64**(4), 046128 (2001).
17. E. Lorenz, *Journal of the Atmospheric Sciences* **20**, p. 130 (1963).
18. A. Oppenheim and R. Schafer, *Digital Signal Processing* (Prentice-Hall, Englewoog Cliffs, NJ, 1975).
19. G. Sugihara and R. May, *Nature* **344**, p. 734 (1990).
20. A. Fraser and H. Swinney, *Phys. Rev. A* **33**, p. 1134 (1986).

Is There a Chaos Occurrence in Athens Exchange? Testing Chaotic Behavior in Bank Stocks and ATHEX Indices

A. Saraidaris and A. Margaris
University of Macedonia, Thessaloniki, Greece

This paper is an attempt to test for nonlinear structure and chaos indicators on the returns of bank stocks listed in Athens Exchange (ATHEX) as well as the indices: ATHEX Composite index, FTSE/ASE 20 and FTSE/ASE mid 40.

Keywords: Chaos; stock returns; Lyapunov exponents.

1. Introduction

Examining chaotic behavior in nonlinear systems we first account for the identification of the deterministic nature of them. Since we identify no stochastic or random behavior of series, we can further proceed to the analysis and quantification of their chaotic nature of evolution. Existence of chaos is usually based on the application of tests like: surrogate data analysis,[1] phase space reconstruction methods,[2] Lyapunov exponents tests,[3] Kolmogorov-Sinai entropy tests,[4] unstable periodic orbit test[5] and further more a variety of tests regarding the etimation of fractal dimensions such as capacity, information, and correlation dimension.[6] In applying those tests a variety of custom made applications as well as known chaos characterization tools such as TISEAN[7] were employed.

By estimating the above indicators one can extract valuable quantitative information for the nature of the chaotic dynamic system.

In terms of bank stock returns and ATHEX indices that is the subject of our investigation, the steps to be followed are:

- Investigation for chaotic characteristics in the behavior of bank stock and indices returns considering for similarities and differences that may derive from the peculiar nature of the individual series.
- Split of the investing period in two subperiods ((a) 02-01-1999 to 31-05-2001 and (b) 01-06-2001 to 16-10-2006), because of their dif-

ferent statistic characteristics (Table 4) and application of the same tests in each one independently, in order to account for the event of the MSCI Greece index classification to a developed Market Index effective from 01-06-2001.

As far as concerns methodology, modeling stock returns series with linear models was outperformed during the last decade by the use of non linear models due to their advantage to be distribution and assumption free. They are capable of recognizing patterns and discovering relationships in the data, whereas regression models require a priori knowledge of the nature of the underlying relationships.

2. Literature Review

Since 1980, several economists have tried to test for non-linearity and in particular for chaos in economic and financial time series.[8] The tests of chaos demand first, the correlation dimension measuring the fractal nature of the possibly underlying attractor and second, the largest Lyapunov exponent which when found to be positive measures the sensitive dependence on initial conditions which is a strong characteristic of a chaotic system.

However, the models to detect chaos suffer from high sensitivity to noise. Brock et al.[9] developed a statistical test for independence known as BDS-test based on correlation integral which can be used as a general specification test. A step forward is the asymptotic distribution of a non parametric neural network estimator of the Lyapunov exponent; this test may be seen as a direct test for chaos. Also, testing nonlinearity and modeling economic and financial time series[10] applied some tests for non linearity and chaos effects in financial and stock series ending to a model with a dynamic chaotic mean process coupled with a variance that follows the GARCH process (Mackey-Glass-GARCH model).

There is a number of studies in the literature that refer to empirical analysis of the efficient market Hypothesis (EMH) and the nonlinear behavior in Athens Stock Exchange.

Stengos and Panas[11] and Theodosiou,[12] conducted some of the few researches prior to 1996 that applied nonlinear methods to explore the dynamic processes of ATHEX stock returns.

Siriopoulos,[13] conducting an empirical research on ATHEX General Index for the period: 1974-1994 by applying BDS test statistic and correlation dimension, concluded that ATHEX return series appear non-linearities which might be generated by a 'semi-chaotic' behavior. Later on, Barkoulas and Travlos[14] investigated the existence of a deterministic nonlinear structure in ATHEX stock returns for the period: 1981-1990. Employing the correlation dimension and Kolmogorov entropy methodologies tested for chaos and found no strong evidences for a chaotic structure in ATHEX returns and concluded that a probable nonlinear stochastic process describes more accurately the behavior of ATHEX stock returns in the examining period.

Panas,[15] employing ARFIMA models and estimating the fractal dimensions provided further evidence of nonlinear characteristics in a sample of ATHEX listed companies' stock returns.

Laopodis[16] tested ATHEX efficient operation for the period: 1985-2001 through the scope of the Greek financial market liberalization and the Morgan Stanley classification of MSCI Greece Index from an emerging to a developed market index. He found a weak-from efficiency long before the announcements of market liberalization were made.

Panagiotidis[17] tested the efficient market hypothesis in ATHEX for the period 01-06-2000 – 14-03-2003 under the fact that Greece officially announced to join the Economic and Monetary Union on 19-06-2000.

Examining the indices FTSE/ASE 20, FTSE/ASE mid 40 and FTSE/ASE small cap, concluded that after the introduction of common currency, there is a strong evidence against the random walk hypothesis and also that lower capitalization index FTSE /ASE small cap is more efficient compared to other two in the sense that past volatility does not help in explaining future returns.

3. The Structure of the Experimental Time Series

The data employed in this study consist of ten Greek Bank daily stock returns and daily returns of the three indices: ATHEX Comprosite index, FTSE/ATHEX 20, FTSE/ATHEX mid 40. The period covered is: 01.01.1999 - 16.10.2006. The time series named as: DLNALPHA for the Alpha Bank, DLNATT for the Bank of Attica, DLNEGNAK and DLNEGNAP for the Egnatia Bank (common stocks and preferred stocks respectively), DLNEUROB for the Eurobank, DLNTELL for the Bank of Greece, DLNTEMP for the Emporiki Bank, DNLTGEN for the General Bank of Greece, DLNETE for the National Bank of Greece and DLNTPEIR for the Piraeus Bank. Regarding the time series associated with the three indices they are characterized by the names: GD for the ATHEX composite index, FTSE20 for the FTSE/ATHEX 20 index and FTSE40 for the FTSE/ATHEX 40 index. Additionally, the suffix P, where found, denotes adjusted (vs historical) stock prices, while the suffixes A and B indicate the corresponding time period.

The data series consist of daily returns on a continuous trading basis. To get equal number of observations for all series of our group we excluded all banks listed in ATHEX later than 02.01.1999 as well as, those banks that were delisted for any reason from ATHEX during the same period. In order to overcome the problem of existence of some no trading days for some of the stocks, we assume for these days the same stock price with the last trading day.

Sample data series consist of first differences (returns) of logarithmic adjusted stock prices (dlnprice). In other words, if we denote as $d(i)$ $(i = 1, 2, \ldots, N)$ the actual time series data - i.e. the stock price levels - then, the used time series $x(i)$ is defined as $x(i) = \log[d(i)] - \log[d(i-1)]$. This data representation has be chosen since from the application of the Augmented Dickey-Fuller unit-root and Phillips-Perron tests steams that stock series should be log differenced to fulfill the term of stationarity. However a quantitative test for the stock prices in levels is also presented.

The selection of this group of individual stocks and Market indices was made in order to account for differences in trading behavior between individual stocks and the market indices.

The interest in the bank stock returns comes first from the fact that in the advanced capital markets bank sector is deemed to lead the evolution in stock market prices and also, bank sector is the sector with the least no trading days, a problem that might hinder the reliability and comparability of the series.

The three indices represent different groups of companies and stock trading characteristics. 1 Their special characteristics are summarized in tables 1 and 4.

Last, from the sample descriptive statistics is apparent that returns distribution deviate from normality

4. Analysis of Chaotic Data

The analysis of the time series is based to the phase space reconstruction technique performed via the method of time delays. The appropriate values for the embedding dimension d were selected by using the following methods: (a) False Nearest Neighbors (FNN), (b) Singular System Approach (SSA) and (c) Γ-Test method. Regarding the optimum value for the time delay τ it was selected by applying the autocorrelation method. The quantitative characterization of the chaotic nature of the time series is based

Table 1. The values of the most important statistical properties for the DLN series

Name	MEAN	ADEV	SDEV	VAR
DLNALPHA	0.000522	0.016571	0.022459	0.000504
DLNATT	0.000270	0.022126	0.031400	0.000986
DLNEGNAK	0.000326	0.020662	0.029713	0.000833
DLNEGNAP	0.000382	0.021649	0.029715	0.000936
DLNETE	0.000747	0.016450	0.023000	0.000529
DLNEUROB	0.000908	0.017018	0.034210	0.001170
FTSE20	0.000442	0.012263	0.017409	0.000303
FTSE40	0.000569	0.013536	0.019290	0.000372
GD	0.000459	0.011696	0.016778	0.000282
DLNTELL	0.000913	0.015362	0.023662	0.000560
DLNTEMP	0.000553	0.017789	0.025031	0.000627
DLNTGEN	0.000213	0.019482	0.027493	0.000756
DLNTPEIR	0.000944	0.016356	0.023667	0.000560

on the Wolf's algorithm. Unfortunately the estimation of the capacity, information and correlation dimension by using the FD3 application was not possible since the number of time series samples was characterized as inadequate by the application for the selected value of the embedding dimension.

The most important statistical properties of the DLN time series for the full period are presented in Table 1. For each time series the mean value, the average deviation (ADEV), the standard deviation (SDEV) and the variance (VAR) values are shown.

By applying the methods described above, the optimum values for the embedding dimension and the time delay were the values $d = 4, 5$ and $\tau = 4$, while, the evolution time parameter of the Wolf's algorithm was set to the value $\tau_e = 5$. However, to ensure for a reasonable and correct calculation of the maximum positive Lyapunov exponent, the combinations $(d, \tau_e) = (5, 3)$, $(d, \tau_e) = (5, 4)$, $(d, \tau_e) = (6, 3)$ and $(d, \tau_e) = (6, 4)$ were used during experiments. The results of those calculations for the DLN time series are shown in Table 2.

Table 3 presents experimental results for the stock price levels time series. The last column represents the value of the estimated Lyapunov exponent for the parameter values $(d, \tau, \tau_e) = (5, 4, 5)$.

Finally, Table 4 presents the values of the main statistical properties and the estimated Lyapunov exponent for the first differences of the logarithmic adjusted stock prices for periods A and B and for parameter values $(d, \tau, \tau_e) = (5, 4, 5)$.

Table 2. The maximum positive Lyapunov exponent for the experimental time series as it has been estimated by using the Wolf's algorithm for various combinations of parameter values

Name	$(d, \tau_e) = (5, 3)$	$(d, \tau_e) = (5, 4)$	$(d, \tau_e) = (6, 3)$	$(d, \tau_e) = (6, 4)$
DLNALPHA	0.005519	0.002817	0.000702	0.001929
DLNATT	0.001623	0.001833	0.001816	0.002133
DLNEGNAK	0.001852	0.001749	0.001609	0.001236
DLNEGNAP	0.002140	0.002385	0.002116	0.002249
DLNETE	0.001692	0.001309	0.001120	0.001589
DLNEUROB	0.001891	0.001550	0.000879	0.001988
FTSE20	0.000545	0.002744	0.001008	0.000525
FTSE40	0.002582	0.002261	0.001592	0.001704
GD	0.000837	0.001933	0.000904	0.001185
DLNTELL	0.000966	0.001081	0.000439	0.001230
DLNTEMP	0.001003	0.001133	0.002971	0.002073
DLNTGEN	0.001916	0.001826	0.002189	0.001725
DLNTPEIR	0.002274	0.001831	0.002537	0.001921

Table 3. The FNN-based estimated embedding dimension together with the statistical properties and the maximum positive Lyapunov exponent for the stock price levels time series.

Name	d	MEAN	SDEV	ADEV	VAR	LYAP
DLNALPHA	4-5	14.409390	5.520228	4.723476	30.472921	0.049116
DLNATT	5	6.534151	3.869586	2.726943	14.973692	0.024922
DLNEGNAK	5	6.133373	4.428127	3.441790	19.608305	0.029151
DLNEGNAP	4-5	5.612581	4.443141	3.384411	19.741501	0.022826
DLNETE	4-5	21.445034	8.555467	7.629788	73.196014	0.038031
DLNEUROB	5	17.139836	5.661896	6.546905	42.861971	0.036970
FTSE20	5	1665.12	478.08	569.70	324563.8	0.046242
FTSE40	4-5	3522.295	2124.151	1510.560	4512019.4	0.013190
GD	4-5	3014.740	1018.017	827.251	1036360.4	0.035328
DLNTELL	5-6	57.154279	25.9586	22.351401	673.850	0.038324
DLNTEMP	5-6	24.993596	13.5434	10.965464	183.425853	0.034630
DLNTGEN	4-5	11.454834	8.082820	5.812678	65.331987	0.038031
DLNTPEIR	4-5	11.370715	5.567555	4.709937	30.997669	0.046868

5. Conclusions

We have applied an investigation of chaotic features to a sample of ATHEX Bank stock and Indices returns for the period: 02-01-1999 – 16-10-2006, as well as for periods (A) 02-01-1999 – 31-05-2001 and (B) 01-06-2001 – 16-10-2006. By employing Lyapunov exponent tests we find a weak though clear evidence of chaos in ATHEX returns series. Regarding the two subperiods A and B, there is a slightly stronger evidence of chaos in the first subperiod

Table 4. The main statistical properties and the estimated Lyapunov exponent for the DLN times series for periods A and B.

PERIOD A					
Name	MEAN	SDEV	ADEV	VAR	LYAP
DLNALPHA	0.000853	0.027787	0.020757	0.000772	0.364777
DLNATT	0.000564	0.036683	0.028077	0.001346	0.368679
DLNEGNAK	0.000422	0.036917	0.027532	0.001363	0.366094
DLNEGNAP	0.000607	0.037716	0.028662	0.001423	0.364023
DLNETE	0.001611	0.027676	0.019909	0.000766	0.367298
DLNEUROB	0.001550	0.051097	0.025416	0.002611	0.016695
FTSE20	0.000888	0.022774	0.016727	0.000519	0.367088
FTSE40	0.001131	0.025732	0.019506	0.000662	0.355423
GD	0.000510	0.020768	0.015711	0.000431	0.355624
DLNTELL	0.001326	0.031020	0.021347	0.000962	0.359179
DLNTEMP	0.002117	0.030561	0.022302	0.000934	0.366241
DLNTGEN	0.000534	0.034047	0.025896	0.001159	0.363799
DLNTPEIR	0.001725	0.030903	0.022489	0.000955	0.348597

PERIOD B					
Name	MEAN	SDEV	ADEV	VAR	LYAP
DLNALPHA	0.000311	0.018276	0.013913	0.000334	0.338722
DLNATT	0.000083	0.027520	0.018332	0.000757	0.242736
DLNEGNAK	0.000265	0.024038	0.016225	0.000578	0.278154
DLNEGNAP	0.000238	0.025025	0.017175	0.000626	0.287828
DLNETE	0.000196	0.019432	0.014269	0.000378	0.361798
DLNEUROB	0.000499	0.015896	0.011742	0.000263	0.367409
FTSE20	0.000158	0.012872	0.009431	0.000166	0.303962
FTSE40	0.000210	0.013682	0.009746	0.000187	0.279681
GD	0.000205	0.011967	0.008690	0.000143	0.308133
DLNTELL	0.000650	0.017429	0.011567	0.000304	0.223455
DLNTEMP	-0.000444	0.020702	0.014890	0.000429	0.354728
DLNTGEN	0.000009	0.022345	0.015376	0.000499	0.273789
DLNTPEIR	0.000446	0.017557	0.012486	0.000308	0.295804

which ends with the classification of MSCI Greece Index to a developed market index.

The evidence of weaker chaos in the second period should be attributed to the change in the nature of ATHEX resulting from likely regulation effects and the two above mentioned major events that provoked the substitution of a great portion of domestic and foreign speculative investment capitals by long run foreign and domestic ones.

The latest can be explained by the common investment practice of international speculative funds in the world emerging capital markets as well as, the experience gained by domestic investors through the real case studies run during the turbulence period. Therefore, during the second period there is a clear adherence of the ATHEX to the movement of the Major World Stock Market indices and primarily Dow Jones and NASDAQ.

For future research is recommended farther investigation of chaos evidence employing Fractals and Neural Networks accounting also for extreme observations in return series.

References

1. J. Theiler, S. Eubank, A. Lougtin, B. Galdrikian and J. Farmer, *Physica D* **58**, 77 (1992).
2. F. Takens, *Detecting Strange Attractors in Turbulence*, in *Dynamical Systems and Turbulence*, eds. D. Rand and L. S. Young, Lecture Notes in Mathematics, Vol. 898 (Springer Verlag, 1981), pp. 366–381.
3. A. Wolf, J. Swift, H. Swinney and J. Vastano, *Physica 16D* , 285 (1985).
4. K. Alligood, T. Sauer and J. Yorke, *Chaos: An Introduction to Dynamical Systems* (Springer Verlag, Pittsburg, 1996).
5. D. Lathrop and E. Kostelich, *Physical Review A* **40**(7), 4028 (1989).
6. L. Liebovich and T. Toth, *Physics Letters A* **141**(8,9), 386 (1989).
7. R. Hegger, H. Kantz and T. Schreiber, *CHAOS* **9**, 413 (1999).
8. W. Brock, D. Hsieh and L. LeBaron, *Nonlinear Dynamics, Chaos and Instability* (MIT Press, Cambridge, 1991).
9. W. A. Brock, W. Dechert, J. Scheinkman and B. LeBaron, *Econometrics Reviews* **15**, 197 (1996).
10. C. Kyrtsou and M. Terraza, *International Review of Financial Analysis* **11**(4), 407 (2002).
11. T. Stengos and E. Panas, *Empirical Economics* **17**, 239 (1992).
12. P. Theodossiou, *Applied Financial Economics* **3**(2), 119 (1993).
13. C. Siriopoulos, *Indian Journal of Quantitative Economics* **11**(1), 76 (1996).
14. J. Barkoulas and N. Travlos, *Applied Financial Economics* **8**, 231 (1998).
15. E. Panas, *Applied Financial Economics* **11**, 395 (2001).
16. N. Laopodis, *Global Finance Journal* **15**, 103 (2004).
17. T. Panagiotidis, *Applied Financial Economics* **15**, 707 (2005).

Some Remarks Concerning the Application of the Variation Method in the Study of Fractal Curves

C. Secrieru[*,‡], J.-M. Nianga[†] and A. Iost[‡,§]

[*]*"Politehnica" University of Timisoara, Faculty of Mechanical Engineering Department Strength of Materials, Timiçoara, 300222 B-dul Mihai Viteazu, nr.1, Romania*

[†]*Pole Structures et Materiaux, Hautes Etudes d'Ingenieur (H.E.I.) 13 rue de Toul, 59000, Lille cedex, France*

[‡]*Equipe CPP, Laboratoire de Metallurgie Physique et de Genie des Materiaux CNRS, UMR 8517, ENSAM, 8 Boulevard Louis XIV, 59546 Lille cedex, France*

*Emails: *claudia.secrieru@gmail.com, [†]jean-marie.nianga@hei.fr, [§]Alain.IOST@lille.ensam.fr*

This paper discusses the pertinence of application of variation method in the study of phenomena which present fluctuations in time. One presents the advantages and the limits of this method. This paper discusses the local and the global aspects of this technique. The variation method applies exclusively to the graphs of continuous functions while making it possible to obtain the Minkowski-Bouligand dimension by the integral of the local oscillation. The sum of the oscillations on a curve does the global aspect of this technique. The variation σ will correspond to an area. Finally one applies this technique for the study of fractal surfaces.

Keywords: Variation method; fractal dimension; local and global aspects.

1. Introduction

In recent years Fractal Analysis is obviously a new direction for understanding the physics of phenomena which presents fluctuations in many fields. It is easier to understand that a measure of the Minkowski-Bouligand type is unsuited to such sets since they present distinct scale properties according to the two axes of the graph. It is thus advisable in this case to find methods more adapted such as for exemple the variation method proposed by Tricot.[1] The cover of a set could be carried out in a homogeneous way by meeting of the "balls" of $_n$ diameter for determining the Minkowski-Bouligand dimension[2] Δ_{MB}. This method is one of most precise to calculate the fractal dimension of the function $z = f(x, y)$. The method can be applied for the self-affine and continuous graphs. We'll refer to the graph of a function of a single variable. The objective is to estimate the fractal dimension of this graph. The fractal dimension can be used to compare the complexity of two curves or two surfaces. It has been proposed that the anisotropy of the fracture process might lead to self-affine rather than self-similar interfaces so we try to apply this algorithm for the fracture surface. The algoritm of the Variation Method is described in the Figure 1.

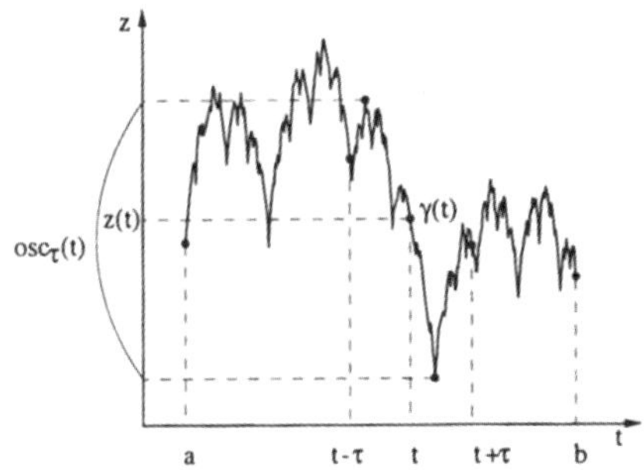

Fig. 1. The oscilation $\mathrm{osc}_\tau\ (t)$ of $z(t)$ used in the variation method

When the abscissa t represents the time, the curve Γ is a trajectory. One defines oscillation of function z in t as:

$$\mathrm{osc}_\tau(f,x) = \left| \max_{|x-t|<\tau} f(t) - \min_{|x-t|<\tau} f(t) \right|$$

One introduces the notion of "variation of a function". The function of $z(t)$ on $[a,b]$ is the integrant of the oscillations. We note it $\mathrm{VAR}_\tau\ (z)$ or VAR_τ .

$$\mathrm{VAR}_\tau = (b-a)\overline{osc}_\tau = \int_a^b \mathrm{osc}_\tau(t)dt$$

$$\mathrm{VAR}_\tau(f,a,b) = \frac{1}{b-a}\int_a^b \mathrm{osc}_\tau(f,x)dx$$

The fractal dimension associated with the graph of the function f defined on the interval $[a,b]$ has the following form:

$$\Delta\,(f,a,b) = \lim_{\tau\to 0}\left(2 - \frac{\log \mathrm{VAR}\,\tau\,(f,a,b)}{\log\tau}\right)$$

One of the advantages of the Variation Method is to substitute for the geometrical measures the integrant of a function with real values, making the digital estimation more precise. By diverse simulations Dubuc et al.[3,4] and Tricot[1] showed that the Variation Method is the best adapted for calculation of fractal dimension of graphs stemming from functions of the type $y = f(x)$. The graph of f is one plane set having the following form:

$$G_f = \{(x, f(x)) : x \in [0, 1]\}$$

2. Aspects of the Method

2.1. *Local aspect of the method*

We analyse the two cases of lengths of a curve:

a) The curve is of finite length
In the case of finite length, the function $z(t)$ is derivable almost everywhere and there exists thus, almost everywhere, an instantaneous speed $v(t)$ having the form:

$$v(t) = \sqrt{1 + z'(t)^2}$$

For the calculation of derivative of z we measure the slope:

$$z'(t) = \lim_{\tau \to 0} \frac{z(t+\tau) - z(t)}{\tau}$$

The length of the curve is defined by:

$$L(\Gamma) = \int_a^b 1 + z'(t)^2 \, dt$$

The larger the oscillation of z in t is, the more the speed of z in t is large.
One considers the curve of Figure 1 and one looks at vibrations around a point. In each point there is the need for making spectral analysis by using the transform of Fourier in the field of frequency.

b) The curve is of infinite length (fractal)
The curve Γ is fractal if:

$$\frac{\mathrm{osc}_\tau(t)}{\tau} \xrightarrow[\tau \to 0]{} +\infty$$

uniformly compared to t.
The curves of infinite length can be classified in two families: those which are locally rectifiable, such spirals and those which are not rectifiable nowhere, whose any under-arc is of infinite length and who can be then called fractals. If the curve is of infinite length $z'(t)$ does not have sense anymore, the derivative concept disappears and

one can only measure local oscillations (around a point). The study of the oscillations in any point of
[a, b] requires the study of the cases where:

$$\begin{cases} t-\tau<a \\ t+\tau>b \end{cases}$$

z is henceforth defined on the interval:

$$z:[a-\tau_0,b+\tau_0]\,,\tau_0>0$$

2.2. *Global aspect of the method*

The Variation Method is essentially deducted from the method of Minkowski.[2] Indeed we always leave a covering of the initial curve. Here, it means making a covering by segments of equal length at given t with osc_σ. What thus interests us it is the sum of the oscillations on a curve Γ .

$$\mathrm{VAR}_\tau=\int_a^b \mathrm{osc}_\tau\,(t)\,dt$$

The variation is going to correspond to an area, similar to a sausage. We deduct the fractal dimension of this calculation of τ variation by:

$$\Delta\left(\Gamma\right)=\lim_{\tau\to 0}\left(2-\frac{\log \mathrm{VAR}\,\tau\,(z)}{\log(\tau)}\right)$$

2.3. *Limits of the method*

The problem major of the method is the sensitivity in the extremes values.[3–5]

3. Application of the Variation Method for the Fracture Surfaces

The assumption of continuity is essential for the study of the fracture profile through the Variation Method. The oscillation is the difference between the largest and the smallest value of z, which represents the local roughness Rt.

The τ oscillation of the two-dimensional function f in x for a surface is defined by:

$$f:\ [a,b]\times[c,d]\to R$$

$$\text{osc}_\tau\,(f,x,y) = \left| \max_{|x-t|<\tau,|y-t|<\tau} f(t) - \min_{|x-t|<\tau,|y-t|<\tau} f(t) \right|$$

By taking the average of the oscillations $\text{osc}_\tau\,(f,x,y)$ on the surface $[a,b]\,x\,[c,d]$ one obtains:

$$\text{VAR}_\tau\,(f,a,b,c,d) = \frac{1}{(b-a)}\frac{1}{(d-c)}\int_a^b\int_c^d \text{osc}_\tau\,(f,x,y)\,dxdy$$

The fractal dimension of a continuous surface is done by:

$$\Delta\,(f,a,b,c,d) = \lim_{\tau\to 0}\left(3 - \frac{\log \text{VAR}\,\tau\,(f,a,b,c,d)}{\log\tau}\right)$$

Observation: the measurement technique for this method is practically independent from a change of amplification.

4. Conclusions

In this paper we discussed the local and the global aspect of the Variation Method for a curve continuous on its domain of definition. An essential condition of the technique is that it can be applied only for self-affine graphs.

We had also studied this algorithm for the fracture profile of a continuous surface. In this case the measurement technique for the Variation Method is practically independent from a change of amplification.

References

1. C. Tricot, *Courbes et dimension fractale* (Springer-Verlag, Berlin, 1999).
2. A. Le Méhauté, *Les géométries fractales* (Hermès, Paris, 1990).
3. B. Dubuc, J. F. Quiniou, C. Tricot and S. W. Zucker, *Physical Review A.* **39**(3), 1500 (1989).
4. B. Dubuc and S. Dubuc, *Siam J. Numer. Analys.* **33**, p. 602 (1996).
5. E. Charkaluk, M. Bigerelle and A. Iost, *Engineering Fracture Mechanics* **61**, 119 (1998).

Study on Bifurcation, Chaos and Chaotic Control of Vehicle Suspension with Nonlinearities under Road Excitation

Y. Sheng*,‡, G.-Q. Wu* and X.-J. Meng*,†

*School of Automobile, Tongji University, 201804, Shanghai, P. R. China
†School of Traffic and Automobile Engineering
Shandong University of Technology, 255049, Zibo, P. R. China
Email: ‡yun.sheng_tj@yahoo.com.cn

In this paper, the nonlinear suspension mathematical model obtained from the actually measured data of a car is applied to simulate the nonlinear spring force and damping force of vehicle suspension system. A dynamic equation of quarter suspension system with one d.o.f is derived. The theories of nonlinear dynamics are applied to study the nonlinear model and to reveal its nonlinear vibration characteristics. The bifurcation behavior is analyzed by using central manifold theorem. The response under road stochastic excitation is obtained by computer simulation. Through the methods of phase trajectory, Poincaré map, time history, power spectrum diagram and Lyapunov exponent, it is revealed that the chaos occurred in the nonlinear suspension system. Furthermore, the speed feedback controller is applied to control the chaos motion of the suspension system. This method is proved to be valid and feasible by numerical simulation.

Keywords: Plasma-surface interaction; impurities; plasma sheath; conservation of iron objects; hydrogen plasma.

1. Introduction

Suspension system is one of the most important parts of the vehicle. The dynamic performance of suspension system directly influences the ride and handling stability of the driving vehicle. Therefore, it is quite necessary to design finer suspension systems in order to improve the quality of vehicles.

Since the disturbance from the road may induce uncomfortable shake and noise in the vehicle body, it is important to study the vibrations of the vehicle. Many studies have been carried out on the dynamic response and the vibration control with linear mechanical model, and various control strategy of active and semi-active suspension, such as optimal control, fuzzy control etc, were advanced based on these researches. However, an automobile is a nonlinear system in practical terms because it consists of suspensions, tires and other components that have nonlinear properties. Therefore, the chaotic response may appear as the vehicle moves over a bumpy road.

Much progress has been made in nonlinear suspension modeling and dynamic characteristic research studying by now.[1–3] Some experts brought forward control strategies of chaos phenomenon and the application of chaos.[4,5] There are also some studies about the chaotic vibration of vehicle suspension system with hysteretic nonlinearity.[6–8] However, the nonlinear spring and damping force of suspension are mainly modeled by the three cubed function of displacement and velocity, and the excitations are mostly sine excitation and multi-frequency excitation in existing studies. The researches on the motion of nonlinear suspension under actual road excitation are almost not involved by now.

In this paper, the nonlinear suspension mathematical model obtained from the actually measured data of a car is applied to simulate the nonlinear spring force and damping force of vehicle suspension system. A dynamic equation of quarter suspension system with one d.o.f is derived. The theory of nonlinear dynamics is applied to study the nonlinear model and to reveal its nonlinear vibration characteristics. The bifurcation behavior possibly existed is analyzed by using central manifold theorem. The response under road stochastic excitation is obtained by computer simulation. Through the ways of phase trace, Poincaré map, time history, power spectrum diagram and Lyapunov exponent, it is revealed that the chaos occurred in the nonlinear suspension system. Furthermore, the speed feedback controller is applied to control the chaos motion of the suspension system.

2. Dynamic Modeling of Nonliner Suspension

The simplified model of nonlinear suspension system with one degree-of-freedom is shown in Figure 1. In Figure 1, K is the spring of nonlinear suspension, C is the nonlinear damper, z is the displacement of the body, and z_0 is the road excitation.

By applying Newton's second law, the differential equation of the heave motion of the vehicle body can be expressed as

$$M\ddot{z} + f_k + f_c = 0 \tag{1}$$

where M is the body mass, f_k is the force of nonlinear spring, f_c is the force of nonlinear damping.

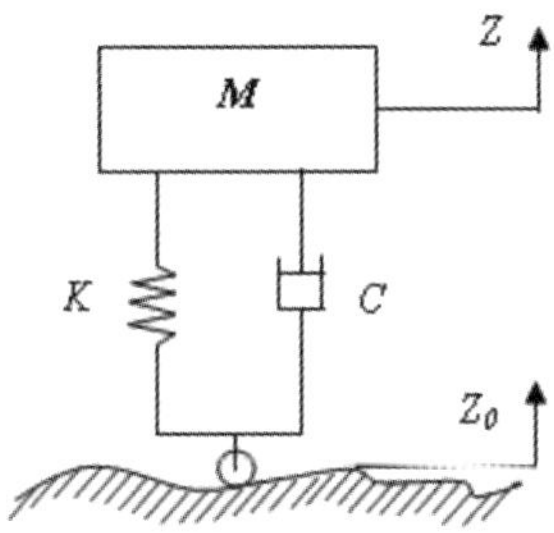

Fig. 1. Quarter vehicle model with one d.o.f.

The nonlinear spring force and damping force were computed through the SPMD (Suspension Parameter Measurement Device) test data as follows:[9]

$$f_s = k_0 + k_1\Delta z + k_2\Delta z^2 + k_3\Delta z^3 \tag{2}$$

$$f_d = c_1\Delta\dot{z} + c_2\Delta\dot{z}^2 \tag{3}$$

where $\Delta z = z - z_0$, $\Delta\dot{z}$ is the relative velocity. The unit of Δz is m and the unit of $\Delta\dot{z}$ is m/s, $k_0 = -2316.4N$, $k_1 = 12394N/m$, $k_2 = -73696N/m^2$, $k_3 = 3170400N/m^3$, $c_1 = 1385.4Ns/m$, $c_2 = 524.28Ns^2/m^2$.

Substituting (2) and (3) into (1), system (1) becomes

$$M\ddot{z}+k_0+k_1(z-z_0)+k_2(z-z_0)^2+k_3(z-z_0)^3+c_1(\dot{z}-\dot{z}_0)+c_2(\dot{z}-\dot{z}_0)^2 = 0 \tag{4}$$

Let $x = z - z_0$, then the equation (4) can be simplified as

$$\ddot{x} = -\ddot{z}_0 - B_0 - B_1x - B_2x^2 - B_3x^3 - C_1\dot{x} - C_2\dot{x}^2 \tag{5}$$

where $B_0 = k_0/M$, $B_1 = k_1/M$, $B_2 = k_2/M$, $B_3 = k_3/M$, $C_1 = c_1/M$, $C_2 = c_2/M$.

3. Bifurcation of Nonlinear Suspension

In this chapter, the center manifold theorem[10–12] is been used to analysis the bifurcation of nonlinear suspension.

Let $\dot{x} = y$, then (5) can be written as

$$\begin{cases} \dot{x} = y \\ \dot{y} = -\ddot{z}_0 - B_0 - B_1x - B_2x^2 - B_3x^3 - C_1y - C_2y^2 \end{cases} \tag{6}$$

System (6) is a nonlinear non-autonomous case. It is known that the nonlinear system is always analyzed by using the equilibrium of autonomous case in most dynamic theories. However, for non-autonomous case, the general method is to seek its average equations without variable t in a period, and analyze the system. But some characteristics of the system might be ignored while using this method. In this paper, we adopt another method to analyze the nonlinear non-autonomous case. This is done as follows.

Write (6) as a matrix form

$$\begin{bmatrix} \dot{x} \\ \dot{y} \end{bmatrix} = \begin{bmatrix} 0 & 1 \\ -B_1 & -C_1 \end{bmatrix} \begin{bmatrix} x \\ y \end{bmatrix} + \begin{bmatrix} 0 \\ -\ddot{z}_0 - B_0 - B_2 x^2 - B_3 x^3 - C_2 y^2 \end{bmatrix} \tag{7}$$

The "equilibrium aggregate" (in the extend sense) is $\{(b, 0)\}$, where b is a real number which satisfies the balanceable equation

$$-\ddot{z}_0 - B_0 - B_1 b - B_2 b^2 - B_3 b^3 = 0 \tag{8}$$

Because b is decided by t, regards b as a variable parameter. Make the coordinate transform

$$\begin{cases} u = x - b \\ v = y \end{cases} \tag{9}$$

Apparently, the equilibrium of system (7) is (0,0). Notice (8), then system (7) becomes

$$\begin{aligned} \begin{bmatrix} \dot{u} \\ \dot{v} \end{bmatrix} &= \begin{bmatrix} 0 & 1 \\ -B_1 - 2B_2 b - 3B_3 b^2 & -C_1 \end{bmatrix} \begin{bmatrix} u \\ v \end{bmatrix} \\ &\quad + \begin{bmatrix} 0 \\ -(B_2 + 3B_3 b)u^2 - B_3 u^3 - C_2 v^2 \end{bmatrix} \end{aligned} \tag{10}$$

where $b = b(t)$ is a variable parameter. (10) has already become an autonomous case. Let

$$f(u, v, t) = -(B_2 + 3B_3 b)u^2 - B_3 u^3 - C_2 v^2 \tag{11}$$

For the computation of the center manifold, the first step is to separate the linear part of the center manifold from the linear part of the stable manifold. The linear system is separated from the system (10) as

$$\begin{bmatrix} \dot{u} \\ \dot{v} \end{bmatrix} = \begin{bmatrix} 0 & 1 \\ -B_1 - 2B_2 b - 3B_3 b^2 & -C_1 \end{bmatrix} \begin{bmatrix} u \\ v \end{bmatrix} \tag{12}$$

Its eigenvalues can be obtained by (12)

$$|\lambda E - A| = \begin{vmatrix} \lambda & -1 \\ B_1 + 2B_2 b + 3B_3 b^2 & \lambda + C_1 \end{vmatrix} = \lambda^2 + C_1 \lambda + B_1 + 2B_2 b + 3B_3 b^2 = 0 \tag{13}$$

By solving (13) we can get

$$\lambda_{1,2} = \frac{-C_1 \pm \sqrt{C_1^2 - 4(B_1 + 2B_2 b + 3B_3 b^2)}}{2}$$

Since the single term of (13) $C_1 \neq 0$, system (12) can't have two zero eigenvalues obviously. If $B_1 + 2B_2 b + 3B_3 b^2 = 0$, system (12) only has one zero eigenvalue, that is $\lambda_1 = 0$. Then another eigenvalue is $\lambda_2 = -C_1$. From (13) we can get the eigenvector $(1, 0)T$ corresponding to $\lambda_1 = 0$ and the other eigenvector $(1, -C_1)T$ corresponding to $\lambda_2 = -C_1$. Therefore the transform matrix is

$$T = \begin{bmatrix} 1 & 1 \\ 0 & -C_1 \end{bmatrix}, \quad T^{-1} = \begin{bmatrix} 1 & \frac{1}{C_1} \\ 0 & -\frac{1}{C_1} \end{bmatrix} \tag{14}$$

Make linear transform by using matrix T.

$$\begin{bmatrix} u \\ v \end{bmatrix} = T \begin{bmatrix} w \\ z \end{bmatrix} \tag{15}$$

Then (10) becomes

$$\begin{bmatrix} \dot{w} \\ \dot{z} \end{bmatrix} = \begin{bmatrix} \lambda_1 & 0 \\ 0 & \lambda_2 \end{bmatrix} \begin{bmatrix} w \\ z \end{bmatrix} + \begin{bmatrix} \frac{1}{C_1} f(w, z, t) \\ -\frac{1}{C_1} f(w, z, t) \end{bmatrix} \tag{16}$$

Where $f(w, z, t)$ is obtained by $f(u, v, t)$ through linear transform (14)

$$f(w, z, t) = -B_3(w + z)^3 - (B_2 + 3B_3 b)(w + z)^2 - C_1^2 C_2 z^2 \tag{17}$$

By using the center manifold theorem mentioned before, we get the dynamic equation of system (17) on the center manifold.

$$\begin{aligned} \dot{w} = \lambda_1 w - \tfrac{B_2+3B_3 b}{C_1} w^2 - \tfrac{2(B_2+3B_3 b)^2 + B_3 C_1^2}{C_1^3} w^3 - \\ \tfrac{9(B_2+3B_3 b)^3 + (5B_3 + B_2 C_2 + 3B_3 b C_2)(B_2+3B_3 b) C_1^2}{C_1^5} w^4 + \cdots \end{aligned} \tag{18}$$

Omit those with three cubed terms and those exceeding three cubed terms, (18) becomes

$$\dot{w} = \lambda_1 w - \frac{B_2 + 3B_3 b}{C_1} w^2 - \frac{2(B_2 + 3B_3 b)^2 + B_3 C_1^2}{C_1^3} w^3 \tag{19}$$

For the bifurcation condition $B_1 + 2B_2 b + 3B_3 b^2 = 0$, the condition of real roots is $\Delta = B_2^2 - 3B_3 B_1 > 0$. Then $b \neq -B_2/3B_3$, thus $B_2 + 3B_3 b \neq 0$ is tenable. The second term of (19) won't be equal to zero. The concrete bifurcation behavior is analyzed as follows.

(1) While $2(B_2 + 3B_3b)^2 + B_3C_1^2 = 0$, then (19) becomes

$$\dot{w} = \lambda_1 w - \frac{B_2 + 3B_3 b}{C_1} w^2 \quad (20)$$

The transcritical bifurcation phenomenon happens.

(2) While $2(B_2 + 3B_3b)^2 + B_3C_1^2 \neq 0$, then the pitchfork bifurcation phenomenon happens.

4. Chaos under Road Stochastic Excitation

4.1. *Modeling stochastic road*

Speed spectrum of the white noise considering the effect of speed and its corresponding expression in time field are always used to model the continuous stochastic road. The statistical characteristic of road surface irregularity which is qua the input of vehicle vibration is mainly described by road power spectrum density. In this paper, we use speed power spectrum density to build the stochastic road model. Based on [10], the speed power spectrum density of road surface irregularity in time-frequency field is

$$G_{\dot{q}}(f) = (2\pi f)^2 G_{\dot{q}}(n_0) n_0^2 u \quad (21)$$

Where f is time frequency (Hz), n_0 is reference space frequency $(n_0 = 0.1m^{-1})$, u is vehicle speed (m/s), $G_{\dot{q}}(n_0)$ is road surface irregularity coefficient (m_3),

Let vehicle speed be constant, and then the spectrum density (21) is a constant too, it can be written as $C = (2\pi n_0)^2 G_{\dot{q}}(n_0)u$. Road surface profile can be generated by the white noise whose power spectrum density is $\sqrt{C}$ through an integrator. It is described as

$$z(t) = \sqrt{C} \int_0 w(t)dt \quad (22)$$

4.2. *Numerical simulation*

Some regular methods, including phase trajectory, Poincaré map, time-history curve, power spectrum diagram and Lyapunov exponents, are selected to study whether there exist chaos phenomena in quarter suspension system with one d.o.f.[4,5,11,12]

Owing to the non-linearity of the differential equation (5), the dynamic responses of the vehicle model are studied numerically with the fourth order Runge-Kutta algorithm provided by MATLAB. In the computation, the absolute error tolerance is less than 10^{-6}. Since numerical integration can give spurious results with regard to the existence of chaos due to insufficiently small time steps,[12] the step size is verified to ensure no such results are generated as a result of time discretization. While computing, C grade road surface is selected, and vehicle speed is set to $u = 15m/s$. The initial conditions are set as $y = 0$ and $\dot{y} = -0.01$. Simulation results can be seen from Figures 2 (left and right) and Figures 3 (left and right).

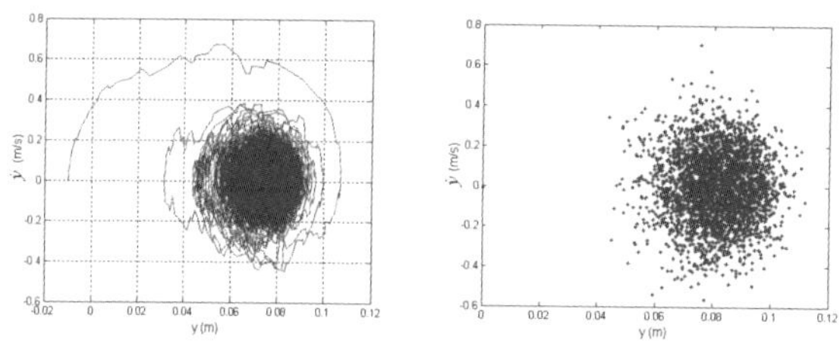

Fig. 2. Phase track (left image) and Poincaré map (right image)

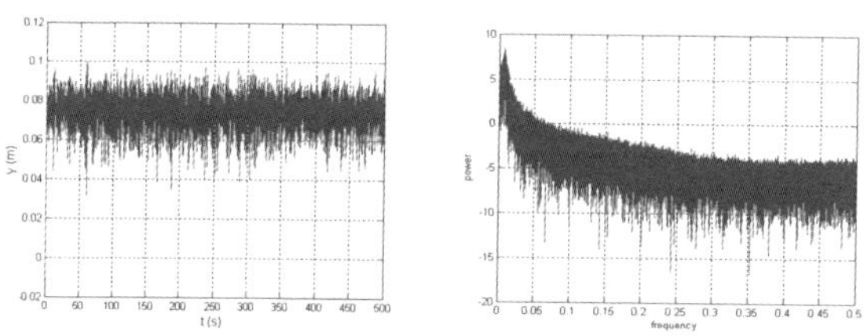

Fig. 3. Time history curve (left image), Power spectrum density (right image)

As is shown in Figure 2 (left), the phase track under road stochastic excitation is disorderly and unsystematic, and it doesn't repeat. From Figure 2 (right) we observe that there are a series of serried points in the Poincaré map. While doing the simulation, we find that the output is very sensitive to the initial conditions. And from Figure 3 (right) we can see the power spectrum density is continuous. Finally, we compute the largest Lyapunov exponent of the system based on,[13] the result is 0.2084, it's positive. Based on these results, we conclude that chaos phenomenon exists in the nonlinear suspension system.

5. Chaotic Control of Nonlinear Suspension

Chaos control, in a broader sense, can be divided into two categories: one is to suppress the chaotic dynamical behavior and the other is to generate or enhance chaos in nonlinear systems. Nowadays, different techniques and methods have been proposed to achieve chaos control. For instance, OGY method, differential geometric method, linear state space feedback, inverse optimal control and output feedback control, among many others.

In this paper, the speed feedback control strategy is used to control the chaos motion of nonlinear suspension system.[14] It is a kind of active controller. The speed feedback controller is described as follows.

Suppose the following autonomic chaos system:

$$\begin{cases} \dot{x} = F(x(t), t) \\ y = Dx \end{cases} \tag{23}$$

where $x = [x_1, x_2, \cdots, x_n]^T$ is the state variant of the system, $D = (d_{ij})_{n \times n}$ is constant matrix, y is the output of the system, $F(x(t), t)$ is a nonlinear function.

Let the speed feedback control

$$u = K\dot{x} \tag{24}$$

where K is the feedback plus matrix.

Add this controller to the right side of the system (23), then:

$$\dot{x} = F(x(t), t) - u \tag{25}$$

The speed feedback control of nonlinear suspension system is simulated by Matlab/Simulink. The results are shown in Figures 4 (left and right) and Figure 5. It is shown in Figure 4 (left) that the phase track is still disorderly and unsystematic, and it doesn't repeat. From Figure 4 (right) we observe that there are a series of scattered points in the Poincaré map. The largest Lyapunov exponent is calculated through the Wolf method,[13] the result is -0.0487 and it's negative. Based on these results we conclude that the motion of nonlinear suspension system is stochastic. At the same time, the vibration acceleration and relative velocity of the body is restrained too.

6. Conclusions

By using the nonlinear spring and damping model obtained from the actual test data of a car, the chaotic response and bifurcation of a nonlinear vehicle model with one d.o.f which is subjected to stochastic road excitation

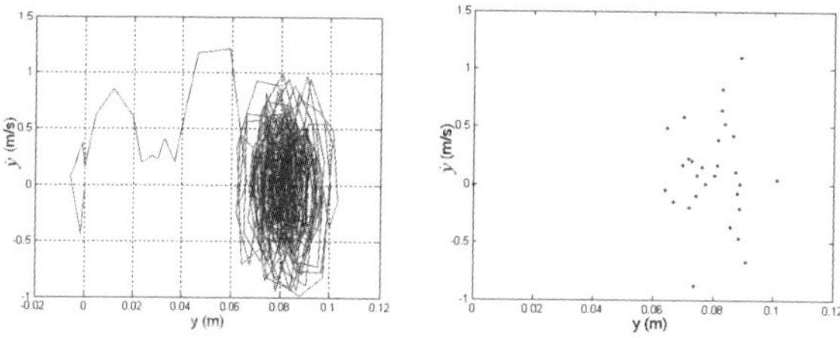

(a) (b)

Fig. 4. Phase Track (left image), Poincaré map (right image)

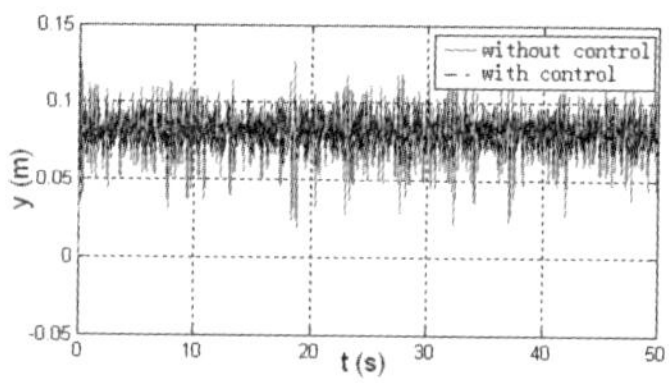

Fig. 5. Time history curve

are studied through numerical simulation. It is found that the chaotic response exists in nonlinear suspension. The theory of nonlinear dynamics is applied to study the nonlinear model and to reveal its vibration characteristics with nonlinear properties. The bifurcation phenomenon is analyzed by using center manifold theorem. Through the ways of phase trace, Poincaré map, time history, power spectrum diagram and Lyapunov exponent, it is revealed that the chaos occurred in the nonlinear suspension system. Furthermore, the speed feedback controller is applied to control the chaos motion of the suspension system. Although the mechanical model of the vehicle is only a simplified one, the results may still be useful in dynamic design of the ground vehicle. The confirmation of the existence of the chaos in this kind of model by experiment is left for further study.

Acknowledgements

The project was supported by Hi-Tech Research and Development Program (863 Program) of China, Support No: 2007AA04Z132

References

1. K. Rebrouck, *SAE paper* **940869**, 1170 (1994).

2. F. H. Besinger, D. Cebon and D. J. Coje, *Vehicle System Dynamics* **24**, 35 (1995).
3. Z. Heping, *Journal of Mechanical Strength* **23**(2), 165 (2001).
4. M. D. Barrett, *Phys. D* **91**(4), 340 (1996).
5. L. Yunjia, *Study Method and Practice of Chaos Vibration* (Tsinghua University Press, 1997).
6. L. Shaohua and Y. Shaopu, *Journal of Vibration, Measurement & Diagnosis* **23**(2), 86 (2003).
7. Y. Shaopu, L. Shaohua and G. Wenwu, *Journal of Vibration, Measurement & Diagnosis* **25**(1), 22 (2005).
8. Q. Zhu and M. Ishitobi, *Journal of Sound and Vibration* **275**, 1136 (2004).
9. C. Kim and P. I. Ro, *Proceedings of the Institution of Mechanical Engineers* **212**, 79 (1998).
10. Y. Zhi-sheng, *Dynamical theory of motor vehicle*, 3 edn. (Mechanical Engineering Press, Beijin, 2003).
11. Z. Qi-chang, W. Hong-li and Z. Zhi-wen, *Theories of Bifurcation and Chaos and Applications* (Tianjin University Press, Tianjin, 2005).
12. B. H. Tongue, *American Society of Mechanical Engineers Journal of Applied Mechanics* **54**, 695 (1987).
13. A. Wolf, J. B. Swift, H. L. Swinney and J. A. Vastano, *Phys. D* **16**, 285 (1985).
14. T. Chaohai, Y. Chunde and L. Yan, *Chaos, Solitons & Fractals* **23**(1), 259 (2005).

Von Karman Streets Chaotic Simulation

C. H. Skiadas

Technical University of Crete, Chania, Crete, Greece
Email: skiadas@asmda.net

In this paper we model and simulate the so-called Von Karman Streets, a characteristic form of turbulent (vortex) flow that appears in small or large systems. Very interesting examples are the large turbulent formations in the sea or in the clouds that have been viewed and photographed from the space (from satellites). The modeling approach is based on the Rotation-Reflection-Translation theory developed in the recent book [1]. The related theory was already applied in various fields and especially in flows.

Keywords: Von Karman Streets; chaotic modeling; chaotic simulation; vortex flow; difference equations; rotation-translation; reflection-translation.

1. Introduction

The famous Von Karman chaotic phenomena are presented in several formations, especially in the sea and are illustrated by photographs taken from space (satellite views).

The Von Karman phenomenon appears when a fluid flowing past an obstacle that appears in the form of a cylinder (A vortex street develops behind a cylinder moving through fluid) or an island in the ocean. In the later case the flow (fluid) lines or trajectories pass from both sides of the island. To simplify the modeling we observe that a mirror image of the flow appears in the other side of the island or a reflection like process appears.

2. Chaotic Modeling

The chaotic theory from its first application of the famous Lorenz[2] model gave rise to a new modeling approach that is the search for the main characteristics of the phenomenon thus neglecting the secondary characteristics. This was applied by Lorenz who showed that a fluid flow model could be modeled by only 3 basic equations instead of the numerous and more complicated equations that where suggested from the fluid flow theory and the Navier-Stokes equations. Summarising the related achievements of several chaotic modeling applications we may observe that the modeling of a chaotic system could result in good approach to the real situations if:

a) The main chaotic-nonlinear characteristic part of the process is retained
b) The number of equations expressing the phenomenon is reduced to a minimum.
c) The "chaotic parameters" selected clearly represent the main nonlinear characteristics of the system.

Another modeling issue is the selection of the form of the chaotic system of equations by means of the selection between a set of differential and a set of difference equations. The differential equation theory is much more studied in details the last centuries instead of the difference theory approach.

However, the difference theory approach looks more essential to model chaotic systems especially from the simulation point of view.[1,3–5] Usually to simulate a set of difference equations we need only a simple "loop" in a computer program to model the iterative process. Instead the differential equations simulation needs more laborious methods.

In our case of modeling the Von Karman Streets we observe that the flow is around a solid form like a cylinder represented by an island in the ocean (The Island is assumed to have a circular form). We assume that a fluid (air or water) flow is directed from left to the right of the coordinates, the island located at the centre of coordinates at $(0, 0)$. The flow (fluid) lines or trajectories past both sides of the island. To simplify the modeling we observe that a mirror image of the flow appears in the other side of the island or a reflection like process appears provided that a symmetric island form is present (In real situations there is not a perfect symmetry present, but a non-perfect mirror image may also provide a chaotic process). To model the phenomenon in the (x, y) plane we observe that:

(1) A fluid is flowing from left to the right with a speed represented by a parameter a. This parameter in geometric notation is noted as the "translation" parameter. The translation parameter is added to keep the direction of the flow. In modeling the process the direction of flow is parallel to the x axis.
(2) When the fluid lines past the island move following the circular form of the obstacle "rotating". Behind the island the rotating lines are joint together having opposite rotation directions thus forming a "reflection" phenomenon. We thus need a rotation equation set.

(3) In some instances the space contraction or expansion is present thus modeled by a space contraction parameter b.

The phenomenon can be modeled by using a reflection-translation equation set that we have introduced in [1]. The system has the form:

$$\begin{aligned} x_{t+1} &= a + b(x_t cos\theta_t + y_t sin\theta_t) \\ y_{t+1} &= b(x_t sin\theta_t - y_t cos\theta_t) \end{aligned} \tag{1}$$

where a is the translation parameter and b is the space contraction ($b < 1$) or expansion ($b > 1$) parameter. The rotation angle selected is of the form: $\theta_t = d/r^k$, d, k are parameters and $k > 0$ and $r = \sqrt{x^2 + y^2}$.

The rotation angle for the reflection-translation scheme must assure that the influence of the obstacle (island) is negligible in long distances, thus r should be in the denominator and the parameter k should be large. The parameter d accounts for the reflection and can be taken as 2π or π.

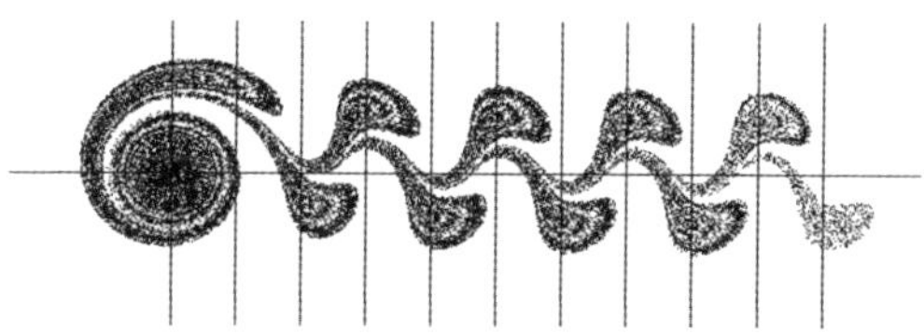

Fig. 1. Von Karman streets simulation (Parameters: $b = 1, d = 2\pi, a = 1, k = 3, t = 10$)

3. Chaotic Simulation

The simulation starts from an original circle of rotating particles with $R = 1$ at $(0, 0)$. The parameters selected are $b = 1, d = 2\pi, a = 1, k = 3, t = 10$. The resulting figure after $t = n = 10 iterations$ is illustrated in figure 1.

Behind the island the flow follows by forming chaotic forms or vortices. The distance between vortices (the frequency of vortex shedding) is equal to a.

Another point is to simulate the reflection-translation case but following a space contraction by means of a space contraction parameter < 1. For the simulation the parameter values of the previous case are retained except for the parameter $k = 4$ and for the space contraction parameter ($b = 0.9$). The resulting form is illustrated in Figure 2. The resulting chaotic forms slowly disappear under the space contraction process.

Fig. 2. Von Karman streets simulation (Parameters: $b = 0.9, d = 2\pi, a = 1, k = 4, t = 10$)

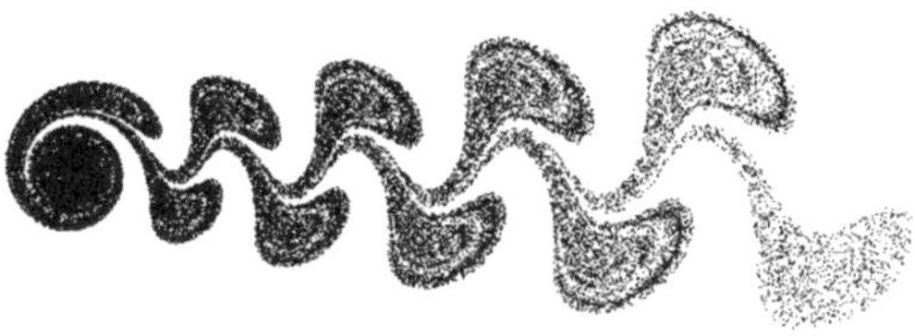

Fig. 3. Von Karman streets simulation (Parameters: $b = 1.1, d = 2\pi, a = 1, k = 3, t = 10$)

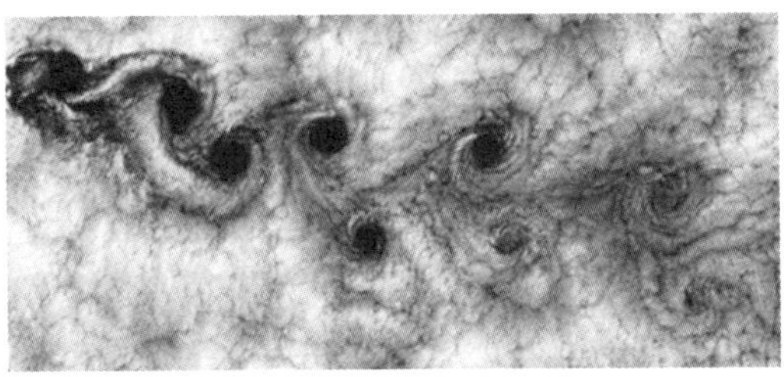

Fig. 4. Von Karman Vortex Street off the Chilean coast near the Juan Fernandez Islands.

The space expansion process ($b > 1$) is illustrated in Figure 3. The parameter values selected are as in the first case except for the space expansion parameter that is $b = 1.1$.

An example that follows the later space expansion case is illustrated in Figure 4. This is Landsat 7 image of clouds off the Chilean coast near the Juan Fernandez Islands (also known as the Robinson Crusoe Islands) on September 15, 1999. It is one of the Classical Von Karman Vortex Streets pictures.

(For more details on von Karman vortices, refer to http://climate.gsfc.nasa.gov/ cahalan, Image and caption courtesy Bob Cahalan, NASA GSFC).

According to Cahalan:[6]

> The island is about 1.5 km in diameter, and rises 1.6 km into a layer of marine stratocumulus clouds. This type of cloud is important for its strong cooling of the Earth's surface, partially counteracting the Greenhouse warming. An extended, steady equatorward wind creates vortices with clockwise flow off the eastern edge and counterclockwise flow off the western edge of the island. The vortices grow as they advect hundreds of kilometers downwind, making a street 10,000 times longer than those made in the laboratory.

The simulation is quite representative of the phenomenon, although it is only a 2-dimentional modeling of a 3-dimentional case. The introduction of a third equation expressing the vertical movements to the z direction would improve the simulation results.

4. Conclusions

In this paper we model and simulate the so-called Von Karman Streets by using difference equations approach modelling. The modeling approach based on the Rotation-Reflection-Translation theory was applied in the so-called Von Karman vortex streets with very promising results.

References

1. C. H. Skiadas and C. Skiadas, *Chaotic Modelling and Simulation: Analysis of Chaotic Models, Attractors and Forms* (Taylor & Francis / CRC Press, New York, 2008).
2. E. N. Lorenz, *J. Atmos. Sci.* **20**, 130 (1963).
3. C. H. Skiadas, *ArXiv.org* **arXiv:nlin/0611036v1**, 1 (2006).
4. C. H. Skiadas and C. Skiadas, *ArXiv.org* **arXiv:nlin/0701012v1**, 1 (2007).
5. C. H. Skiadas, Exploring and simulating chaotic advection: A difference equations approach, in *Recent Advances in Stochastic Modeling and Data Analysis*, ed. C. Skiadas (World Scientific, Singapore, 2007).
6. B. Cahalan, *NASA GSFC* `http://climate.gsfc.nasa.gov/cahalan` (1999).

Quantum Chaos: Degree of Reversibility of Quantum Dynamics of Classically Chaotic Systems

V. V. Sokolov* and O. V. Zhirov

Budker Institute of Nuclear Physics, Novosibirsk, 630090, Russia; CNISM, CNR-INFM, and Center for Nonlinear and Complex Systems, Università degli Studi dell'Insubria, Via Valleggio 11, Como, 22100, Italy
*Email: *V.V.Sokolov@inp.nsk.su*

G. Benenti

CNISM, CNR-INFM, and Center for Nonlinear and Complex Systems, Università degli Studi dell'Insubria, Via Valleggio 11, Como, 22100; Istituto Nazionale di Fisica Nucleare, Sezione di Milano, Via Celoria 16, Milano, 20133, Italy
Email: Giuliano.Benenti@uninsubria.it

G. Casati

CNISM, CNR-INFM, and Center for Nonlinear and Complex Systems, Università degli Studi dell'Insubria, Via Valleggio 11, Como, 22100; Istituto Nazionale di Fisica Nucleare, Sezione di Milano, Via Celoria 16, Milano, 20133, Italy
and
Department of Physics, National University of Singapore Singapore 117542, Republic of Singapore
Email: Giulio.Casati@uninsubria.it

We present a quantitative analysis of the reversibility properties of classically chaotic quantum motion by relating the degree of reversibility to the rate at which a quantum state acquires a more and more complicated structure during its time evolution. This complexity can be characterized by the number $\mathcal{M}(t)$ of harmonics of the (initially isotropic, i.e. $\mathcal{M}(0) = 0$) Wigner function, which are generated by the time t. We show that, in contrast to the classical exponential increase, this number can grow after the Ehrenfest time t_E not faster than linearly. It follows that if the motion is reversed at some arbitrary moment T immediately after applying an instant perturbation with intensity described by the parameter ξ, then there exists a critical perturbation strength, $\xi_c \approx \sqrt{2}/\mathcal{M}(T)$, below which the initial state is well recovered, whereas reversibility disappears when $\xi \gtrsim \xi_c(T)$. In the classical limit the number of harmonics proliferates exponentially with time and the motion becomes practically irreversible. The above results are illustrated in the example of the kicked quartic oscillator model.

Keywords: Quantum chaos; instability; complexity; reversibility.

1. Introduction

Extreme sensitivity of the phase space trajectories to initial conditions and system's parameters, which is characterized by positive Lyapunov exponent or, in more general terms, positive *algorithmic complexity*, is the very essence of the classical dynamical chaos. Such an exponential instability of classical motion results in extraordinary complexity of random and unpredictable classical trajectories. In computer simulations of classical motion this leads to rapid loss of memory and practical irreversibility even though the exact equations of motion are reversible. Any, however small, imprecision such as computer round-off errors, is magnified by the exponential instability of the classical orbits to the extent that the memory of the initial conditions is effaced. In other words, due to this instability smaller and smaller scales of the classical phase space are explored during evolution of a phase space distribution exponentially fast with time. These fine details are lost due to the finite accuracy of numerical simulations, and therefore the reversibility of the evolution is destroyed.

In contrast, almost exact reversion is observed in numerical simulations of the quantum motion of classically chaotic systems, even in the regime in which statistical phenomena such as deterministic diffusion take place.[1,2] Qualitatively, this crucial difference is explained by a much simpler structure of quantum states as compared to the complexity of random classical trajectories.[3,4] Unfortunately, the concept of complexity formulated in terms of the exponential instability of the phase trajectories cannot be immediately transferred to quantum mechanics, where the very notion of trajectory is irrelevant and there is no quantum analogue to the Lyapunov exponent. At first glance, there is no means to measure comparative complexity of classical and quantum states of motion.[5]

However, individual classical trajectories are, in essence, of minor interest if the motion is chaotic. They all are alike in this case and rather behavior of manifolds of them carries really valuable information. Therefore the methods of the phase space and the Liouville form of the classical mechanics become the most adequate. It is very important that, opposite to the classical trajectories, the classical phase space distribution and the Liouville equation have direct quantum analogs. Hence, a comparison between classical and quantum dynamics can be made by studying the evolutions in time of the classical and quantum phase space distributions expressed in similar canonical action I and angle θ variables and both ruled by linear equations.

The exponentially fast structuring of the system's phase space on finer and finer scales, which is the paramount property of the classical chaotic dynamics, is restricted in quantum case by the quantization of the phase space. This makes quantal phase-space quasi-distribution, e.g. the quantum Wigner function, "simpler" in comparison with its classical counterpart. In particular, while in the case of chaotic motion the mean number $\mathcal{M}(t)$ of θ-Fourier components of the classical phase-space distribution grows exponentially in time, similar growth in the case of the quantum Wigner function turns out, as we will show below, to be impossible. We notice in this connection that this number also increases much slower (only power-like) if the classical motion is regular.

It is intuitive that the degree of reversibility of a motion should depend on the ratio with which complexity of the state grows. Nevertheless, a rigorous link between the intuitively expected different degrees of reversibility of quantum and classical motion and the structure developed by the phase-space distributions during dynamical evolution has never been established before. We aim to clarify this problem.

2. The Phase Space Approach

Let $\hat{H} \equiv H(\hat{a}^\dagger, \hat{a}; t) = H^{(0)}(\hat{n} = \hat{a}^\dagger \hat{a}) + H^{(1)}(\hat{a}^\dagger, \hat{a}; t)$ be the Hamiltonian of a generic nonlinear system with a bounded below discrete energy spectrum $E_n^{(0)} \geq 0$, which is driven by a time-dependent force of such a kind that the classical motion exhibits a transition from integrable to chaotic behavior when the strength of the driving force is increased. Here $\hat{a}^\dagger, \hat{a}$ are the bosonic, creation-annihilation operators: $[\hat{a}, \hat{a}^\dagger] = 1$. In our analytical study and numerical simulations we considered as an illustrative example the kicked quartic oscillator with Hamiltonian $\hat{H}^{(0)}(\hat{n}) = \hbar\omega_0\hat{n} + \hbar^2\,\hat{n}^2$ driven by periodic kicks $\hat{H}^{(1)}(\hat{a}^\dagger, \hat{a}) = -\sqrt{\hbar}\,g(t)(\hat{a}+\hat{a}^\dagger)$ where $g(t) = g_0\sum_s \delta(t-s)$. In our units, the time and parameters $\hbar, \omega_0$ as well as the strength of the driving force are dimensionless. The period of the driving force is set to one. The classical dynamics of such a non-linear oscillator becomes chaotic when the kicking strength g_0 exceeds a critical value $g_{0,c} \approx 1$. The angular phase correlations decay in this case exponentially and the mean action grows diffusively with the diffusion coefficient $D = g_0^2$.

We use the method of c-number α-phase space borrowed from the quantum optics (see for example [6–8]). It is, basically, built upon the basis of the coherent states $|\alpha\rangle = \hat{D}\left(\frac{\alpha}{\sqrt{\hbar}}\right)|0\rangle$ obtained from the ground state $|0\rangle$ of the unperturbed Hamiltonian with the help of the unitary displacement operator $\hat{D}(\lambda) = \exp(\lambda\,\hat{a}^\dagger - \lambda^*\hat{a})$. Here α is a complex phase space variable independent of the effective Planck's constant $\hbar$.

The Wigner function W in the α-phase plane is defined by the following Fourier transformation

$$W(\alpha^*, \alpha; t) = \frac{1}{\pi^2\hbar}\int d^2\eta\, e^{(\eta^*\frac{\alpha}{\sqrt{\hbar}} - \eta\frac{\alpha^*}{\sqrt{\hbar}})}\mathrm{Tr}\left[\hat{\rho}(t)\,\hat{D}(\eta)\right], \tag{1}$$

where $\hat{\rho}$ is the density operator and the integration runs over the complex η-plane. The Wigner function is normalized to unity, $\int d^2\alpha W(\alpha^*, \alpha; t) = 1$ and is real though, in general, not positive definite. It satisfies the evolution equation

$$i\frac{\partial}{\partial t}W(\alpha^*, \alpha; t) = \hat{\mathcal{L}}_q W(\alpha^*, \alpha; t), \tag{2}$$

with the Hermitian "quantum Liouville operator" $\hat{\mathcal{L}}_q$. This equation reduces in the case $\hbar = 0$ to the classical Liouville equation with respect to the canonical pair $\alpha, i\alpha^*$ with the classical Hamiltonian function being given by the diagonal matrix elements $H_c(\alpha^*, \alpha; t) = \langle\alpha|\hat{H}^{(N)}(\hat{a}^\dagger, \hat{a})|\alpha\rangle$ of the normal form $\hat{H}^{(N)}$ of the quantum Hamiltonian operator. In other words, this function is obtained from the quantum Hamiltonian by substituting $\hat{a} \to \alpha/\sqrt{\hbar}$, $\hat{a}^\dagger \to \alpha^*/\sqrt{\hbar}$.

We define the harmonic's amplitudes $W_m(I)$ as the Fourier components of the Wigner function with respect to the angle variable θ introduced by the canonical transformation $\alpha = \sqrt{I}\,e^{-i\theta}$. The normalization condition reduces then to $\int_0^\infty dI\, W_0(I; t) = 1$ whereas the amplitudes of other harmonics are expressed in terms of the matrix elements $\langle n+m|\hat{\rho}|n\rangle$ of the density operator along the mth collateral diagonal as

$$\begin{gathered} W_m(I; t) = \tfrac{2}{\hbar}\, e^{-\frac{2}{\hbar}I}\sum_{n=0}^{\infty}(-1)^n\sqrt{\tfrac{n!}{(n+m)!}}\times \\ (4I/\hbar)^{\frac{m}{2}}\, L_n^m\,(4I/\hbar)\,\langle n+m|\hat{\rho}(t)|n\rangle, \quad m \geq 0, \end{gathered} \tag{3}$$

with L_n^m Laguerre polynomials and $W_{-m} = W_m^*$. The inverted relation reads

$$\begin{gathered} \langle n+m\big|\hat{\rho}(t)\big|n\rangle = (-1)^n\, 2\sqrt{\tfrac{n!}{(n+m)!}}\times \\ \int_0^\infty dI\, e^{-2\frac{I}{\hbar}}\,(4I/\hbar)^{\frac{m}{2}}\, L_n^m\,(4I/\hbar)\; W_m\,(I; t)\,. \end{gathered} \tag{4}$$

These formulae allow us to freely translate all relations given below from the language of the density matrix into the language of the Wigner function and back.[9,10]

3. Reversibility and Peres Fidelity

Aiming to connect the reversibility of the motion with the complexity of the Wigner function, we follow the approach developed in [11]. We consider first the forward evolution $\hat{\rho}(t) = \hat{U}(t)\hat{\rho}(0)\hat{U}^\dagger(t)$ of a simple initial (generally mixed) state $\hat{\rho}(0)$ up to some time $t = T$. An instantaneous Hermitian perturbation $\xi\hat{V}\,\delta(t-T)$ with the intensity ξ is then applied which transforms the state $\hat{\rho}(T)$ into $\hat{\tilde{\rho}}(T,\xi) = \hat{P}(\xi)\hat{\rho}(T)\hat{P}^\dagger(\xi)$. The resulting transformation $\hat{P}(\xi) = e^{-i\xi\hat{V}}$ is unitary. For example this transformation is equivalent to the global rotation $W(I,\theta;T) \to W(I,\theta+\xi;T)$ by the angle ξ in the phase plane if the operator $\hat{V} = \hat{n}$. In particular, we use below an infinitesimal perturbation of such a kind to reveal complexity of the Wigner function at the instant T.

The new state $\hat{\tilde{\rho}}(T,\xi)$ serves as the initial condition for the backward evolution $\hat{U}(-T) = \hat{U}^\dagger(T)$ during the same time T, after which the reversed state $\hat{\tilde{\rho}}(0|T,\xi) = \hat{U}^\dagger(T)\hat{\tilde{\rho}}(T,\xi)\hat{U}(T) = \hat{P}(\xi,T)\hat{\rho}(0)\hat{P}^\dagger(\xi,T)$ is finally obtained. Here $\hat{P}(\xi,T) \equiv e^{-i\xi\hat{V}(T)}$, with $\hat{V}(T) \equiv \hat{U}^\dagger(T)\hat{V}\hat{U}(T)$ being the Heisenberg evolution of the perturbation during the time T. At last, to characterize the degree of reversibility we consider the distance between the reversed $\hat{\tilde{\rho}}(0|T,\xi)$ and the initial $\hat{\rho}(0)$ states, as measured by the Peres fidelity[12]

$$F_{\rm rev}(\xi;T) = \frac{{\rm Tr}[\hat{\tilde{\rho}}(0|T,\xi)\hat{\rho}(0)]}{{\rm Tr}[\hat{\rho}^2(0)]} = \left.\frac{{\rm Tr}[\hat{\tilde{\rho}}(t,\xi)\hat{\rho}(t)]}{{\rm Tr}[\hat{\rho}^2(t)]}\right|_{t=T} = F\,(\xi;t)|_{t=T}\,. \tag{5}$$

The fidelity is bounded in the interval $[0,1]$ and is the closer to unity the more similar are the initial and reversed states. The second equality in (5) is a consequence of the unitary time evolution. The fidelity $F(\xi;T)$ measures the *complexity* of the Wigner function at the moment $t = T$ (see below). Both the functions $F_{\rm rev}(\xi;T)$ and $F(\xi;T)$ are numerically identical but expressed in different variables. The relation (5) plays the key role in our analyses. It allows us not only to relate the degree of reversibility to the complexity of the state at the reversal time T but also *to establish a strong restriction on the upgrowth* of the number $\mathcal{M}(t)$ of harmonics of the Wigner function. The crucial point is that while in classical mechanics the number of Fourier components has no direct physical meaning, in quantum mechanics the number of the components of the Wigner function at any given time is related to the *degree of excitation* of the system (see for example eq. (10) below) and therefore unrestricted exponential growth of this number is not allowed.[13–16]

4. Peres Fidelity and the Number of Harmonics

First of all, let us establish a connection between fidelity and complexity of the Wigner function at an arbitrary moment t. To do this we make the instant rotation in the phase plane $e^{-i\xi\hat{n}}$ at this moment and utilize then the second form of fidelity in Eq. (5). This yields[9,10]

$$F(\xi;t) = 1 - 2\sum_{m=1}^{\infty} \sin^2(\xi m/2)\, \mathcal{W}_m(t) = 1 - \frac{1}{2}\xi^2 \langle m^2\rangle_t + O(\xi^4), \tag{6}$$

where $\langle m^2\rangle_t = \sum_{m=0}^{+\infty} m^2\, \mathcal{W}_m(t)$ and

$$\mathcal{W}_{m\geqslant 0}(t) = \frac{(2-\delta_{m0})\sum_{n=0}^{\infty}\left|\langle n+m|\hat{\rho}(t)|n\rangle\right|^2}{\sum_{m=-\infty}^{\infty}\sum_{n=0}^{\infty}\left|\langle n+m|\hat{\rho}(t)|n\rangle\right|^2}. \tag{7}$$

We define then the number of harmonics of the Wigner function at the time t as $\mathcal{M}(t) = \sqrt{\langle m^2\rangle_t}$. This is in line with [14–16]. The set of positive definite quantities $\mathcal{W}_m$ is normalized to unity, $\sum_{m=0}^{+\infty} \mathcal{W}_m = 1$, and can be, therefore, given a probabilistic interpretation. According to (6) the number of harmonics at arbitrary time t is found as

$$\mathcal{M}^2(t) \equiv \langle m^2\rangle_t = -\frac{d^2F(\xi;t)}{d\xi^2}\bigg|_{\xi=0}. \tag{8}$$

This formula remains valid also in classical limit provided that the fidelity is calculated with the help of the corresponding classical phase-space distribution function.

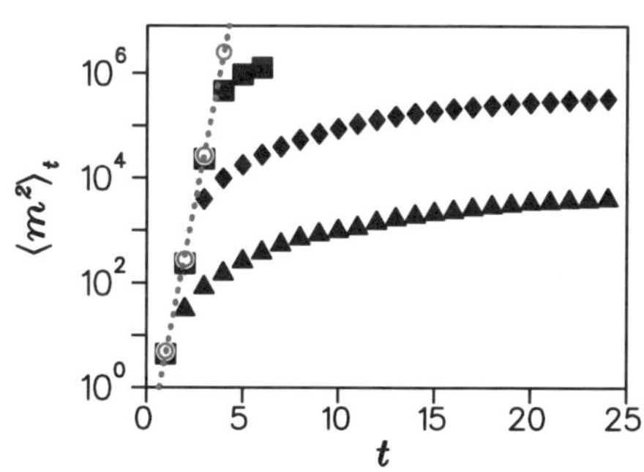

Fig. 1. Root-mean-square $\langle m^2\rangle_t$ versus time t at $g_0 = 1.5$. Squares, diamonds and triangles correspond to $\hbar = 0.01, 0.1$ and 1. Empty circles refer to classical dynamics and the dashed line fits these data.

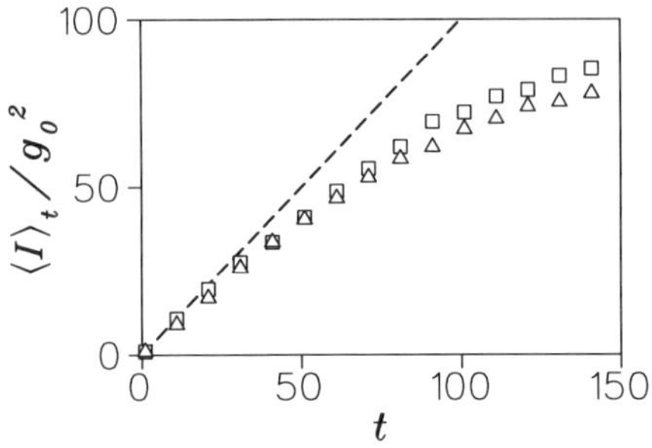

Fig. 2. Quantum diffusion: mean value $\langle I \rangle_t / g_0^2 = \hbar \langle n \rangle_t / g_0^2$ in function of time t. Squares and triangles correspond to $(\hbar, g_0)=(1,2)$ and $(2,3)$. The straight line shows the classical diffusion law $\langle I \rangle_t = g_0^2 t$.

The time evolution of $\mathcal{M}^2(t)$ is illustrated for different values of the effective Planck constant $\hbar$ in Fig.1. The initial state is chosen to be a pure ground state $\hat{\rho}(0) = |0\rangle\langle 0|$ which corresponds to the isotropic Wigner function $W(\alpha^*, \alpha; 0) = 2\, e^{-2|\alpha|^2}$ with the size 1/2. This size is kept constant throughout all calculations whereas the quantum Liouville equation is solved for a number of decreasing values of the Planck's constant thus approaching the classical dynamics. It is clearly seen that the exponential increase of $\langle m^2 \rangle_t$ takes place only up to the Ehrenfest time scale $t_E \propto \ln \hbar$, consistently with the findings reported in [14–16]. Then, a much slower power-law increase follows.

To reveal the reason for such a behavior we will analyse now the fidelity $F_{\rm rev}(\xi; t)$ expressed in terms of the density matrices (or, equivalently, the Wigner functions) at the initial time of the forward – backward cycle of evolution. This yields[9,10] $F_{\rm rev}(\xi; t) = |f(\xi, t)|^2$ where

$$f(\xi, t) = \langle 0|e^{-i\xi\hat{n}(t)}|0\rangle = \mathrm{Tr}\left[e^{-i\xi\hat{n}}\hat{\rho}(t)\right] = \sum_{n=0}^{\infty} w_n(t)\, e^{-i\xi n} \qquad (9)$$

with $w_n(t) \equiv \langle n|\hat{\rho}(t)|n\rangle = \left|\langle n|\hat{U}(t)|0\right|^2$ being the excitation number probability distribution. In such a way we relate the behavior of fidelity to evolution of the excitation numbers. Equating now the ξ^2 terms in expansions of the both possible representations of fidelity we arrive at the following *significant exact relation* between the number of harmonics and the root-mean-square deviation of the action

$$\langle m^2 \rangle_t = 2\,\chi_2(t), \quad \chi_2(t) \equiv \langle 0|\hat{n}(t)^2|0\rangle - \langle 0|\hat{n}(t)|0\rangle^2 . \qquad (10)$$

Thorough numerical study[9] convince us that after proper averaging over strong irregular fluctuations (*coarse graining*) the excitation number distribution w_n decays exponentially with n,

$$w_n(t) \approx \frac{1}{\langle n\rangle_t + 1}\left[\frac{\langle n\rangle_t}{\langle n\rangle_t + 1}\right]^n, \quad \langle n\rangle_t \equiv \langle 0|\hat{n}(t)|0\rangle\,. \tag{11}$$

It follows then[9,10] that after the Ehrenfest time

$$\mathcal{W}_m(t) = (2-\delta_{m0})\sum_{n=0}^{\infty} w_{n+m}(t)\, w_n(t) \approx \frac{2-\delta_{m0}}{2\langle n\rangle_t + 1}\left[\frac{\langle n\rangle_t}{\langle n\rangle_t + 1}\right]^m, \tag{12}$$

and

$$\langle m^2\rangle_t = 2\,\chi_2(t) \approx 2\langle n\rangle_t^2\,. \tag{13}$$

The time dependence of the mean action $\langle I\rangle_t$ (the deterministic quantum diffusion) is shown in Fig. 2. Thus the relation (13) implies that the number of harmonics grows after the Ehrenfest time not faster than linearly.

5. Reversibility Versus the Number of Harmonics

One can readily show now that for any finite $\xi \ll 1$ fidelity equals in the approximation (11)

$$F_{\rm rev}(\xi;T) = F(\xi;t=T) = \frac{1}{1+\frac{1}{2}\xi^2\langle m^2\rangle_T}\,. \tag{14}$$

More than that, in fact this formula remains valid for any time including the times $T < T_E$.[9] This formula directly relates reversibility of the motion to complexity of the Wigner function at the reversal T.

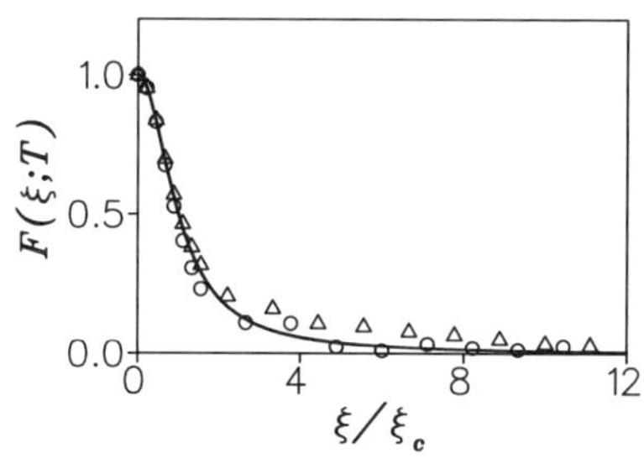

Fig. 3. Fidelity $F(\xi;T)$ versus the scaled variable $\xi/\xi_c(T)$. Data correspond to: (a) $\hbar = 1, g_0 = 2$; circles: $T = 10$; triangles: $T = 50$. The full curve shows the theoretical prediction Eq. (14). The deviations on the tail of the curve are due to fluctuations neglected in (14).

According to Eq. (14) a crossover takes place near the *critical value* $\xi_c(T) \equiv \sqrt{2}/\mathcal{M}(T)$, from good, $F(\xi;T) \approx 1$, to broken, $F(\xi;t) \approx (\xi/\xi_c(T))^2 \ll 1$, reversibility. Our numerical simulations (Fig. 3) nicely confirm this formula.

6. Summary

We have established a quantitative relation between complexity of the Wigner function and degree of reversibility of motion of a classically chaotic quantum system. We have analytically proved that the number of harmonics $\mathcal{M}(t)$ of this function, which can serve as a natural measure of the complexity, increases after the Ehrenfest time *not faster than linearly* in striking contrast with classical dynamics, where the number of harmonics of phase space distribution proliferates exponentially. We have shown that if a quantum motion has been perturbed at some moment T by an external force with intensity ξ and then reversed, its initial state is recovered with the accuracy $\sim (\xi/\xi_c)^2$ as long as the strength is restricted to the interval $0 < \xi < \xi_c(T) = \sqrt{2}/\mathcal{M}(T)$. This interval decreases with the time T at most linearly beyond the semi-classical domain but shrinks exponentially due to the classical exponential instability when this domain is approached.

References

1. D. L. Shepelyansky, *Physica D* **8**, 208 (1983).
2. G. Casati, B. V. Chirikov, I. Guarnery and D. L. Shepelyansky, *Phys. Rev. Lett.* **56**, p. 2437 (1986).
3. V. M. Alekseev and M. V. Jacobson, *Phys. Rep.* **75**, p. 287 (1982).
4. J. Ford, *Phys. Today* **36**, p. 40 (1983).
5. T. Prosen, *J. Phys. A* **40**, p. 7881 (2007).
6. R. J. Glauber, *Phys. Rev.* **131**, p. 2766 (1963).
7. G. S. Agarwal and E. Wolf, *Phys. Rev. D* **2**, p. 2161 (1970).
8. G. S. Agarwal and E. Wolf, *Phys. Rev. D* **2**, p. 2187 (1970).
9. V. V. Sokolov, O. V. Zhirov, G. Benenti and G. Casati, *Phys. Rev. E* **78**(046212) (2008).
10. V. V. Sokolov and O. V. Zhirov, *Europhys. Lett.* **84**, p. 30001 (2008).
11. K. S. Ikeda, in *Quantum Chaos: Between Order and Disorder*, eds. G. Casati and B. V. Chirikov (Cambridge University Press, 1995).
12. A. Peres, *Phys. Rev. A* **30**, p. 1610 (1984).
13. B. V. Chirikov, F. M. Izrailev and D. L. Shepelyansky, *Sov. Sci. Rev. C* **2**, p. 209 (1981).
14. Y. Gu, *Phys. Lett. A* **149**, p. 95 (1990).
15. A. K. Pattanayak and P. Brumer, *Phys. Rev. E* **56**, p. 5174 (1997).
16. J. Gong and P. Brumer, *Phys. Rev. A* **68**, p. 062103 (2003).

Multi Channel Wave Transfer in Sparsely Connected Systems

R. Sprik
Van der Waals-Zeeman Instituut, Universiteit van Amsterdam
Amsterdam, The Netherlands
Email: sprik@science.uva.nl

Wave transport in finite systems connected through multiple channels is strongly influenced by inter channel correlations. Such correlations are induced by scattering and restrictions in the complex medium in which the waves propagate. Examples can be found in electron conduction in mesoscopic systems with multiple leads and in electromagnetic wave transfer in multi-antenna wireless communication. The influence of the correlations on the overall wave transport can be uncovered by analyzing the eigenvalue distributions of the transfer matrix between the ingoing and outgoing channels. A model system is studied experimentally in the microwave region where the correlation between sets of transmitting and receiving antennas are induced by connecting complex cavities through a limited number of leads. The eigenvalue distribution of this correlated multi channel system is described using random matrix theory.

Keywords: Random matrix modeling; eigenvalue distribution; classical wave transport; coupled chaotic systems.

1. Introduction

Wave transport in a strongly scattering environment occurs in many systems. Examples are found in classical wave transport of light, electromagnetic waves, sound and elastic waves.[1] Also in quantum mechanical problems such as electronic transport in small devices multiple scattering is frequently encountered. In the past decades the effects of multiple scattering on wave transport has been studied intensely both experimentally and theoretically. E.g. scattering theory developed for quantum field problems has been successfully transferred to solve problems in classical and quantum mechanical wave scattering and lead to understanding of wave scattering related phenomena such as weak and strong localization.

In recent developments of wireless communication in strongly scattering indoor (e.g. office spaces) and outdoor (e.g. cities) environments the scattering is exploited to improve the transfer.[2] The advantage of multiple antennas on transmit and receive side relies on scattering induced by the environment. The theoretical framework to analyze the wave transfer in strongly scattering and noisy systems has been developed in the past decade as an important topic in information theoretical studies[3] starting with the work of Shannon in 1948.[4]

The details of the underlying mechanisms of scattering on transport in wireless communication or of the Hamiltonian of a (quantum) mechanical system are not required to model a system. A powerful approach is provided by random matrix theory (RMT).[5] The random matrix approach was started by Wigner[6] to describe distributions of quantum mechanical eigenvalues of complex systems and found early applications in the field of mathematical statistics by Wishart.[7] In the information theory of wireless communication RMT has been applied with success to a variety of multichannel situations without and with correlations between channels.[8] The elements in the transfer matrix $\mathbf{H}$ describing the phase and amplitude at the receiving antennas due to a signal at the transmitting antennas can be modelled by statistically defined distributions.[9] The eigenvalues of $\mathbf{HH}^\dagger$, with $\mathbf{H}^\dagger$ the Hermitian adjoint of $\mathbf{H}$, are an essential ingredient in estimating the transfer capacity of a multi channel connection. For example for random matrices with independent and identically distributed elements Marčenko and Pastur[10] derived general analytical expression for the eigenvalue distributions applicable to transmissions between rectangular antenna arrays. Once the properties of the eigenvalue distributions are known, an estimate of the transfer capacity and the fluctuations can be made. Finding the eigenvalues λ_i^2 of $\mathbf{HH}^\dagger$ is equivalent to finding the singular values λ_i of $\mathbf{H}$, which is sometimes more convenient. The generalized Shannon expression for the transfer capacity is:[9] $C_{\text{shannon}} = -\log\det[\mathbf{I} + \eta\mathbf{HH}^\dagger]$, with η the signal to noise ratio in the system, $\mathbf{I}$ the unit matrix and depends intricately on the eigenvalue distribution.

Correlation between channels decreases the transfer capacity considerably. The effect of correlation induced by constrictions in space between transmitter and receiver in the form of keyholes or waveguides has been the topic of mostly theoretical[11–13] and some experimental studies.[14,15] Such constrictions are found typically in transfer between floors in buildings and urban environments.

In this article a controllable model system with classical waves in two coupled complex and dissipative closed cavities will be used to study the influence of correlations between channels on wave transport. The experiments described in Sections 2 and 2.2 enable a systematic variation of the induced correlation and the influence on resonances in the wave transport. The results are analyzed with a random matrix approach in Section 3.

2. Wave Transfer in Sparsely Coupled Systems

2.1. *Experiments*

The microwave experiment in the 0.5-2 GHz band (i.e. 30-15 cm wavelength) uses two almost chaotic cavities (see Fig. 1). Within each of the cavities the reflections from the walls and from the enclosed antennas causes an almost chaotic transport of the waves from multiple antennas to a limited number antennas connected to outgoing channels. These outgoing channels are then launched into the second cavity and propagate to the receiving antennas.

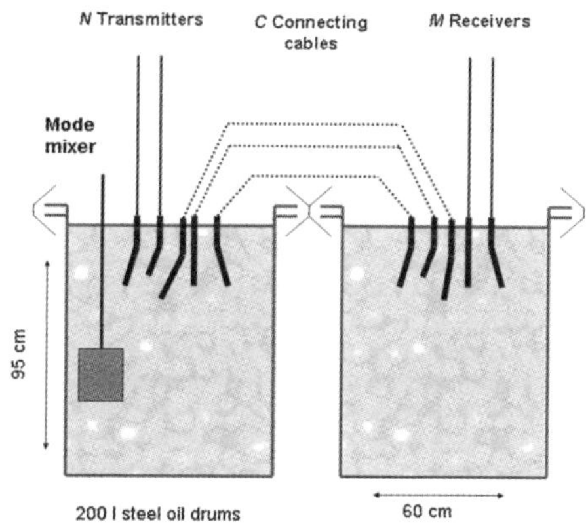

Fig. 1. Layout of a multi-channel microwave experiment with microwaves in the 0.5-2 GHz band. N transmitting antennas in a closed cavity are sending the signal through C connecting cables to a second closed cavity with M receiving antennas.

The receiver and transmitter antennas are 2.4 GHz omni-directional antennas on a swivel mount used in e.g. computer WiFi connections and are fitted into the lids of two 200 liter steel oil drums. Some of the antennas at the transmitter side are connected by coax cables directly to antennas at the receiver side. The overall quality factor of the drum cavity is $Q \approx 300$. The transfer as function of frequency between the M transmitting and N receiving antennas is determined by measuring the S-parameters S_{21}, S_{11}, S_{12}, and S_{22} with an Agilent 4396B network analyzer in the 0.5GHz - 1.8GHz range. The S-parameters between two ports are defined by the ratio of the voltage of the reflected and the transmitted wave with respect to the voltage of the incoming wave in a 50Ω terminated system. E.g. S_{21} characterizes the transmitted and S_{11} the reflected wave for a signal injected into port 1 and received at port 2. In a dissipative system $1 - |S_{12}|^2 + |S_{11}|^2$ gives the dissipation and in a reciprocal system $S_{21} = S_{12}$. A homemade coaxial multiplexing switch selects all possible transmit-receive combinations to obtain the multi-channel scattering matrix S_{ij} with $i = 1, ..., N$ and $j = 1,, M$ and gives the full $M \times N$ complex transfer matrix H of the system as function of frequency. The connecting paths for the different switch selections are identical and all unused antennas are terminated by a 50Ω impedance. Calibration of the S-parameter measurement is performed by fitting the 12 parameter correction model.[16] In the reference measurements the antenna connections are terminated with a short, an open, a 50Ω load, and a direct through connection (SOLT calibration model). The calibration procedure eliminates the influence of the leads and switch contacts and gives the true multiport S-parameters between the antennas. Averaging over configurations is achieved by the rotation of a copper plate inside the oil drum ('mode mixer' see Fig. 1). The set up is under computer control to enable signal averaging over configurations and perform multiplexing of a bi-directional measurement of an antenna array with a maximum of 8 transmitters and 8 receivers.

2.2. *Experimental results*

Fig. 2 shows the typical observed variations in the matrix elements for the uncorrected S-parameters in a 3 transmitter × 3 receiver configuration with 2 connecting cables between the drums.

After correction the data is used to extract the statistics of the elements in $\mathbf{H}(\omega)$ and are shown in Fig. 3. The distribution shows a uniform coverage in phase, i.e. $\langle H(\omega)\rangle = 0$. The distribution of the amplitude is not Gaussian but is characteristic for the fading behavior in frequency for a set of well separated narrow resonances with resonance frequencies distributed over the frequency interval and with varying strengths and widths as shown in Fig. 2).

The distribution of the singular values of $\mathbf{H}(\omega)$ are given in Fig. 4 for different number of connecting cables ($C = 1, 2$, and 3). The distribution for $C = 1$ show a strong peak at 0 and a suppressed probability close to zero. The distributions with $C > 1$ look very similar. The distributions show a distinct structure that can be associated with the direct line of sight contact between antennas.

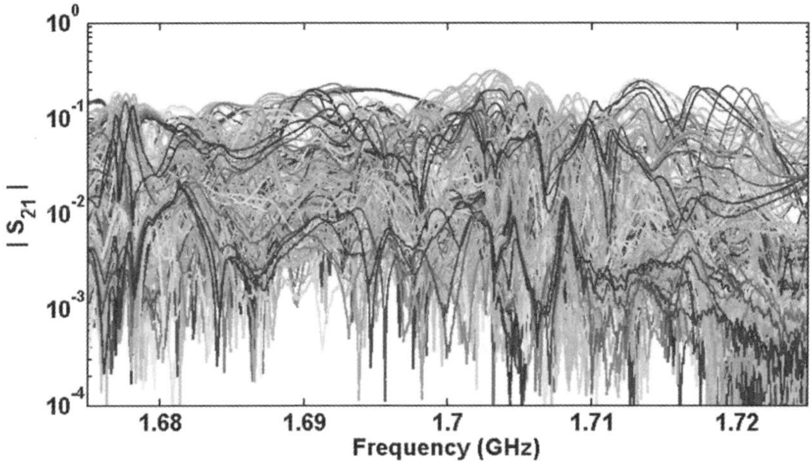

Fig. 2. The different curves combine all the uncorrected S_{21} measurements with a $N = 3$, $C = 2$, and $M = 3$ configuration for 21 different positions of the 'mode mixer' in 50MHz band around 1.7GHz.

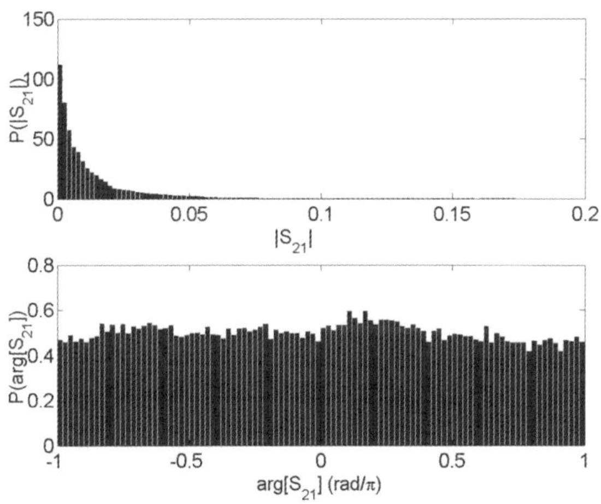

Fig. 3. Distribution of amplitude and phase angle of the corrected transmission parameter S_{21} between the different transmit and receive antennas with $C = 1$.

3. Modelling the Results with Random Matrices

3.1. *Random matrix model for coupled systems*

The transfer matrix $\mathbf{H}$ between the antennas in drum 1 to the receiving antennas in drum 2 will be modelled by a product of random matrices:

$$\mathbf{H} = \mathbf{TCX}. \tag{1}$$

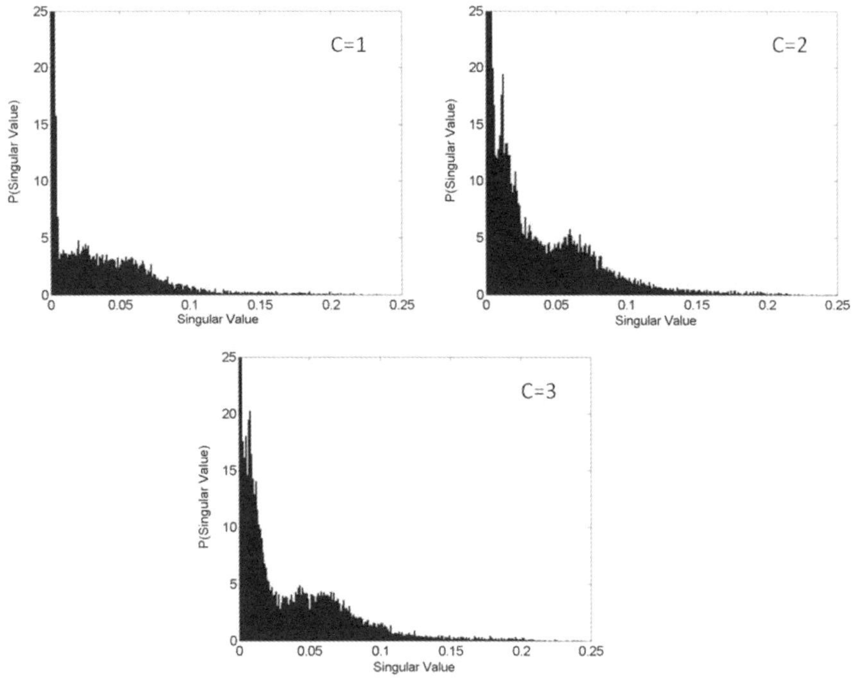

Fig. 4. Singular value distributions of $\mathbf{H}(\omega)$ with $N = M = 3$ receiving and transmitting antennas and $C = 1, 2, 3$ interconnections averaged over all 'mode mixer' positions and frequencies.

Here matrix $\mathbf{T}$ is a $N \times C$ dimensioned and matrix $\mathbf{X}$ a $C \times M$ dimensioned random matrix with identical (independent) distributed elements. The matrix $\mathbf{T}$ represents the transfer between the M antennas on the transmitter to the C antennas connected to the cables between the transmitter and receiver side. Likewise $\mathbf{X}$ represents the connection from the C cables to the M receiving antennas. The deterministic connection matrix $\mathbf{C}$ is a diagonal $C \times C$ matrix representing the transfer through the C connecting cables between the transmitter and the receiver side.

3.2. *Independent transmit and receive channels*

The simplest approach is to assume that the random elements in $\mathbf{T}$ and $\mathbf{X}$ are independent identical complex distributions with random phase and equal amplitude (Gaussian distribution). Simulation results for the matrix elements of $\mathbf{H}$ and the eigenvalue distribution of $\mathbf{H}\mathbf{H}^\dagger$ are shown in Fig. 5 for 5000 realizations and increasing number of links.

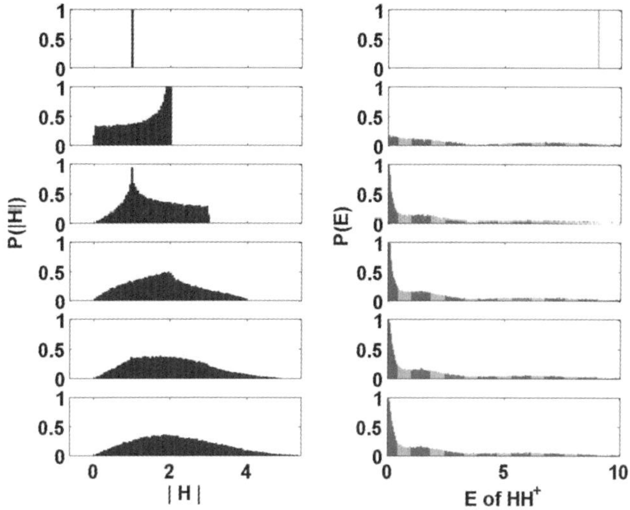

Fig. 5. Random matrix simulation of the eigenvalue distribution of $\mathbf{HH}^\dagger$ and matrix elements for the random transfer matrix $\mathbf{H}$ with $M = 3$ transmitters and $N = 3$ receivers connected through C restricting links. The elements of $\mathbf{T}$ and $\mathbf{X}$ are independently identically distributed complex numbers with random phase and unit amplitude. Coupling matrix $\mathbf{C}$ is the unit matrix for $C = 1$ to 6 from top to bottom.

The rank of the matrix $\mathbf{H}$ is an indication of the number of available channels: rank$(\mathbf{H}) \leq \min(\mathbf{M}, \mathbf{C}, \mathbf{N})$. Hence, if the number of connecting cables C is less than the number of transmitting or receiving antennas, the interconnects are restricting the transfer. When the rank$(\mathbf{C}) >$ max(rank$(\mathbf{M}, \mathbf{N})$) the distributions quickly converge to the same behavior for the elements of $\mathbf{H}$ and for the eigenvalue distribution.

Müller[17] derived the asymptotic eigenvalue distribution for products of random matrices in the limit of very large matrix sizes as a generalization of the Marčenko-Pastur result.[9] The current examples are not in that limiting regime yet. However, the convergence to the asymptotic limit is in general approached fast for very small systems already.[8] Including the fading behavior as function of frequency indicated by Fig. 2 and Fig. 3 does not influence the asymptotic behavior for uncorrelated matrix elements.

3.3. *Correlations on the transmit and receive side*

In the experiment there are different causes for correlations between channels. The connecting cables restrict the transfer to a finite number of channels and cause correlation as shown in the previous paragraph. In addition, the antennas on the receiver and transmitter side are placed close enough to be correlated due to the finite extent of coherent regions in the wave pattern inside the drum. Analysis of the signals from the transmitters on one side to the connecting antennas in the same drum show indeed correlation inside each of the drums. Using the statistics of the transfer within the drums in combination with the effects of the extra links will not be pursued in detail here. Given the limited number of configurations included in the estimates of the distributions, it is difficult to sample enough configurations to be able to use the random matrix approach.

4. Conclusions

The random matrix theory is not describing the observed distribution in detail, but outlines the consequence of the sparse connections between two strongly scattering spaces. Improvement of the description can be obtained by averaging over more antenna configurations and choosing a geometry where near field and direct coupling is suppressed.

References

1. P. Sheng, *Introduction to Wave Scattering, Localization, and Mesoscopic Phenomena* (Academic Press, 1995).
2. S. H. Simon, A. L. Moustakas, M. Stoytchev and H. Safar, *Physics Today* **38** (september 2001).
3. T. M. Cover and J. A. Thomas, *Elements of Information Theory* (John Wiley & Sons Inc, 1991).
4. C. Shannon, *Bell Syst. Tech. J.* **27**, p. 379 and 623 (1948).
5. C. Beenakker, *Rev. Mod. Phys.* **69**, 731 (1997).
6. E. Wigner, *Ann. Math.* **67**, 325 (1958).
7. J. Wishart, *Biometrika* **20**, 35 (1928).
8. A. Tulino and S. Verdù, *Found. Trends Commun. Inf. Theory* **1**, 1 (2004).
9. V. A. Marčenko and L. A. Pastur, *Math. USSR-Sbornik* **1**, 457 (1967).
10. E. Telatar, *Eur. Trans. Telecom.* **10**, 585 (1999), AT & T-Bell Technical Memorandum (1995, see: http://mars.bell-labs.com).
11. R. Sprik, *Phys. Rev. E* **72**, 037602 (2005).
12. D. Chizhik, G. Foschini, M. Gans and R. Valenzuela, *IEEE Trans. Commun.* **1**, 361 (2002).
13. S. Loyka and A. Kouki, *IEEE Trans. on Comm.* **50**, 1886 (2002).
14. D. Chizhik, J. Ling, P. W. Wolniansky, R. A. Valenzuela, N. Costa and K. Huber, *IEEE Journal on Sel. Areas in Comm.* **21**, 321 (2003).
15. P. Almers, F. Tufvesson and A. F. Molisch, *IEEE Comm. Lett.* **7**, 373 (2003).
16. Agilent application note AN 1287-3, Applying Error Correction to Network Analyzer Measurements, 2002 (see: http://www.agilent.com).
17. R. Müller, *IEEE Trans. Inform. Theory* **9**, 2495 (2002).

Prediction of Dominant Frequencies of CFOA-Based Sprott's Sinusoidal and Chaotic Oscillators

B. Srisuchinwong*, C.-H. Liou and T. Klongkumnuankan

Sirindhorn International Institute of Technology, Thammasat University
Pathum-Thani, Thailand 12000
*Email: *banlue@siit.tu.ac.th*

Prediction of dominant frequencies of two examples of current-feedback op amp (CFOA)-based Sprott's chaotic oscillators is presented through the use of simple Fourier analysis. In addition, two examples of CFOA-based Sprott's sinusoidal oscillators exhibiting periodic sinusoids are also demonstrated. In the Sprott's sinusoidal oscillators, the fundamental frequencies may be approximated by the prediction of the dominant frequencies.

Keywords: Sinusoidal oscillators; chaotic oscillators; dominant frequencies; fundamental frequencies; current-feedback op amps.

1. Introduction

Chua's circuit[1,2] is one of the best-known chaotic oscillators based on three first-order ordinary differential equations (ODEs), but is relatively difficult to scale to arbitrary frequencies. Sprott[3] has alternatively proposed chaotic oscillators that can be easily scaled to different frequencies based on a single third-order ODE in a simple "jerk function" such as

$$\frac{\partial^3 x}{\partial t^3} + A\frac{\partial^2 x}{\partial t^2} + \frac{\partial x}{\partial t} = G(x) \tag{1}$$

where $G(x)$ is a nonlinear function and $A = 0.6$, using op amps operating at a low frequency of around 1.5 kHz. Recently, high-frequency Sprott's chaotic oscillators have been implemented using current-feedback op amps (CFOAs) and a dominant frequency where the strongest peak of the power spectrum is centred has been relatively high and observed at 1.2 MHz by Srisuchinwong and Liou.[4] In this paper, prediction of the dominant frequencies of two examples of CFOA-based Sprott's chaotic oscillators is presented through the use of simple Fourier analysis. In addition, two examples of CFOA-based Sprott's sinusoidal oscillators exhibiting periodic sinusoids are also demonstrated using $A = 0.85$ and $A = 1.184$. In the Sprott's sinusoidal oscillators, the fundamental frequencies may be approximated by the prediction of the dominant frequencies.

2. Two Examples of CFOA-Based Sprott's Chaotic Oscillators

2.1. *Circuit description*

Figure 1 shows a high-frequency Sprott's chaotic oscillator using CFOAs AD844/AD. The CFOA U_0 forms an integrator U_0 where the zero-dB crossing (ZdB) frequency $\omega_0 = 1/\tau_0$ and $\tau_0 = R_0C_0$. The CFOA U_2 forms an integrator U_2 where the ZdB frequency $\omega_2 = 1/\tau_2$ and $\tau_2 = R_2C_2$. An expected integrator U_1 between U_0 and U_2 is replaced with a simpler passive R_1C_1 filter. A routine analysis in Fig. 1 reveals that the resulting jerk function at node N_1 is

$$K_3\frac{\partial^3 x}{\partial t^3} + K_2\frac{\partial^2 x}{\partial t^2} + K_1\frac{\partial x}{\partial t} = G(x) \tag{2}$$

where

$$\begin{aligned} K_1 &= \frac{\tau_0}{R_2} \\ K_2 &= \frac{\tau_0\tau_2}{R_2}(1 + \frac{R_1}{R_X}) \\ K_3 &= \frac{\tau_0\tau_X\tau_2R_1}{R_XR_2} \end{aligned} \tag{3}$$

where $R_X = (R_0R_2)/(R_0 + R_2)$ and $\tau_X = R_XC_1$. As (1) = (2), therefore $K_1 = 1$, $K_2 = A = 0.6$ and $K_3 = 1$. For simplicity, let $R_2 = R_1 = R_0 = R_T$, consequently $C_2 = A/(3R_T)$, $C_1 = 3/(AR_T)$ and $C_0 = R_2/R_0 = 1$. If $R_T = 1\ k\Omega$, then $C_2 = 0.0002\ F$, $C_1 = 0.005\ F$ and $C_0 = 1\ F$. In practice, slight modification through a scaling factor m = 100 and arbitary scaling factors S_0, S_1 and S_2 yields that $C_2 = AS_2m/(3R_T)$, $C_1 = 3S_1m/(AR_T)$ and $C_0 = S_0/m$. Equation (3) suggests that $K_1 = f(C_0)$ and $K_3 = f(C_0, C_1, C_2)$, therefore practical values are $K_1 = S_0/m$, and $K_3 = S_0S_1S_2m^2/m = S_0S_1S_2m$, as summarized in Table 1(a).

2.2. *Nonlinear components G(x)*

Two examples of CFOA-based chaotic oscillators are demonstrated by two different types of nonlinear components G(x). For example, G(x) can be either $G_1(x) = |x| - 2$ or $G_2(x) = -1.2x + 2sgn(x)$, implemented in a

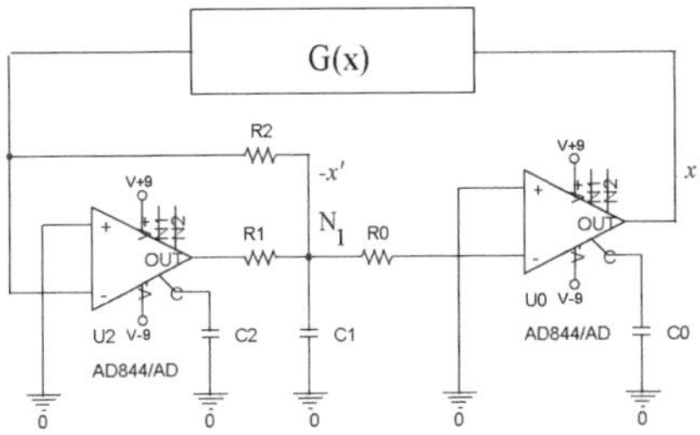

Fig. 1. High-frequency CFOA-based Sprott's sinusoidal and chaotic oscillators.

form of current using CFOAs AD845/AD. As shown in Fig. 2(a), $G_1(x)$ is implemented using CFOA U_3, resistors $R_3 = 3.333\ k\Omega$, $R_4 = 1.667\ k\Omega$, $R_5 = 5\ k\Omega$ and $R_6 = 45\ k\Omega$. As shown in Fig. 3(a), $G_2(x)$ can be implemented using CFOAs U_4 and U_5, resistors $R_7 = R_8 = 1\ k\Omega$, $R_9 = 8.333\ k\Omega$ and $R_{10} = 45\ k\Omega$.

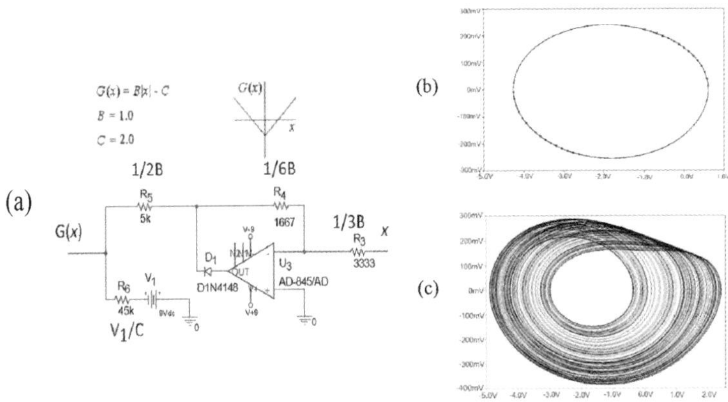

Fig. 2. (a) Nonlinear component $G(x) = G_1(x) = |x| - 2$, (b) Periodic attractor using $A = 0.85$, (c) Chaotic attractor using $A = 0.6$.

3. Prediction of Dominant Frequencies

3.1. *Use of nonlinear component $G_1(x)$*

Equation (2) may be analyzed through Fourier analysis using $G(x) = G_1(x) = |x| - 2$, therefore,

$$K_3(j\omega)^3X + K_2(j\omega)^2X + K_1(j\omega)X - F[G_1(x)] = 0 \tag{4}$$

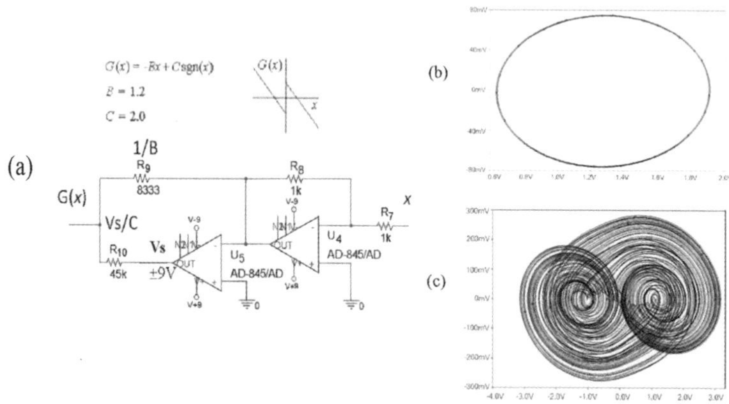

Fig. 3. (a) Nonlinear component $G(x) = G_2(x) = -1.2x + 2sgn(x)$, (b) Periodic attractor using $A = 1.184$, (c) Chaotic attractor using $A = 0.6$.

where $F[G_1(x)]$ is the Fourier transform of $G_1(x)$ and $X = F[x]$. By setting the real part = 0, and the imaginary part = 0, therefore

$$-\omega^3 K_3 X + K_1 \omega X = 0 \tag{5}$$

$$\omega = \sqrt{\frac{K_1}{K_3}} = \sqrt{\frac{1}{m^2 S_1 S_2}} \tag{6}$$

A dominant frequency f_0 may be defined at the frequency where the strongest peak of the power spectrum is centered. For G(x) = G_1(x), the dominant frequency $f_0 = \omega/(2\pi)$.

3.2. *Use of nonlinear component $G_2(x)$*

Equation (2) may be analyzed through Fourier analysis using $G(x) = G_2(x) = -1.2x + 2\,\mathrm{sgn}(x)$, therefore,

$$K_3(j\omega)^3 X + K_2(j\omega)^2 X + K_1(j\omega)X - F[G_2(x)] = 0 \tag{7}$$

where $F[G_2(x)]$ is the Fourier transform of G_2(x) and $X = F[x]$. By setting the real part = 0, and the imaginary part = 0 , therefore

$$-\omega^3 K_3 X + K_1 \omega X = 0 \tag{8}$$

$$\omega = \sqrt{\frac{K_1}{K_3}} = \sqrt{\frac{1}{m^2 S_1 S_2}} \tag{9}$$

For $G(x) = G_2(x)$, a dominant frequency $f_0 = \omega/(2\pi)$.

4. Simulation Results

4.1. *Two examples of CFOA-based Sprott's sinusoidal oscillators*

The CFOA-based Sprott's chaotic oscillator shown in Fig. 1 can be simulated as a Sprott's sinusoidal oscillator, where A is no longer 0.6. Two examples of CFOA-based Sprott's sinusoidal oscillators are demonstrated using $G_1(x)$, A = 0.85, and parameters as shown in Table 1(b), or using $G_2(x)$, A=1.184, and parameters as shown in Table 1(c).

Table 1. Summaries of parameters,calculation and simulation results

Parameters	Ideally m=1	Practically m=100. S_0, S_1, S_2 are Scaling Factors. $R_T = 1\ k\Omega$.				
		(a)	Sprott's Sinusoidal Oscillators		Sprott's Chaotic Oscillators	
			(b)	(c)	(d)	(e)
			$G_1(x)$	$G_2(x)$	$G_1(x)$	$G_2(x)$
			A=0.85	A=1.184	A=0.6	A=0.6
		S_0=1 S_1=1 S_2=1	$S_0 = \frac{10^{-6}}{400}$ S_1=S_0 S_2=S_0	$S_0 = \frac{10^{-6}}{1000}$ S_1=$\frac{10^{-6}}{768}$ S_2=S_0	$S_0 = \frac{10^{-6}}{400}$ $S_1 = S_0$ S_2=S_0	$S_0 = \frac{10^{-6}}{800}$ $S_1 = S_0$ S_2=S_0
C_0	1	$\frac{S_0}{m}$	25 pF	10 pF	25 pF	12.5 pF
C_1	$\frac{3}{AR_T}$	$\frac{3S_1 m}{AR_T}$	0.88 nF	0.33 nF	1.25 nF	0.63 nF
C_2	$\frac{A}{3R_T}$	$\frac{AS_2 m}{3R_T}$	70.8 pF	39.5 pF	50 pF	25 pF
K_1	1	S_0/m	25×10^{-12}	10×10^{-12}	25×10^{-12}	12.5×10^{-12}
K_3	1	$S_0 S_1 S_2$m	1.56×10^{-24}	1.3×10^{-25}	1.56×10^{-24}	1.95×10^{-25}
f_0(Calculation)			636.62 kHz	1.39 MHz	636.62 kHz	1.27 MHz
f_0(Simulation)			553.3 kHz	1.14 MHz	543.33 kHz	1.24 MHz
f_0(Error)			13.2 %	17.9 %	14.7 %	2.4 %
THD			2.6 %	0.3 %		

Two resulting Lissajous Figures as the periodic attractors in the (x, dx/dt) plane are shown in Fig. 2(b) for $G_1(x)$ and A = 0.85, and in Fig. 3(b) for $G_2(x)$ and A = 1.184. The fundamental frequency of such sinusoidal oscillators may be approximated by the prediction of the dominant frequency f_0, described in equations (6) and (9). As shown in Table 1(b), the calculated and simulated results of the dominant frequency f_0 are 636.62 kHz and 553.3 kHz, respectively, i.e. an error of 13.2 %, whilst the total harmonic distortion (THD) is measured approximately 2.6%. As shown in Table 1(c), the calculated and simulated results of the dominant frequency f_0 are 1.39 MHz and 1.14 MHz, respectively, i.e. an error of 17.9%, whilst the THD is measured approximately 0.3%. As an example, Fig. 4(a) shows the power spectrum and the resulting sinusoidal waveform (in the inset)

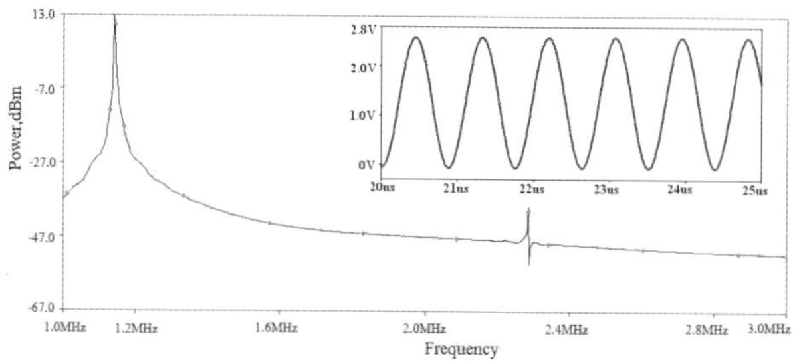

(a) Power spectrum of the Sprott's sinusoidal oscillator and the sinusoidal waveform (in the inset) using $G(x) = G_2(x) = -1.2x + 2\,\mathrm{sgn}(x)$ and A = 1.184.

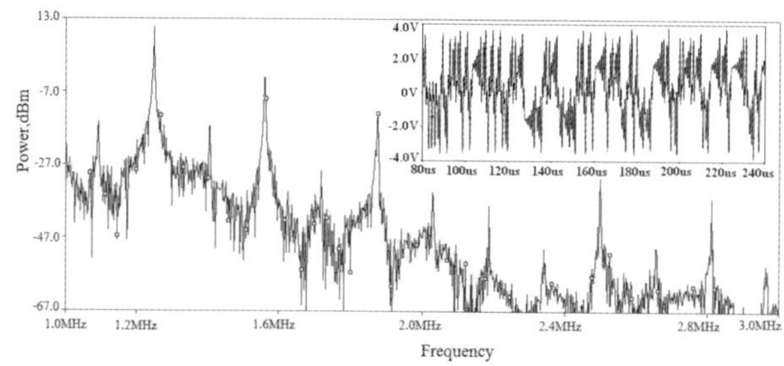

(b) Power spectrum of the Sprott's chaotic oscillator and the chaotic waveform (in the inset) using $G(x) = G_2(x) = -1.2x + 2sgn(x)$ and A = 0.6.

Fig. 4. Power spectra for Sprott's sinusoidal oscillator and chaotic oscillator

of the Sprott's sinusoidal oscillator at the fundamental frequency of 1.14 MHz.

4.2. *Two examples of CFOA-based Sprott's chaotic oscillators*

For $A = 0.6$, two examples of CFOA-based Sprott's chaotic oscillators shown in Fig. 1 are demonstrated using $G_1(x)$ and parameters shown in Table 1(d), or using $G_2(x)$ and parameters shown in Table 1(e). Two resulting chaotic attractors in the $(x, dx/dt)$ plane are shown in Fig. 2(c) for $G_1(x)$, and in Fig. 3(c) for $G_2(x)$. As shown in Table 1(d), the calculated and simulated results of the dominant frequency f_0 are 636.62kHz and 543.33 kHz, respectively, i.e. an error of 14.7%. As shown in Table 1(e), the calculated and simulated results of the dominant frequency f_0 are 1.27 MHz and 1.24 MHz, respectively, i.e. an error of 2.4%. As an example,

Fig. 4(b) shows the power spectrum and the resulting chaotic waveform (in the inset) of the Sprott's chaotic oscillator at the dominant frequency of 1.24 MHz.

5. Conclusions

Prediction of the dominant frequencies of two types of CFOA-based Sprott's chaotic oscillators has been presented through the use of simple Fourier analysis. In addition, two types of CFOA-based Sprott's sinusoidal oscillators exhibiting periodic sinusoids have been demonstrated at relatively low THD. In the Sprott's sinusoidal oscillators, the fundamental frequencies may be approximated by the prediction of the dominant frequencies.

Acknowledgements

This work is supported by Telecommunications Research and Industrial Development Institute (TRIDI, NTC), Thailand.

References

1. T.Matsumoto, *IEEE Transactions on Circuits and Systems* **CAS-31**, 1055 (1984).
2. B. Srisuchinwong and W. San-Um, *Proceedings of the Asia-Pacific Conference on Communications 2007 (APCC 2007)* **1**, 47 (2007).
3. J. Sprott, *Physic Letters A* **266**, 19 (2000).
4. B. Srisuchinwong and C.-H. Liou, *International Symposium on Signals Circuits and Systems - ISSCS 2007* **1**, 97 (2007).

Dynamics of the Coin Tossing

J. Strzałko, J. Grabski, A. Stefanski, P. Perlikowski and T. Kapitaniak*

Division of Dynamics, Technical University of Lodz
Stefanowskiego 1/15, 90-924, Lodz, Poland
*Email: *tomaszka@p.lodz.pl*

The basins of attraction of heads and tails in the realistic coin tossing are considered. We give evidence that from the point of view of dynamics the outcome of the coin tossing is predictable. In a limit theoretical case we identify the chaotic process and argue that its existence explains why the coins behave in practice as perfect randomizers.

1. Introduction

It is commonly known that the toss of a coin is random and this statement is fundamental in the probability theory.[1–3] On the other hand, the dynamics of the tossed coin is described by deterministic equations, with no external source of random influence,[4–8] so one can expect predictability in the results. It is possible to construct a mapping of the initial conditions: position, configuration, momentum, and angular momentum at the beginning of the coin motion to a final observed configuration, i.e., the coin terminates with its heads (tails) side up or on its edge. The initial conditions which are mapped onto heads create heads basin of attraction while these mapped onto tails create tails basin of attraction.[4] The boundary which separates heads and tails basins consists of initial conditions mapped onto the coin standing on the edge. The structure of these boundaries has the significant impact on the problem of the coin tossing predictability, i.e., smooth basin boundaries allow predictability while fractal boundaries can lead to unpredictability.[9,10] However, the precise structure of the heads-tails basin boundaries for a realistic model of a coin tossing is unknown. Here, we show that the heads-tails basin boundaries are smooth so the outcome of the coin tossing is predictable. We have found that the increase of the number of impacts in the period when the coin bounces on the floor makes the basin boundaries more complex and in the limit of infinite number of impacts the behavior of the coin is chaotic and the basins of heads and tails become intermingled.[11–15] Our results demonstrate that although the coin tossing is predictable it can also be considered as an approximation of the random process and can serve as the foundation for understanding the behavior of physical (mechanical) randomizers.[16–18] We anticipate our results to be a new point in the discussion of the nature of random processes.[16,19]

2. The Coin Model

A coin can be modeled as a rigid body, namely a cylinder with a radius r and height h as shown in Figure 1. The center of the coin mass C is located at the distance $\xi_C \neq 0$, $\eta_C \neq 0$, $\zeta_C \neq 0$ from the geometrical center B. Any arbitrary position of a rigid body with respect to the fixed reference frame $Oxyz$ can be described by a combination of the position of the origin of the local reference frame $x'y'z'$ and the orientation (angular position) of this frame $\xi\eta\zeta$ (see [20,21]). The local reference frame $x'y'z'$ is rigidly attached to the body and its axes are parallel to the xyz frame and $\xi\eta\zeta$ is the frame embedded and fixed in the body. As the origin of the local frames it is convenient to choose the geometric center of the body model B.

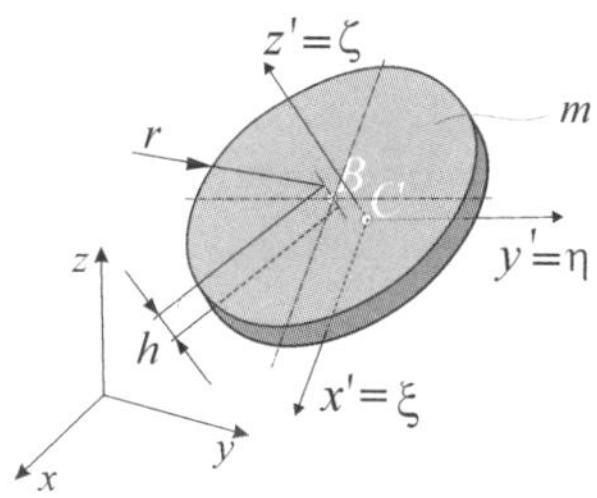

Fig. 1. 3-dimensional model of the coin and its orientation in the space.

In our studies, we consider the following motion of the coin. We assume that the coin is thrown at the height z_0 with the following initial conditions $\Phi_0 = \{x_0, y_0, z_0, \dot{x}_0, \dot{y}_0, \dot{z}_0, \psi_0, \upsilon_0, \phi_0, \omega_{\xi 0}, \omega_{\eta 0}, \omega_{\zeta 0}\}$, i.e, the mass centre initial position is given by $\{x_0,\ y_0\ ,\ \mathrm{z}_0\}$, its initial velocity $\{\dot{x}_0, \dot{y}_0, \dot{z}_0\}$, coin initial orientation $\{\psi_0, \theta_0, \phi_0\}$ and initial angular velocity $\{\omega_\xi 0, \omega_\eta 0, \omega_\zeta 0\}$. After a free fall when z coordinate of one of the coin points is equal to zero, the coin collides the parallel base (floor). It is assumed that at the collision, a portion of the coin energy is dissipated, i.e., the collision is described by the restitution coefficient $\chi < 1$. At the contact between the coin and the floor the developed friction described by the friction coefficient μ_{fr} is assumed.[22] After the collision, the coin mass center moves to the height z_1 in which the total mechanical energy of the coin $E = T + V$ is equal to its total energy in the moment after the collision $E' = T' + V'$ minus the energy dissipated due to the air resistance. Next, the coin moves on until it collides with the floor again. The calculations are terminated when after n-th collision the total mechanical energy of the coin $E = T + V$ is smaller than the potential energy at the coin center level equal to its radius r, i.e., $E < mgr$, where g is a gravitational acceleration, as this condition disables the change of the coin face during further motion.

The equations of motion describing the tossed coin are Newton's equations, with no external source of random influence, i.e., the fluctuations of air, thermodynamic or quantum fluctuations of the coin. If the outcome of the long sequence of the coin tossing is to give a random result, it can only be because the initial conditions vary sufficiently from toss to toss. The flow given by the equations of motion maps all possible initial conditions into one of the final configurations. The set of initial conditions which is mapped onto heads configuration creates heads basin of attraction while the set of initial conditions mapped onto tails configuration creates tails basin of attraction. The boundary, which separates heads and tails basins, consists of initial conditions mapped onto the coin standing on the edge configuration. For infinitely thin coin, this set is a set of zero measure and thus with probability one the coin ends up either heads or tails. For the finite thinness of the coin this measure is not zero but the probability of edge configuration to be stable is low.

Assume that one can set the initial conditions Φ_0 with uncertainty ϵ. If the ball B in the phase space centered at Φ_0 contains only the points which go to one of the final states, the outcome is predictable and repeatable. If in the ball B there are points leading to different final states (denote the set of points leading to heads as H and the set points leading to tails as T), then the result of tossing is not predictable. One can calculate the probability of heads (tails) as $P(\text{heads}) = \mu(H)/\mu(B)$ $(P(\text{tails}) = \mu(T)/\mu(B))$ where μ is a measure of the sets H, T and B.

The possibility that heads-tails basin boundaries are fractal,[9] particularly riddled[11–13,15] or intermingled[11,14] ones is worth investigating. Near a given basin boundary, if the initial conditions are given with the uncertainty ϵ, then a fraction $f(\epsilon)$ of initial conditions give the unpredictable outcome. In the limit $\epsilon \to 0$, $f(\epsilon)\tilde{\epsilon}^{\alpha}$ where $\alpha < 1$ for fractal and $\alpha = 1$ for smooth boundary. Fractal basins boundaries are discontinuous (an uncountable sequence of disjoint stripes) or continuous (a snowflake structure).[10] From the point of view of the predictability of the coin toss the possibility of the occurrence of intermingled basins is the most interesting. Let us briefly explain the term of the intermingled basins of attraction. Let A be an attractor with a basin of attraction $\beta(A)$. A basin $\beta(A)$ which has a positive Lebesque measure but does not contain any neighborhood of the attractor A is called a riddled basin, i.e., for any point in the riddled basin of the attractor, a ball in the phase space of arbitrarily small radius has a nonzero fraction of its volume in some other (say D) attractor basin. The basin of the attractor D may or may not be riddled by the basin $\beta(A)$. If the basin $\beta(D)$ is also riddled by the basin $\beta(A)$ such basins are called the intermingled ones. In the case of the tossed coin the intermingled basins of attraction between heads and tails will mean, that in any neighborhood of the initial condition leading to heads, there are initial conditions which are mapped to tails, i.e., there does not exist an open set of initial conditions which is mapped to one of the final states or infinitely small inaccuracy in the initial conditions makes the state of the coin tossing unpredictable.

In our numerical calculations we consider the following coin data: $m=$ 20 g, $r=$ 1.25 cm, $h=$ 0.2 cm (former Polish 1 PLN coin made of a light aluminum based alloy) and $\xi_C = 0.1$ cm , $\eta_C = 0.1$ cm , $\zeta_C =$ -0.02 cm. We considered the air resistance acting on the coin in both tangential and normal directions and described by by the following coefficients $\lambda_n = 0.8$, $\lambda_\tau = 0.2$. The friction between the coin and the floor during the impact is described by friction coefficient $\mu_{fr} = 0.1$.

3. Results and Discussion

Figure 2(a-d) shows the basins of attraction of heads and tails calculated for various coin models. The dark regions correspond to heads and the white ones to tails. The case of the coin terminating on the soft floor (restitution coefficient $\chi = 0$) is shown in Figure 2(a). The case, which allows the bouncing of the coin on the floor surface ($\chi = 0.6$), is shown in Figure 2(b). The structure of the basins boundaries for the case without bouncing on the floor is similar to the structure in the Keller model.[7] One can notice, that the structure of the basins boundaries is more complex (looks like fractal or intermingled) when the coin is allowed to bounce on the floor as it can be seen in Figure 2(b). To check the possibility, that these basins are fractal (or intermingled), the appropriate enlargements are presented in Figure 2(c,d). It can be seen that apart from the graininess due to the finite number of points, the boundaries are smooth (see Figure 2(d). Under further magnification no new structure can be resolved, i.e., no evidence of intermingled or even fractal basin boundaries is visible. The same conclusion has been reached in the studies of simple one-dimensional models.[4,5] Figure 2(a-d) has been calculated by numerically integrating equations of motion. We fixed all initial conditions except two namely: the position of the coin mass center z_0 and the angular velocity $\omega_{\xi 0}$. We check that the similar structures of the basins boundaries are observed when different initial conditions are allowed to vary, so the presented in Figure 2(a-d) the two-dimensional sections of the phase space, are good indication of what happens in the entire phase space.

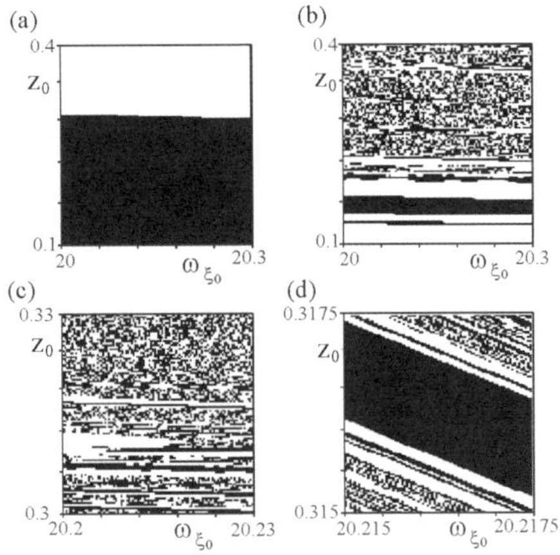

Fig. 2. Basins of attraction of heads (black) and tails (white); (a) coin lands on the soft surface, (b-d) coin bounds on the floor, (c,d) enlargements of (b).

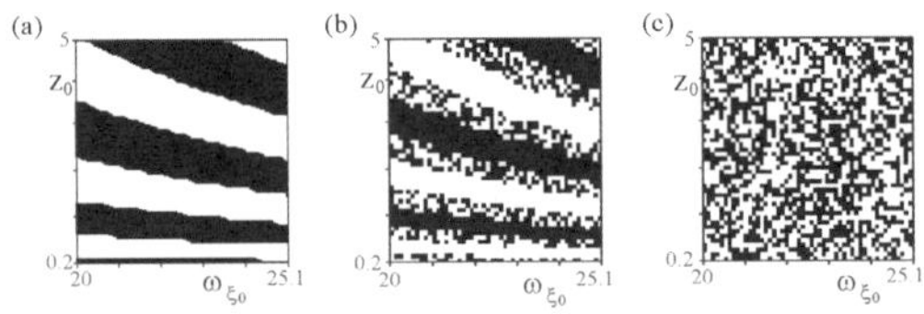

Fig. 3. Basins of attraction indicating the face of the coin which is up after the n-th collision: (a) n=0, (b) n=3, (c) n=10, heads and tails are indicated in black and white respectively.

This allows us to state our main result: for any initial condition $\mathbf{\Phi}_0$

there exists such $\epsilon > 0$ that the ball with radius ϵ centered at $\mathbf{\Phi}_0$ is mapped into the set H or T. In other words, if one can settle the initial condition with appropriate accuracy, the outcome of the coin tossing procedure is predictable and repeatable.

Acknowledgements

This study has been partially supported by the Polish Department for Scientific Research (DBN) under project No. N N501 0710 33.

References

1. W. Feller, *An Introduction to Probability: Theory and Examples* (Wiley, New York, 1957).
2. E. T. Jaynes, *Probability Theory: The Logic of Science* (Cambridge University Press, Cambridge, 1996).
3. J. E. Kerrich, *An Experimental Introduction to the Theory of Probability* (J. Jorgensen, Copenhagen, 1946).
4. V. Z. Vulović and R. E. Prange, *Physical Review A* **33**(1), p. 576 (1986).
5. P. Diaconis, S. Holmes and R. Montgomery, *SIAM Rev.* **49**, p. 211 (2007).
6. T. Mizuguchi and M. Suwashita, *Progress of Theoretical Physics Supplement* **161**, p. 274 (2006).
7. J. B. Keller, *Americam Mathematical Monthly* **93**, p. 191 (1986).
8. J. Strzalko, J. Grabski, A. Stefanski, P. Perlikowski and T. Kapitaniak, *Physics Reports* **469**, 59 (2008).
9. C. Grebogi, S. W. McDonald, E. Ott and J. A. Yorke, *Phys. Lett. A* **99**, p. 415 (1983).
10. B. B. Mandelbrot, *Fractal Geometry of Nature* (Freeman, San Francisco, 1982).
11. J. C. Alexander, J. A. Yorke, Z. You and I. Kan, *Int. J. Bifur. Chaos* **2**, p. 795 (1992).
12. J. C. Sommerer and E. Ott, *Nature* **365**, p. 136 (1993).
13. E. Ott, J. C. Alexander, I. Kan, J. Sommerer and J. Yorke, *Physica D* **76**, p. 384 (1994).
14. T. Kapitaniak, *Phys. Rev. E* **53**, p. 53 (1996).
15. T. Kapitaniak, Y. Maistrenko, A. Stefanski and J. Brindley, *Phys. Rev. E* **57**, p. R6253 (1998).
16. H. Poincare, *Calcul de Probabilites* (George Carre, Paris, 1896).
17. E. Hopf, *Journal of Mathematical Physics* **13**, p. 51 (1934).
18. E. Hopf, *Jahresbericht Deutsche Mathematische Verlag* **46**, p. 179 (1936).
19. J. Ford, *Physics Today* **40**, p. 3 (1983).
20. H. Goldstein, *Classical Mechanics* (Addison-Wesley, Reading, 1950).
21. J. E. Marsden and T. Ratiu, *Introduction to Mechanics and Symmetry* (Springer, New York, 1994).
22. J. I. Nejmark and N. A. Fufajev, *Dynamics of Nonholonomic Systems* (American Mathematical Society, 1972).

Parametric Rolling Behaviour of a Containership

I. G. Tigkas* and N. Pallikaropoulos†

*National Technical University of Athens, Greece
Email: itigkas@naval.ntua.gr
†HALCOR S. A., Metal Works, Greece
Email: npallika@gmail.com

Parametric roll excitations of a containership are investigated by a "Mathieu-type" roll model. The results of analytical closed-form formulas for the investigated ship are compared with the results obtained by the numerical technique of continuation. The aim of the study is twofold; firstly to extract the parametric roll boundaries of the investigated containership and secondly to validate the analytical expressions developed by international organisations.[1,2] The continuation technique implemented in our study is able to unveil in detail the dynamic portrait of the investigated dynamical system and to identify all associated bifurcations. The method of continuation in nonlinear roll dynamics not only predicts the instability roll boundaries and the steady amplitudes of roll oscillation by following the branches of stable solutions, but is also able to track the branches of unstable solutions and trace the system's bifurcations.

Keywords: Parametric rolling; continuation; containership; bifurcation analysis; Mathieu-type; bifurcation; limit-point of cycles.

1. Introduction

Parametric rolling is a "Mathieu-type" low-frequency resonance induced by the dynamic fluctuations of roll restoring. When a ship encounters a train of regular head or astern steep waves, where each wave's length equals approximately the ship's length, she may experience large heeling angles at the crest of those waves. These roll angles are caused from the reduced stability of the ship when her midships is poised at the crest of the waves, as discussed previously. Each time the vessel lands on the wave's trough, a strong and sudden upright moment opposite to the heeling angle forces the ship to the upright condition. This upright moment produces a dynamic effect and therefore on the next wave crest the vessel heels at an opposite roll angle. This phenomenon induces a parametric oscillation that it is either being damped when the vessel passes out of the wave train, or builds-up to the undesirable consequence of sudden and large heeling angles.

The phenomenon of parametric rolling was firstly addressed in the 1950s.[3,4] It was not until recently, that the dynamics leading to parametric rolling have been clarified.[5–8] Unfortunately, the International Maritime Organization (IMO) has not yet imposed stability criteria in order to safeguard ships from parametric rolling, but only recently issued a "Guidance to the Master" ([9,10]).

Nevertheless, the multi million dollar accidents owed to parametric oscillations (e.g. M/V APL China with estimated losses 100 million USD, 1998) have stimulated the International Towing Tank Committee (ITTC) and a well recognised classification society to develop a technical guide[1] of how to avoid parametric rolling for containers from the early design stage. Both guidelines predict the boundaries of parametric roll excitation by utilising analytical "closed-form" expressions. The method of prediction is based upon perturbation-type elaborations of a generic roll model embedded with a "Mathieu-type" oscillator.

In order to check the validity of the analytical expressions and to construct the boundaries of parametric roll excitations, the method of "continuation" or else "path-following" is implemented. Continuation is able to follow the branches of stable and unstable solutions with the same ease independently of whether they are steady or periodic.

The implementation of continuation analysis has already been accomplished by some of the authors in ship manoeuvrability studies.[11,12] The authors have also addressed the problem of parametric rolling with continuation techniques following a more practical approach dedicated to ship designers and naval architects.[13] Further details and discussion regarding design criteria can be extracted from that study.

2. Analytical Formulation

The roll motion can be mathematically modelled in a single degree of freedom by using the following expression:

$$\ddot{\phi} + 2\zeta\omega_0\dot{\varphi} + \omega_0^2\left[1 - h\cos\left(\omega_e t\right)\right]\phi - c_3\omega_0^2\phi^3 - c_5\omega_0^2\phi^5 = 0 \qquad (1)$$

where, ϕ is the roll angle, ζ is the damping factor, ω_0 is the natural roll frequency, ω_e is the wave encounter frequency, c_3, c_5 are the roll restoring coefficients and h is obtained from metacentric height's GM fluctuation where it can be defined as $h = \frac{GM_{\max} - GM_{\min}}{2\overline{GM}}$.

In the ABS guideline the corresponding equation is identical to (1), if $\overline{GM}$ is interpreted as $\frac{GM_{\max} + GM_{\min}}{2}$ and the 5th-order restoring term is omitted.

The ITTC's instability boundaries are addressed to small perturbations from the upright state, hence the nonlinear restoring terms do not affect the value of h. The formula estimating the principal resonance is outlined below:

$$h = \sqrt{\left(2 - \frac{\omega_e^2}{2\omega_0^2}\right)^2 + 4\zeta^2 \frac{\omega_e^2}{\omega_0^2}} \tag{2}$$

The formula shown above is the 1st-order approximation of the Mathieu function that truly defines the boundary.

Despite the fact that the inception of instability is a 'linear' phenomenon, the roll amplitude that arises from the loss of stability is dominantly influenced by nonlinearities. The formulas used in ITTC guideline for predicting the steady parametric rolling originate from the method of harmonic balance:[5]

$$A^2 = \frac{4}{3\,c_3}\left[\left(1 - \frac{1}{a}\right) \mp \sqrt{\frac{h^2}{4} - \frac{4\zeta^2}{a}}\right] \quad \text{for} \quad c_5 = 0 \tag{3}$$

$$A^2 = -\frac{3\,c_3}{5\,c_5} \pm \sqrt{\left(\frac{3\,c_3}{5\,c_5}\right)^2 - \frac{8}{5c_5}\left(-1 + \frac{1}{a} \pm \sqrt{\frac{h^2}{4} - \frac{4\zeta^2}{a}}\right)} \text{for } c_5 \neq 0 \tag{4}$$

where A is the parametric roll amplitude and $a = 4\omega_0^2/\omega_e^2$.

The $\pm$ sign is attributed to the phenomenon of the coexistence of multiple stable and unstable solutions for the same value of the frequency parameter a.

Conversely, the guide developed by ABS is derived by the method of multiple scales, but after a conversion of the symbols used leads to identical results. A difference occurs on how the value of a is calculated. In ABS guide the frequency parameter is not calculated at the natural frequency, but at a modified one, based upon the metacentric height.

The boundaries of parametric roll excitation at the principal resonance are calculated for a post-panamax containership and are illustrated in Figure 1. The characteristics of the ship used for the computation are: $\zeta = 0.061275$ at $\omega_0 = 0.2076 s^{-1}$ and hardening restoring $c_3 = -0.99861$.

The extra boundary curve drawn in the figure is due to the following expression:[5]

$$h = \frac{4\zeta}{\sqrt{a}}, \quad a \leq 1 \tag{5}$$

From Figure 1 we can depict that the system undergoes a behaviour rich in dynamic phenomena. For example the borderline denoted by (a), explains the transition to the parametric roll oscillatory state through progressively increasing roll excitation amplitudes. This birth and the then gradual increase in the amplitude of the limit cycles is a clear phenomenon of a supercritical-type bifurcation. Therefore, a progressively larger variation in the GM is required to initiate parametric roll excitations for increasing values of a above unity.

An even more interesting behaviour is observed on the left-hand side of Figure 1. In this case two lines denoted by (b) and (c) constitute the borders of stability for the investigated dynamical system. The line (b) remains the natural continuation of the loci of the borderline (a). The main difference is that the limit cycles of the parametric roll oscillatory motion do not evolve smoothly and gradually by an increasing h, but they arise through a sudden "jump". The line (c) on the other hand, is mistakenly not included in the two guidelines. They represent the "fold" of the limit cycles that are generated by their birth at borderline (b). Hence unstable limit cycles are born in borderline (b), where they fold back and turn into stable ones at borderline (c). This "limit-point of cycles" (LPC) bifurcation point, is also referred in the literature as "fold of cycles" or "saddle-node of periodic orbits" bifurcation. When investigating the global stability of the system the borderline (c) should not be omitted. The system's behaviour in the region between the two borderlines (b) and (c) is unclear since co-existence of the two states is apparent. Despite the fact that the ship may remain in the "upright" condition, a certain strong perturbation in the environmental conditions may drive the vessel out of the safe basin and initiate the phenomenon of parametric rolling.

3. The Investigated Containership

A post-panamax containership was the case of our study, where her principal characteristics are summarised in Table 1. It is worth noting, as shown in Figure 2, the change in the wetted area between the crest and the trough at a sample wave for a characteristic containership. A characteristic variation in the roll righting lever GZ of a containership is shown in Figure 3(a).

By using least-squares fits in the true still-water GZ, we can obtain the restoring coefficients for various orders of polynomials. The range of validity of the fit depends clearly upon the order of the polynomials used to model the roll restoring. Figure 3(b) on the other hand, represents a typical time-history of the roll angle leading to parametric roll oscillations.

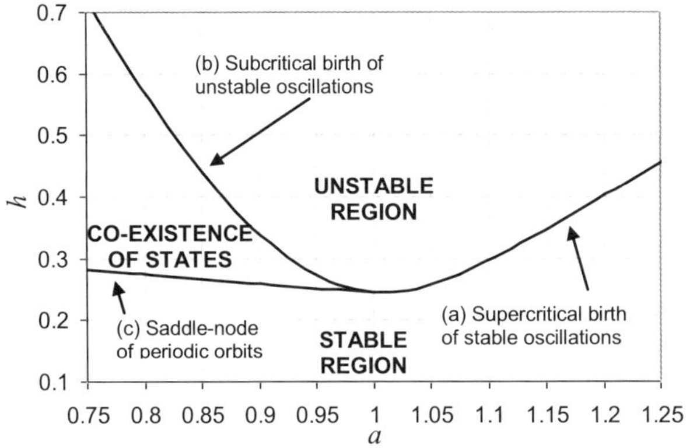

Fig. 1. The stability boundaries of parametric rolling.

Table 1. Particulars and other data of the investigated containership.

Length (L):	288.87 m	Δ:	114,956 t
Beam (B):	42.8 m	TEU:	6,660
Mean draught (T):	14.0 m	GM:	1.08 m
Depth to upper deck (D):	24.4 m	ω_0:	0.2076 rad/s
Block coefficient (C_b):	0.647	ζ:	0.061275

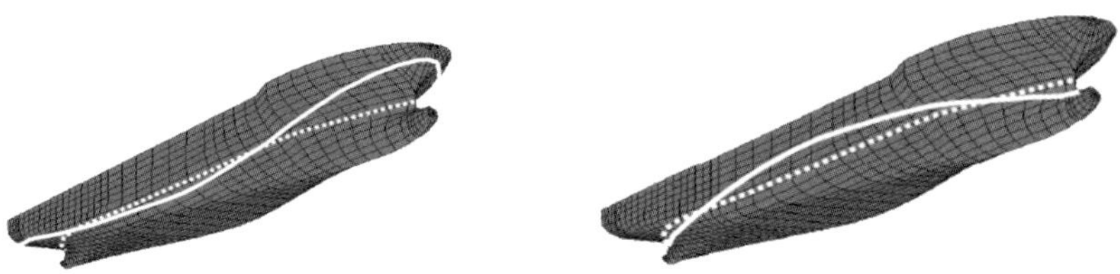

Fig. 2. Change in the wetted area of a containership between the trough and the crest of a sample wave.[7]

Table 2. Roll restoring coefficients

	c_3	c_5	range of validity
Cubic hardening	-0.99861	-	20 deg
Quintic softening	-3.12037	8.287963	44 deg

4. Dynamic Analysis using the Continuation Method

Before the roll dynamical system undergoes a continuation analysis, it has firstly to be transformed into its autonomous canonical form using state

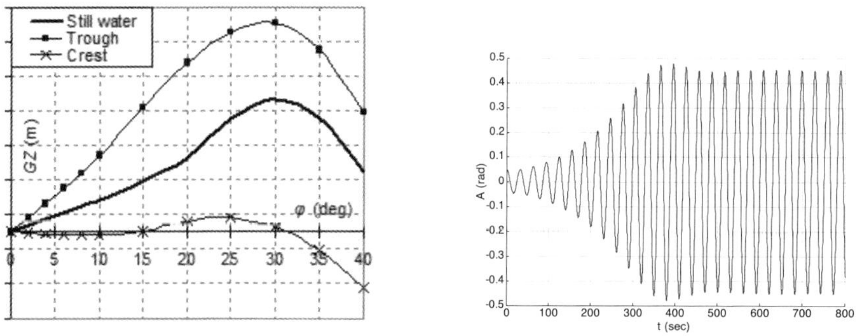

(a) Characteristic GZ change of a containership between still water, crest and trough.

(b) Time-history of a characteristic parametric roll excitation for the investigated containership.

Fig. 3.

space representation:

$$\dot{\mathbf{x}} = f(\mathbf{x}, \mathbf{k}; t) \tag{6}$$

where, $\mathbf{x}$ and $\mathbf{k}$ are the vectors of the variables and of the controlling parameters respectively. The values of a, h, and ζ may comprise the governing parameters of the system. The vector of variables on the other hand requires a special treatment. Therefore, by using certain dummy variables, equation (1) can be transformed into its state space representation by the following set of first order differential equations:

$$\begin{aligned} \dot{\phi} &= p \\ \dot{p} &= -2\zeta\omega_0\, p - \omega_0^2\,(1 - hy)\,\phi + c_3\omega_0^2\phi^3 + c_5\omega_0^2\phi^5 \\ \dot{x} &= x + y\sqrt{4\omega_0^2/a} - x\left(x^2 + y^2\right) \\ \dot{y} &= -x\sqrt{4\omega_0^2/a} + y - x\left(x^2 + y^2\right) \end{aligned} \tag{7}$$

where, $x = \sin(\omega_e t)$, $\quad y = \cos(\omega_e t)$.

The two last added variables are used in order to detach the time dependence of the system that is responsible for the periodic force excitation. Continuation analysis cannot be achieved otherwise unless these two dummy variables, which comprise a nonlinear oscillator, are used. These last two dummy variables produce asymptotically the desired stable solutions.

The nonlinear continuation analysis is achieved through MatCont algorithm.[14] Therefore with this automated procedure, we follow the branches of solutions and we construct the following bifurcation diagrams.

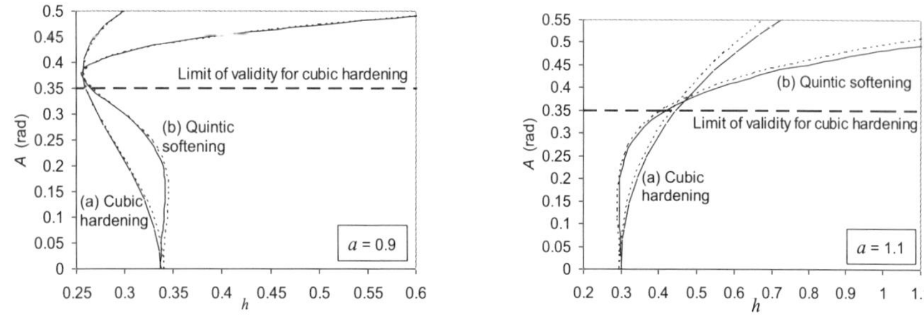

(a) (b)

Fig. 4. Comparisons of numerical (continuous line) and analytical (dotted) parametric roll amplitude (A) predictions for two scenarios with different values of a.

In Figures 4(a) and 4(b) we are able to identify the values of the parametric roll amplitudes corresponding to different values of the GM variation parameter h, for a wide range of frequency parameter a. This is achieved for the two cases with the 3rd and the 5th order polynomials for nonlinear roll restoring. A good agreement between the analytical and the numerical results is observed; meaning that that the analytical expressions are capable of successfully representing the associated dynamics of the Mathieu-type roll model with both 3rd and 5th order restoring. A 3-d diagram and its 2-d projection for certain conditions are shown in Figure 5(a) and 5(b), where an eminent subcritical type of response is present.

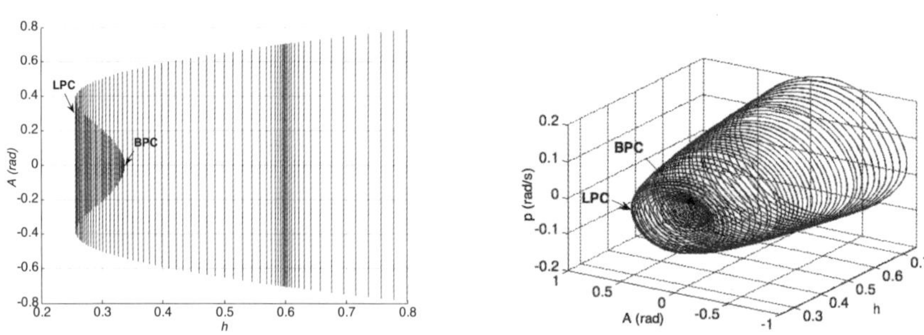

(a) A 3-d representation of the evolution of the limit cycles. ($a = 0.9$)

(b) A projection focused on the effect of the GM parameter h upon the roll excitation amplitude A. ($a = 0.9$)

Fig. 5.

A cusp point of cycles (CPC) appearing in the co-dimensional plane, as shown in Figure 6(a), is responsible for the separation of the qualitative behaviour from supercritical to subcritical. This is evident at decreasing values of a below unity.

A codimension-2 continuation was finally initiated from the bifurcations in Figures 5(a) and 5(b). The results in Figure 6(b) are therefore obtained be varying simultaneously both frequency parameter a and GM parameter h simultaneously. The dynamical portrait of the system is hence fully disclosed and the boundaries of stability are clearly formed. One could falsely expect that the analytical formulas based upon perturbation-type methods cannot successfully model the system' behaviour especially at large angles of roll. The comparison showed that despite minor quantitative differences at high values of a, the analytical expressions can be trusted for depicting the boundaries of stability for the investigated system. One should then consider the validity of the single degree Mathieu-type roll model, for expressing the actual ship dynamics.[13]

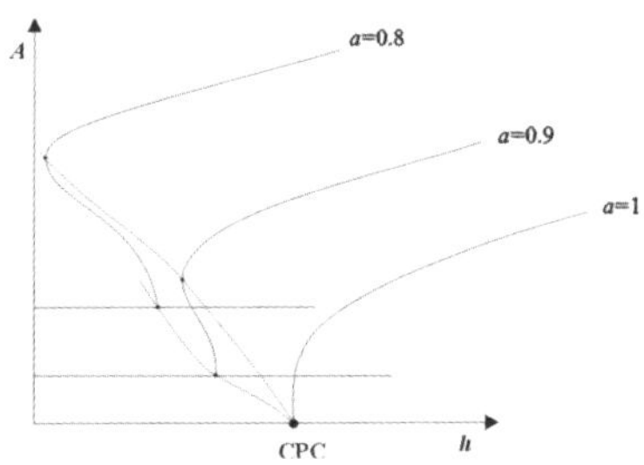

(a) A "Cusp point of cycles" separates the supercritical behaviour from the subcritical one.

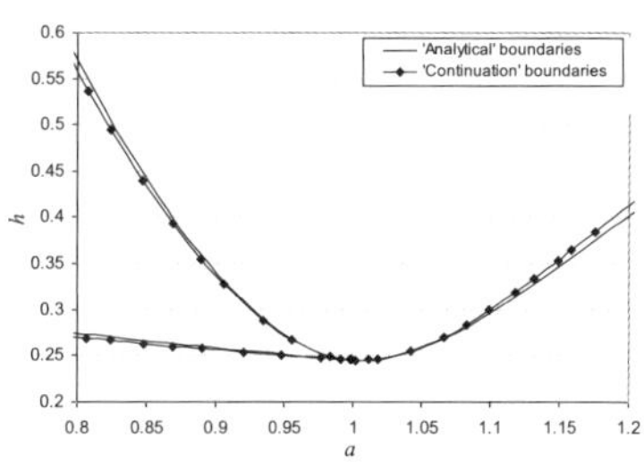

(b) Comparison of the boundaries of stability obtained by analytical formulas and numerical continuation.

Fig. 6.

5. Conclusions

The analytical expressions were tested through the numerical continuation technique and it was found that they are able to successfully represent the behaviour of the simple Mathieu-type roll system.

The importance of the locus of the saddle-nodes of the parametric oscillations is emphasised for an accurate global analysis of the system's behaviour.

Numerical continuation can be proven as a valuable tool for the investigation of the parametric roll boundaries in ship dynamics, since it expedites and automatise the process and on the same way it clearly unveils the system's global dynamic portrait. More complicated mathematical models (firstly transformed into a set of O.D.E.s) may be accommodated in order to depict more detailed and accurate results.

Acknowledgments

The authors acknowledge Professor K. Spyrou for his kind supervision throughout this current effort. Ioannis Tigkas would like to thank "Alexander S. Onassis" Public Benefit Foundation for his PhD scholarship.

References

1. American Bureau of Shipping, ABS Guide for the assessment of parametric roll resonance in the design of container carriers (2004).
2. ITTC, Report of the stability in waves committee, in *Proceedings 24th International Towing Tank Conference*, (2005).
3. O. Grim, *Schiffstechnik* **1**, 10 (1952).
4. J. Kerwin, *International Shipbuilding Progress* **2**, 597 (1955).
5. K. J. Spyrou, *Oceanic Engineering International* **9**, 11 (2005).
6. W. N. France, M. Levadou, T. W. Treakle, J. R. Paulling, R. K. Michel and C. Moore, *Marine Technology* **40**, 1 (2003).
7. Y. S. Shin, V. L. Belenky, J. R. Paulling, K. M. Weems and W. M. Lin, Criteria for parametric roll of large containerships in longitudinal seas, in *SNAME Annual Meeting*, *SNAME Annual Meeting* (Washington D.C., USA, 2004).
8. M. Neves, and C. Rodriguez, *Ocean Engineering* **34**, 1618 (2007).
9. IMO, Revision of the intact stability code, sub-committee on stability and load lines and on fishing vessels safety (May 2006).
10. IMO, Revised guidance to the master for avoiding dangerous situations in adverse weather and sea conditions, resolution msc.1/circ.1228 (2007).
11. K. J. Spyrou, I. Tigkas and A. Chatzis, *Journal of Ship Research* **51**, 160 (2007).
12. K. J. Spyrou and I. Tigkas, Nonlinear dynamics of ship steering behaviour under environmental excitations, in *Proccedings of the IUTAM Symposium on Fluid-Structure Interaction in Ocean Engineering*, ed. E. Kreuzer, IUTAM Bookseries, Vol. 8 (Springer, 2007).
13. K. J. Spyrou, I. Tigkas, G. Scanferla, N. Pallikaropoulos and N. Themelis, *Oceanic Engineering* **35**(11-12), 1235 (2008).
14. A. Dhooge, W. Govaerts, Y. A. Kuznetsov, W. Mestrom, A. M. Riet and B. Sautois, *MATCONT and CL_MATCONT: Continuation toolboxes for MATLAB*, tech. rep., Gent (Belg.) and Utrecht (Neth.) Universities (2003).

Chaos Synchronization of Genesio System

Ö. Umut
Abant İzzet Baysal University
Bolu, Turkey

This paper presents chaos synchronization of two identical Genesio systems by using three methods; Pecora and Carroll, active control and linear generalized synchronization. We demonstrate that coupled Genesio systems can be synchronized. Numerical simulations are given to verify the effectiveness of these methods.

Keywords: Genesio system; chaos synhronization; Pecora and Carroll method; active control; linear generalized synchronization.

1. Introduction

Chaos synchronization is an important problem in nonlinear science. In the last few decades, synchronization has received a great deal of interest among scientists from various fields.[1,2] The various types of chaos synchronization, whose description may require different theoretical frameworks were found in natural systems and specified.

The idea of synchronizing two identical chaotic systems that from different initial conditions was introduced by Pecora.[1,3] Several schemes can be used to achieve such type of synchronization, for example Pecora and Carroll method,[1,4] active control,[5,6] adaptive control,[7,8] etc. Identical synchronization has clear application to secure communication due to special features of chaotic systems, such as being extremely sensitive to tiny variations of initial conditions and system parameters.

The practical applications of chaos synchronization show the limitation of identical synchronization: parameter mismatch will probably destroy the manifold of this synchronization. Consequently, generalized synchronization was introduced,[9,10] Generalized synchronization can give much richer dynamics than identical ones.

This work considers the synchronization problems of two Genesio systems. We will show that it can be achieved by the following methods: Pecora and Carroll, active control and linear generalized synchronization.

The Genesio system proposed by Genesio and Tesi,[11] is one of paradigms of chaos since it captures many features of chaotic systems. It includes a simple square part and three simple ordinary differential equations that depend on three positive parameters. The dynamic equations of the system is given by

$$\begin{aligned}\dot{x} &= y \\ \dot{y} &= z \\ \dot{z} &= -ax - by - cz + x^2\end{aligned} \tag{1}$$

where x, y and z are state variables and a, b and c are positive control parameters. This system is dissipative for $c > 0$ and has a chaotic attractor when $a = 6.0$, $b = 2.92$ and $c = 1.2$ (see Figure 1). Therefore, the state trajectories $x(t)$, $y(t)$ and $z(t)$ of the system in (1) are globally bounded for all $t \geq 0$ and continuously differentiable with respect to time. Consequently, there exists a positive constant M such that $|x(t)| \leq M < \infty$, $|y(t)| \leq M < \infty$ and $|z(t)| \leq M < \infty$ hold for all $t \geq 0$.

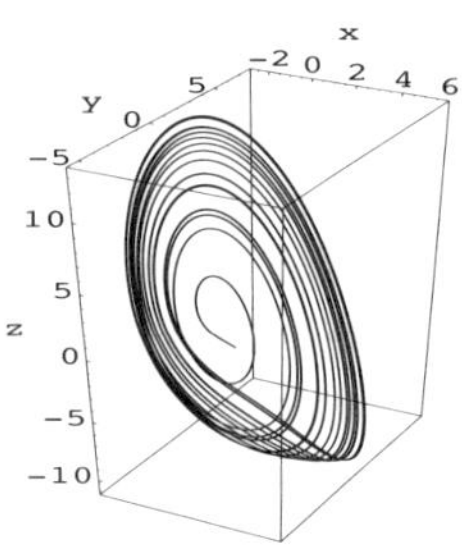

Fig. 1. Genesio attractor

2. Synchronization using Pecora and Carroll Method

In this section, we will apply the most popular method for constructing synchronizing systems, which was introduced by Pecora and Carroll (PC).[1,4] Let us consider an autonomous chaotic dynamical system

$$\dot{u} = f(u)$$

where $u = (u_1, u_2, \ldots, u_n)^T$ is an n-dimensional state vector with f defining a vector field $f\colon \mathbb{R}^n \to \mathbb{R}^n$. The method of Pecora and Carroll decomposes the dynamical system $\dot{u} = f(u)$ into two subsystems

$$\dot{v} = g(v, w) \quad \text{and} \quad \dot{w} = h(v, w)$$

where $v = (v_1, v_2, ..., v_m)^T$ and $w = (w_{m+1}, w_{m+2}, ..., w_n)^T$ such that any second subsystem (response)

$$\dot{w}' = h(v, w')$$

with the same driving v but with different variable w' synchronizes ($w \to w'$ as $t \to \infty$) with the original w-subsystem. The v-system and the w'-system are called the *drive* and the *response* system, respectively. Let us build a PC drive-response configuration with a drive system given by the Genesio system (with three state variables denoted by subscript 1) and with a response system given by the subspace containing the (y, z) variables. We will use the chaotic signal x_1 to drive the response subsystem whose variables are denoted by subscript 2. The drive system is described by

$$\begin{aligned} \dot{x}_1 &= y_1 \\ \dot{y}_1 &= z_1 \\ \dot{z}_1 &= -ax_1 - by_1 - cz_1 + x_1^2 \end{aligned} \tag{2}$$

The response system is given by

$$\begin{aligned} \dot{y}_2 &= z_2 \\ \dot{z}_2 &= -ax_1 - by_2 - cz_2 + x_1^2 \end{aligned} \tag{3}$$

Define the synchronization error by

$$e_y = y_2 - y_1, \quad \text{and} \quad e_z = z_2 - z_1$$

Then subtracting system (2) from system (3), we obtain the error system

$$\begin{aligned} \dot{e}_y &= e_z \\ \dot{e}_z &= -be_y - ce_z \end{aligned} \tag{4}$$

The linear system of synchronization error (4) has two complex eigenvalues with negative real part

$$\lambda_{1,2} = -\frac{c}{2} \pm \frac{1}{2}\sqrt{c^2 - 4b} \qquad (\lambda_2 = \bar{\lambda}_1, \quad \text{since} \quad c^2 < 4b)$$

This implies that the zero solution of system (4) satisfies

$$\| e_y \| \to 0 \quad \text{as} \quad t \to \infty \quad \text{and} \quad \| e_z \| \to 0 \quad \text{as} \quad t \to \infty$$

Hence the zero solution of the error system (4) is asymptotically stable and then the response system (3) with x-drive configuration achieves synchronization.

Remark 1. Only one choice induces the appearance of chaos synchronization namely (y, z) driven by x. For the other possible choices (x, z) driven by y or (x, y) driven by z, the Pecora and Carroll scheme did not succeed in achieving synchronization.

3. Synchronization using Active Control

Here, the synchronization of two identical Genesio systems is achieved using active control.[5,6] We now, assume that we have two Genesio systems and that the drive system with the subscript 1 is to control the response system with subscript 2. The drive and response systems are defined as follows:

$$\begin{aligned} \dot{x}_1 &= y_1 \\ \dot{y}_1 &= z_1 \\ \dot{z}_1 &= -ax_1 - by_1 - cz_1 + x_1^2 \end{aligned} \tag{5}$$

and

$$\begin{aligned} \dot{x}_2 &= y_2 + u_1(t) \\ \dot{y}_2 &= z_2 + u_2(t) \\ \dot{z}_2 &= -ax_2 - by_2 - cz_2 + x_2^2 + u_3(t) \end{aligned} \tag{6}$$

We have introduced three control functions $u_1(t)$, $u_2(t)$ and $u_3(t)$ in the latter system. Our goal is to determine these functions $u_1(t)$, $u_2(t)$ and $u_3(t)$. In order to estimate the control functions, we subtract (5) from (6) and define the error system as

$$\begin{aligned} \dot{e}_1 &= e_2 + u_1(t) \\ \dot{e}_2 &= e_3 + u_2(t) \\ \dot{e}_3 &= -(a - 2x_1)e_1 - be_2 - ce_3 + e_1^2 + u_3(t) \end{aligned} \tag{7}$$

where $e_1 = x_2 - x_1$, $e_2 = y_2 - y_1$, $e_3 = z_2 - z_1$. System (7) is called the *error system* and e_1, e_2, and e_3 are called the *error states.*

The synchronization problem for Genesio system is to achieve the asymptotic stability of the zero solution of the error system (7).

We define the active control functions $u_1(t)$, $u_2(t)$ and $u_3(t)$ as follows:

$$\begin{aligned} u_1 &= V_1(t) \\ u_2 &= V_2(t) \\ u_3 &= (a - 2x_1)e_1 - e_1^2 + V_3(t) \end{aligned} \tag{8}$$

This implies

$$\begin{aligned} \dot{e}_1 &= e_2 + V_1(t) \\ \dot{e}_2 &= e_3 + V_2(t) \\ \dot{e}_3 &= -be_2 - ce_3 + V_3(t) \end{aligned} \tag{9}$$

The error system (7) to be controlled is a linear system with a control input $V_1(t)$, $V_2(t)$ and $V_3(t)$ as functions of e_1, e_2, and e_3. As long as these feedbacks stabilize the system, e_1, e_2, and e_3 converge to zero as time t tends to infinity. This implies that two Genesio systems are synchronized with feedback control. There are many possible choices for the control $V_1(t)$, $V_2(t)$ and $V_3(t)$. We choose

$$\begin{pmatrix} V_1(t) \\ V_2(t) \\ V_3(t) \end{pmatrix} = C \begin{pmatrix} e_1 \\ e_2 \\ e_3 \end{pmatrix} \tag{10}$$

where C is a 3×3 constant matrix. In order to make sure the closed loop system will be stable, the proper choice of the elements of the matrix C is such that the feedback system must have all of the eigenvalues with negative real parts.

Let the matrix C is chosen in the form

$$C = \begin{pmatrix} -1 & -1 & 0 \\ 0 & -1 & -1 \\ 0 & b & c-1 \end{pmatrix}$$

For this particular choice, the closed loop system (7) has the eigenvalues

$$\lambda_1 = -1, \quad \text{and} \quad \lambda_{2,3} = \frac{-(2-c) \pm \sqrt{c^2 - 4b}}{2}, \qquad (c < 2 \ \text{ and } \ c^2 < 4b)$$

With the choices $c < 2$ and $c^2 < 4b$, the error states e_1, e_2, and e_3 converge to zero as time t tends to infinity and hence the synchronization of two Genesio systems is achieved.

4. Linear Generalized Synchronization of Genesio System

A dynamical system can be decomposed into two parts

$$\dot{x} = Ax + \Psi(x) \tag{11}$$

where A is an $n \times n$ constant matrix and $\Psi : \mathbb{R}^n \to \mathbb{R}^n$. We assume that the drive system transmit the signal $\Psi(x)$ to the response system and consider the following unidirectional synchronization scheme:

$$\begin{aligned} \dot{x} &= Ax + \Psi(x) \quad \longleftarrow \quad \text{drive system} \\ \dot{y} &= Ay + \Lambda\Psi(x) \longleftarrow \quad \text{response system} \end{aligned} \tag{12}$$

Theorem 4.1 (Yang and Chua, 1999). *If the matrix Λ commutes with A, then the two dynamical systems in* (12) *are in a state of generalized synchronization via the following generalized synchronization transformation*

$$y(\infty) = H(x) = \Lambda x \tag{13}$$

if and only if A is negative definite.

Proof. Let $z = y - \Lambda x$, then the stability of the generalized synchronization between two dynamical systems in (12) via the generalized synchronization transformation $y = H(x) = \Lambda x$ is equivalent to that of the origin of the following system:

$$\begin{aligned} \dot{z} &= [Ay + \Lambda\Psi x] - [\Lambda Ax + \Lambda\Psi(x)] \\ &= Ay - \Lambda Ax \\ &= A(y - \Lambda x) \quad \text{since } \Lambda \text{ commutes with } A \\ &= Az \end{aligned}$$

Therefore $z = 0$ is asymptotically stable if and only if A is negative definite. □

The Genesio system can be decomposed into two parts as

$$\dot{x} = Ax + \Psi(x) \tag{14}$$

where

$$A = \begin{pmatrix} 0 & 1 & 0 \\ 0 & 0 & 1 \\ -a & -b & -c \end{pmatrix}, \qquad x = \begin{pmatrix} x_1 \\ y_1 \\ z_1 \end{pmatrix}, \qquad \Psi(x) = \begin{pmatrix} 0 \\ 0 \\ x_1^2 \end{pmatrix}$$

The matrix A is negative definite if and only if all eigenvalues of A have negative real parts. By Routh-Hurwitz criteria this provided when $bc > a$

Now, if the driven system is

$$\dot{y} = Ay + \Lambda\Psi(x) \tag{15}$$

where the matrix Λ commutes with A, then the drive Genesio system and the response Genesio system are in a state of generalized synchronization.
Simulation 1. We choose $\Lambda =$ constant, say 2, then

$$\Lambda = \begin{pmatrix} 2 & 0 & 0 \\ 0 & 2 & 0 \\ 0 & 0 & 2 \end{pmatrix}$$

and the response system becomes

$$\begin{aligned} \dot{x}_2 &= y_2 \\ \dot{y}_2 &= z_2 \\ \dot{z}_2 &= -ax_2 - by_2 - cz_2 + x_1^2 \end{aligned} \tag{16}$$

The state variables of the drive system and the response system are related as follows

$$\begin{aligned} x_2 &= 2x_1 \\ y_2 &= 2y_1 \\ z_2 &= 2z_1 \end{aligned}$$

Simulation 2. We now, choose

$$\Lambda = A = \begin{pmatrix} 0 & 1 & 0 \\ 0 & 0 & 1 \\ -a & -b & -c \end{pmatrix}$$

then the response system becomes

$$\begin{aligned} \dot{x}_2 &= y_2 \\ \dot{y}_2 &= z_2 + x_1^2 \\ \dot{z}_2 &= -ax_2 - by_2 - cz_2 - cx_1^2 \end{aligned} \tag{17}$$

The relation between the state variables of the drive and response system is

$$\begin{aligned} x_2 &= y_1 \\ y_2 &= z_1 \\ z_2 &= -ax_1 - by_1 - cz_1 \end{aligned}$$

5. Numerical Results

In this section numerical simulations are given to verify the effectiveness of Pecora and Carroll method, active control method and linear generalized synchronization to achieve synchronization of two Genesio systems. All simulations are carried out using MATHEMATICA. The drive system and the response system are taken as Genesio systems with same parameters but different initial conditions. We select the parameters of Genesio systems as $a = 6$, $b = 2.92$ and $c = 1.2$ to ensure the existence of the chaotic behavior (Figure 1).

Figure 2(a) displays the time response of the drive state y_1 and the response state y_2, while Figure 2(b) displays z_1 and z_2 under Pecora and Carroll method where the initial values of the drive system (2) and the response system (3) are $x_1(0) = 3$, $y_1(0) = -4$, $z_1(0) = 2$, $y_2(0) = 5$ and $z_2(0) = -5$, respectively. The corresponding initial errors are $e_y(0) = 9$ and $e_z(0) = -7$. Figures 3(a)-3(b) show the synchronization errors e_y and e_z of systems (2) and (3), respectively, as functions of time, tend to zero asymptotically. From Figures 2(a)-3(b) we observe that the synchronization of the drive and response of two identical Genesio system occured after $t = 10$.

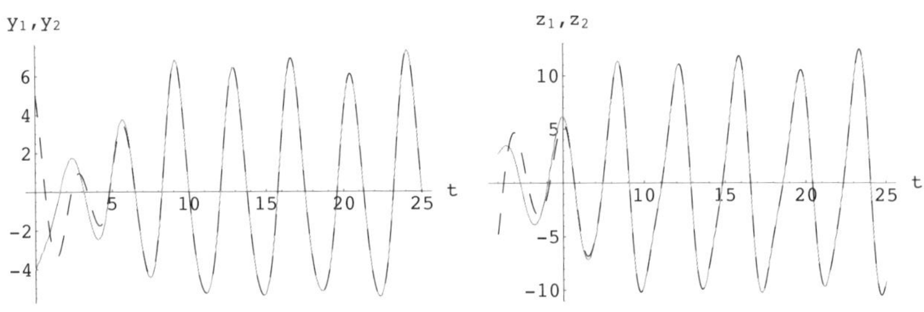

(a) y_1, y_2 (b) z_1, z_2

Fig. 2. Time response of the drive state (gray) and the response state (dash).

The numerical simulations of the drive and response systems (5) and (6), respectively, are presented to show the effect of chaos synchronization by using active control method. The control functions are chosen as in (8) and (10) and the initial states of the drive and response systems are taken as $x_1(0) = 3$, $y_1(0) = -4$, $z_1(0) = 2$, $x_2(0) = 3.4$, $y_2(0) = -5$, $z_2(0) = 2.3$,

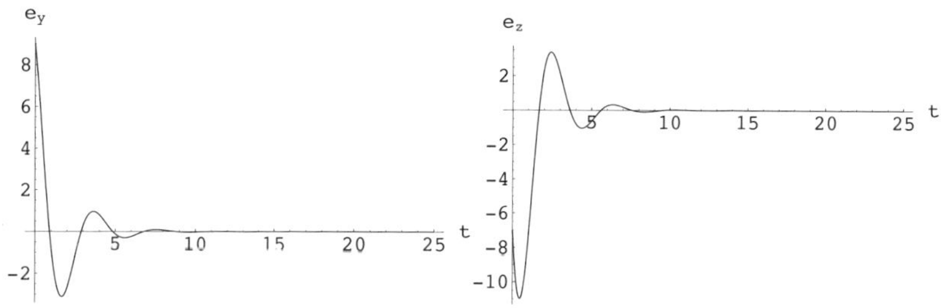

(a) e_y (b) e_z

Fig. 3. The synchronization error as a function of time t

respectively. Then the initial values for the error states are $e_x(0) = 0.4$, $e_y(0) = -1$ and $e_z(0) = 0.3$. Figures 4(a), 4(b) and 4(c) display the time responses of the drive and response states x_1x_2, y_1y_2 and z_1z_2, respectively. Figures 5(a)- 5(c) show the synchronization errors e_x, e_y, and e_z, respectively as functions of time, tend to zero asymptotically. We observe that the synchronization of drive and response of two identical Genesio systems occured after $t = 10$.

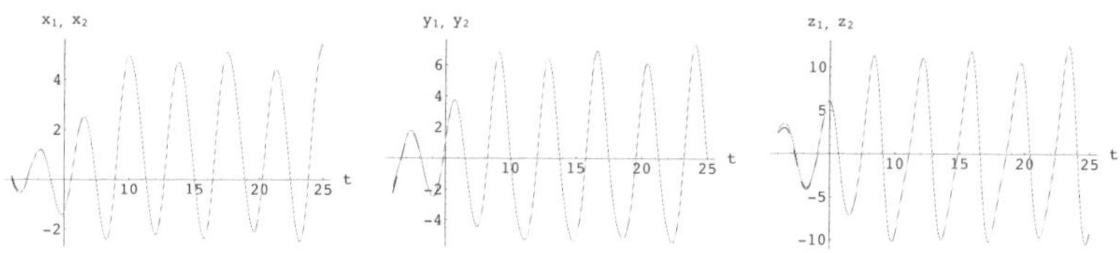

(a) x_1, x_2 (b) y_1, y_2 (c) z_1, z_2

Fig. 4. Time response of the drive states, (gray) and the response states,(dash)

We present two simulations for the linear generalized synchronization. Drive and response trajectories are shown by black and gray lines, respectively, in Figures 6(a)-7(c).

In the first simulation, we take $\Lambda = \text{diag}\,(2, 2, 2)$. The initial values of the drive and the response systems, (14) and (16), respectively are taken as $x_1(0) = 3$, $y_1(0) = -4$, $z_1(0) = 2$, $x_2(0) = 6$, $y_2(0) = -8$ and $z_2(0) = 4$. Figures 6(a)-6(c) display the projections of the synchronized attractors in the coupled Genesio systems onto xy, xz and yz-planes, respectively.

In the second simulation, we take $\Lambda = A$. The initial values of the drive

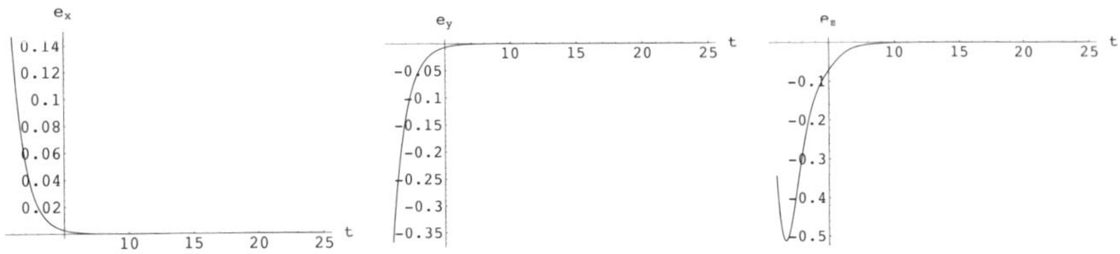

(a) e_x (b) e_y (c) e_z

Fig. 5. The synchronization errors as a function of time t

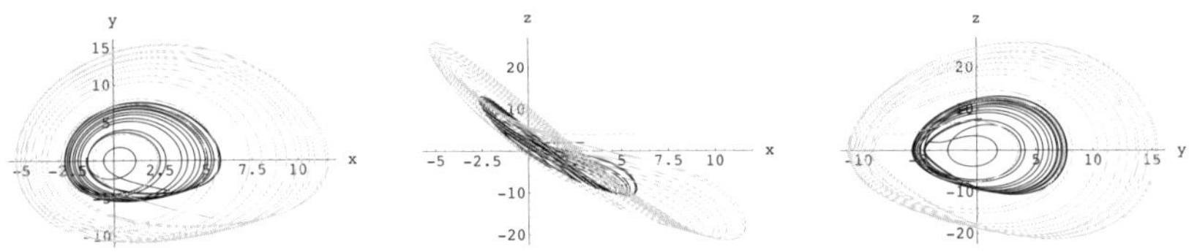

(a) Onto xy-plane (b) Onto xz-plane (c) Onto yz-plane

Fig. 6. Projections of the synchronized attractors

and response systems (14) and (17), respectively are taken as $x_1(0) = 3$, $y_1(0) = -4$, $z_1(0) = 2$, $x_2(0) = -4$, $y_2(0) = 2$, $z_2(0) = -8.72$. Figures 7(a)-7(c) display the projections of the synchronized attractors in the coupled Genesio systems onto xy, xz and yz-planes, respectively.

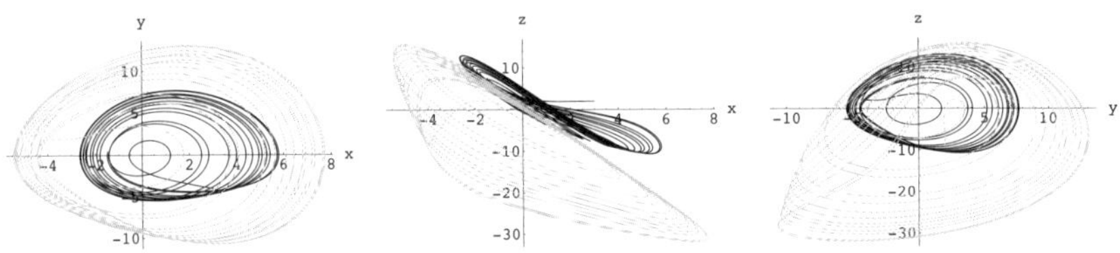

(a) Onto xy-plane (b) Onto xz-plane (c) Onto yz-plane

Fig. 7. Projections of the synchronized attractors

6. Conclusions

This paper addresses chaos synchronization of Genesio system by using Pecora and Carroll method, active control and linear generalized synchronization. A perfect linking of the chaotic trajectories is obtained in the applications of Pecora and Carroll and active control methods. The hooking of the outputs of drive and response systems is achieved when linear general synchronization is used. The conditions sufficient for the occurence of chaos synchronization of coupled Genesio systems are derived by using Lyapunov stability theory. Numerical simulations are performed to verify these results.

References

1. T. L. Carroll and L. M. Pecora, *Phys. Rev. Lett.* **64**(8), 821 (1990).
2. K. M. Cuomo and A. V. Oppenheim, *Phys. Rev. Lett.* **71**, 65 (1993).
3. T. L. Carroll and L. M. Pecora, *IEEE Trans. Circuits Syst.* **38**, 453 (1991).
4. H. N. Agiza, *Nonlinear Analysis* **58**(1–2), 11 (2004).
5. E. W. Bai and K. E. Lonngren, *Chaos, Solitons and Fractals* **8**(1), 51 (1997).
6. H. N. Agiza and M. T. Yassen, *Phys. Lett. A* **278**(4), 191 (2001).
7. A. S. Hegazi, H. N. Agiza and M. El-Dessoky, *Int. J. Bifur. Chaos* **12**(7), 1579 (2002).
8. T. Liao and S. Lin, *J. Franklin Institute* **336**(6), 925 (1999).
9. N. F. Rulkov, M. M. Sushchik, L. S. Tsimring and H. D. I. Abarbanel, *Phys. Rev. E* **51**, 980 (1995).
10. T. Yang and L. O. Chua, *Int. J. Bifur. and Chaos* **9**(1), 215 (1999).
11. R. Genesio and A. Tesi, *Automatica* **28**, 531 (1992).

Processes under Outbursts of Mountain Lakes and Model for Risk Assessment

A. N. Valyaev
Nuclear Safety Institute of Russian Academy of Sciences, Moscow, Russia

S. A. Erochin and T. V. Tusova
Institute of Water Problems and Hydro Energy of National Academy Sciences Bishkek, Kyrgyzstan Republic

The mountain water objects such as large lakes and huge artificial water reservoirs play a very important role for human existence and life in many of the earth's regions. They are very dangerous and complex objects, where natural and man-made catastrophes may result to their outbursts with huge negative irreversible damages on environment and population. Also their large hydro constructions and dams are very attractive for directed terrorist attacks and realization of wide scale global ecologic and water terrorism. Also we present the model for their risk and damage assessments and some of its applications. Results will be presented for the classification of these lakes on their danger degree, management of their risks, development of counter measures for prediction/prevention of the lakes catastrophes and reducing possible negative damages.

1. Motivation

All water objects (WO), such as rivers, lakes, seas and oceans, the artificial water constructions and the different water resources have essential and important political, economic and ecological significance for any country as the source of life and existence of all living organisms. The constantly growing intensive industrial and agricultural activity made worse the WO situation: building of hydro-electric stations (HES) with their huge water reservoirs has broken the natural balance of river flow, clean and drinking water resources have been reduced and the variety and amount of pollution has been increased. As a result the epidemical situation and the population's health have worse. The growing demographic processes stimulated the settlement of new water formed and water carried territories with the additional destruction and reduction of water sources In Eastern Europe and Central Asia the political and economical changes, connected with the creation of new countries from former USSR, resulted to non-sufficient and/or often absent WO ecological monitoring, especially for WO located at transboundary territories. Non-satisfactory or non-sufficient systems of hydrological control of river flows and lakes, radiological and geochemical monitoring created new negative problems both at regional state and global international levels. In addition the absence of the total management system for WO and the system for the population's information and notification greatly decrease the safety level of WO exploitation.

An especially complex and stressful situation has been formed in the countries of Central Asia and Caucasus, where in addition to the above mentioned factors the problem of WO existence, saving and its right safety exploitation has the special significance and actuality. It is connected with the following main moments:[1–3]

(1) The most part of WO is located in mountainous seismically active and dangerous regions with high probability of natural catastrophes and cataclysms such as earthquakes, landslips, mudflows and others, that simulate the negative structural changes in the earth surface crust. As a result some water sources may disappear and amounts of ground and underground water storages will be greatly depleted with the large additional pollution of water in WO.
(2) Under the existence deficit of the natural and artificial water sources here the hot and drought climate causes increased water consumption on different industrial, agricultural and social needs.
(3) It is necessary to provide the WO right management and its safety exploitation in conditions of its specifically complex construction and the large cumulative mechanical and corrosive wear of their elements: high dams with huge water mass in artificial reservoirs; the existence of mistakes in WO projecting and older exploitation norms, developed in general more than 40 years ago in old USSR; significant drifts in rivers and other natural and artificial water arteries; non-sufficient amounts and volumes of defense and nature protection works or often their total absence. In addition, the huge water masses in artificial reservoirs press on the earth crust surface and stimulate frequent and intensive earthquakes, observed now in the regions of the Kazakhstan and Kyrgyzstan Republics.[1,2]
(4) At the mountain territories of some Central Asia and Caucasus countries the high power operated and non-operated industrial objects (nuclear power plants and nuclear reactors with their water

heat sinks, nuclear fuel and weapon production plants with the numerous huge uranium tailing storages, tanks for saving of high radioactivity materials, rocket fuel and others) are often located near natural and industrial WO. All above mentioned objects are very attractive especially for so called "terrorists of new generation", who wish to realize global ecological catastrophes with the use of nuclear or radioactive weapons, for example, to create "dirt nuclear bombs."[3]

(5) All above mentioned factors result to the high risks for realization of man-made catastrophes, including some possible elements of ecological and water terrorism. Opportunity to perform deliberate attacks of terrorists using explosives may cause such catastrophes and stimulate natural calamities.[4] It is easier to implement such attacks there due to the intersection of main lines, an available border with current centers of international terrorism, located in Chechnya, Afghanistan and some others so called zones of "frozen conflicts." Consequences from terrorist attacks will be followed by major human and huge material losses in these densely populated regions, and extremely negative irreversible environmental effects of global scale, such as pollution of the world's oceans.[4]

For example, at the Kyrgyzstan territory there are about 100 mountain lakes with separated surface square value of more than 1 km^2, and about 70 large and dangerous lake outbursts have happened during last 50 years. That is why the prediction of WO behavior and decreasing of their risks are very important today.

2. Physical Processes under Outbursts of Mountain WO and their Risk Assessments

Mountain WO are high complex dynamic non-equilibrium critical systems, that function in conditions depending on many and often non-controlled natural and man-made factors (responses). The direct physical experiments on such open systems and their natural modeling are impossible. Let us consider any mountain lake. It is represented as a space-temporal dynamic object, described by its own natural characteristics and containing its own different particularities and defects, that are varied under the above mentioned responses. These responses may cause three possible kinds of temporal lake transformations at the following stages:

(1) slow evolution of the lake processes without collapse

(2) slow accumulation of lake defects till their collapse values
(3) instant catastrophic collapse of the lake

The most interesting are the two last collapse stages with, for example, outburst of lake or dam destruction (resulting in earthquake, blast or rock slip in lake), when the values of the above mentioned responses exceed some critical thresholds. As a result, huge break-through water waves will be created and their propagation over vast territories will result in great economic and ecologic damages. The parameters of this wave and flood size may be estimated for the separate lake and its high mountain location using a special model. Then it is possible to assess the above mentioned damages using our methods.[1,2] We use a probabilistic approach and present these stages by corresponding stochastic processes.

Today we have very insufficient information on these thematic problems. Any theoretical and mathematical description of lake outburst phenomena is absent at all and the next consideration is only our first step in this direction.

The mountain lake may be considered as a complex non-equilibrium dynamic system with chaotic behavior, described by the following parameters: 1) Its water volume; 2) shape of its water surface; 3) maximum depth; 4) water temperature; 5) height of localization under sea level; 6) origination of lake; 7) its dam parameters; 8) some local environmental characteristics.

The main external responses on lakes are the following: 1) earthquakes; 2) rock slip; 3) mudflows; 4) underground tectonic processes at its bottom 5) environmental temperature changes. These responses may cause three possible kinds of temporal transformation of some lake parameters, corresponding to three above-mentioned stages. We assume that lake outbursts can be realized by the following three mechanisms:

(1) Dam destruction
(2) Subterranean outburst along intra-glacial and intra-moraine tunnels
(3) Coincidence of dam destruction and subterranean outburst

The third case may happen for example as a result of a high intense earthquake.

We use a probabilistic approach and then the total probability (risk) of outburst (P_t) is equal to:

$$P_t = 1 - (1 - P_d)(1 - P_s) \qquad (1)$$

where P_d and P_s are probabilities, corresponding to the first and the second outburst ways. Lake dams may be natural or man-made ones and constructed from different material, including ice for high mountain lakes. The main dam parameters includ its material characteristics and the shape of its construction: height, width and length. The environmental dam characteristics are the following: relief of surround territory, soil, rock stones, glaciers, water, medium and material (ice) temperatures. Any dam is a non-equilibrium dynamic object, where irreversible processes of defect accumulation and evolution with dam partial material destructions are constantly taking place. For ice dams their mass value variation is often caused by ice-water temperature phase transformation.

Dam total or partial destruction can occur as a result of the following processes:

(1) Earthquake response on dam,
(2) intense mechanical shock response on dam, caused by rock slip or mudflow,
(3) growing of lake water level with additional pressure on dam,
(4) slow accumulation of mechanical defects in dam material till threshold values.

Next we describe mechanisms of these processes and their influence on the corresponding probabilities.

(1) Dam seismic stability may be estimated by a safety coefficient of current situation (S), depending on dam parameters, dam material interactions with solid foundation and water medium, vertical and horizontal accelerations, caused by intense earthquakes.[5] Probability of this dam destruction is presented by the following function:

$$P_1 = f_1(F, S, T_d, t) \tag{2}$$

where F is seismic earthquake level (force) by the Richter scale; T_d and t are current dam temperature and time.
The value of $S = 1$ corresponds to the limit of equilibrium of dam state and $S = 0$ to total loss of equilibrium. We consider that S value may describe and reflect the total dam state as a whole. The S value is estimated using the special monitoring results on current situation for a separated dam. Some methods of space monitoring may be effectively used for observation of lake and dam current states.[6,7]

(2) As a first approximation we assume that the probability of dam destruction with the second way (intense shock response) is proportional

to the mass value (m) of rock slip or mudflow resulting from direct contact mechanical response (or shock) on the dam body, its duration (t_s) and space localization of this response on dam construction, characterized by the $\vec{R}_s$ vector:

$$P_2 = f_2(S, m, t_s, \vec{R}_s, t) \tag{3}$$

(3) Growing of lake water level may occur from: (a) ice melting in lake, dam or in surrounded glaciers; (b) rock slip or mudflow penetration into lake; (c) advance of glacier.Assuming that these processes result in an increase of water level on ΔH_m value, we write down:

$$P_3 = f_3(S, \Delta H_m, T_w, T_i, T_m, t) \tag{4}$$

where T_w, T_i and T_m, are current lake water, ice and medium temperatures. For realization of overdammed outburst the lake water level has to exceed the dam height. It is possible for example, if mass of rock slip or mudflow will be more than some threshold value.
We also notice that influence of the global earth climate warming on risk value of lake outburst may be taken into account by the corresponding increasing of T_w and T_m values.

(4) Slow and constant generation of mechanical defects (such as cracks, pores, craters and others) and their accumulation till their threshold values (corresponding to full loss of dam stability and its total fracture) are constantly taking place in the dam material. Their space distributions over dam volume are also essential. All these factors affect the risk value of dam destruction. We assume that their influence may be reflected in variation of the dam current value S. Then the corresponding probability may be written as:

$$P_4 = f_4(S, t) \tag{5}$$

Assuming, that the dam destruction may occur, only if one from above described events are realized, we obtain the following formula for the assessment of P_d total probability of dam destruction:

$$P_d = 1 - (1 - P_1)(1 - P_2)(1 - P_3)(1 - P_4) \tag{6}$$

For the assessment of P_s in (1) we may notice the following. As for the subterranean lake outburst along intra-glacial and intra-moraine tunnels, it is necessary to point out that today this phenomenon is not yet researched in detail. But we have our own experimental investigations on the Kyrgystan high mountains lakes[8–10] and now use them for this risk assessment. Mountain moraine-glacial lakes are formed from atmospheric precipitation in melted snow water form, "young" ice of glacier beetling over lake (meteoric waters) and water from melted "aged" buried ice, lying in adjoining rocks of lake edges and bottom (waters, leached from rocks).[8] Waters from first source are characterized by extremely low uranium (U) content and balanced ratio of its even isotopes ($C = (0,39 \pm 0,02)$ppb, $\gamma = 1,02 \pm 0,02$) [9]. Waters from melted buried ice are noted for higher U content and non-equilibrium γ ratio (C range $= 2,48 \pm 0,08$–$28 \pm 0,2$ppb; γ range $= 0,862 \pm 0,007$–$1,16 \pm 0,02$). The higher the contribution of buried ice in lake feeding, the more water in moraine-glacial lakes deflects from glacial melted water by corresponding C and γ values, which allows the determination of main feeding sources. The latter testifies about lake instability state and increasing of its outburst probability. The presented method is confirmed by our inspection of five Kyrgyz lakes:[9] maximum C, γ deflections from the same values in meteoric waters were in Teztor-1 lake ($C = 23 \pm 2$ppb; $\gamma = 0,93 \pm 0,01$) and Atjailoo lake ($C = 1,69 \pm 0,05$ppb; $\gamma = 0,89 \pm 0,01$). After that Teztor-1 had several outbursts; the last was in 2004. Atjailoo outburst was in 1997 and today it is in a constantly growing outburst stage. The three other lakes are stable and their uranium-isotopic parameters are close to meteoric waters: for Teztor-2 lake $C = 1,14 \pm 0,03$ppb; $\gamma = 1,00 \pm 0,02$; for Kashkasu and Tuuktor lakes $C = 1,39 \pm 0,05$ppb; $\gamma = 1,06 \pm 0,01$. For this reason we assume that:

$$P_d = f_5(C, \gamma, t) \tag{7}$$

Also we notice the following connection between the possible kinds of lake outburst and its origination. There are three variants of lake outbursts, connected with their origination:

(1) Surface overflow. It is typical for lakes of obstructed type, much less possible for glacial and moraine lakes, and even less possible for moraine-glacial and moraine-nogging types of lakes.

(2) Subterranean outburst along intra-glacial and intra-moraine tunnels. Tunnels bring out water beyond bounds of moraine-glacial complex down to a valley. This outburst is typical for lakes of glacial and moraine-glacial types, rarely for moraine-nogging and moraine types.

(3) Compound outburst. Outburst starts subterranean way, and during a process of outburst underground canal of drain widens so much that its roofing can't bear the overlying load and descends. The outburst comes to the surface; usually in this case the volume of the outburst greatly increases. The third variant of outburst is most

typical for moraine-glacial and moraine-nogging lakes, rarely for obstructed lakes. These observations show more a complex nature of lake outburst. Some discussion of these results is presented in the Conclusions section of this article.

3. Break-Through Water Wave Propagation

Let us consider the most catastrophic processes after mountain lake outburst, caused by the total dam destruction and the following generation of a huge break-through water wave. At first this wave will create an intense shock mechanical response on all natural and artificial objects along its propagation. Then the wave will be transformed in a water space submergence zone. The evaluation of the parameters of this catastrophic response will include the following:

(a) maximum possible height and speed of break-through wave propagation;
(b) estimated time of coming of wave crest (front) onto selected object or point of territory;
(c) boundaries of possible submergence zone in the vicinity;
(d) maximum depth of submergence for every definite locality and time of its submergence;
(e) to point out all main objects, that will be under wave response;

As for the last case, the following explanations have to be done. Some objects may be only under response of break-through wave front for the shot time without the next submergence for long time. It will only result in mechanical total or partial destruction of these objects with subsequent water wash-out of soil and artificial constructional elements. This situation will be analyzed in according to the results, caused damages and consequences.

For estimates of the wave parameters and calculations we shall use computer modeling taking into account the real profiles of local earth's crust and mountains valley (including its rock and soil materials), and other water reservoirs, such as lakes, small rivers and other natural objects.

Let us consider the case, when the water object is a lake or another huge water reservoir. The object has a river, flowing through or out of it, and one or few natural or man-made dams separate this river from the WO. Then the possible scenarios of examined catastrophic flooding, related with dam destruction, will be elaborated on the basis of method of the geographic analogous.[11] The height of breakthrough wave is determined as a sum of upper and lower waves of destroyed dams, height of the seasonal high-water (flood) according to data of water register in certain parts of rivers. This method allows to determine the quantitative parameters of breakthrough wave and consequences of damages under the various order of dam total or partial destruction or damage. The calculation results will be usually presented on a special map, which allows to determine the extension of damage and to develop the measurements on prevention of the negative consequences in accordance with distinguished hazard zones.

In the case where the wave propagation is located at different industrial zones, where liquid and water objects, such as tailing storages, heat sinks, artificial water reservoirs and others contain different hazard pollutants (toxins, radionulclides, heavy metals, toxins and others), we have to take into account the influence of these objects on wave propagation and their additional hazard response on the environment after wave response. Our analysis will expanded to include the following:

To analyze the possible scenes of realization of situations on pollutant migration from these objects, for example, from tailing storage:

(a) constant pollutant migration without damage of tailing storage dams;
(b) similar migration with the partial damage of tailing storage dams, for example, under landslide or earth flow;
(c) pollutant migration under total dam destroy;
(d) pollutant migration as a result of: (1) partial flowage; (2) total flowage.

Under realization of the last two scenarios the following two cases of development of catastrophic situation are possible:

(1) all tailings are washed off by a river flow over a period of a few days;
(2) all tailings are washed off by a river flow instantly.

The last situation is the most extreme and dangerous, because it will cause the maximum pollution with maximum losses both for environment and population. For all cases it is necessary to take into account the following kinds of possible damages caused by:

(1) people victims and hazards to human health;
(2) pollution of wide scale territories with subsequent losses in forest, agricultural and fish industries;
(3) wide pollution of buildings and constructions;
(4) pollutant migration in basins of the large rivers.

Under radiological risk assessments it is necessary to take into account the possible chemical nuclear reactions and transformations of pollutants in soil, water and air. For example, transport calculation will be done for decay chain $^{234}U{>}^{226}Ra{>}^{238}U{>}^{230}Th$.

4. Assessment of Damages

We use our method for assessment of damages, that includes the following.[1,2] Let us consider the common case of any object exploration for the fixed time interval under the following assumptions: (1) at initial state the object is in normal (non-accidents) exploitation; (2) the different kinds of accidents may be occurred as noticed $i = 2, 3, \ldots, m$, where m is the total number of possible accidents ($m = 1$ corresponds to the normal regime); (3) every accident may create the different kinds of damages. Assuming that j is the kind of damage with a_j value. Then $j = 1, 2, \ldots, n$, where n is the total number of possible kinds of damages; (4) realization of i accident creates the damage of j kind with P_{ij} probability, thus the matrix of risks (probabilities) is determined. Then the total vector of limited losses $\vec{a}_{\text{lim}}$ may be determined by the next formula:

$$\vec{a}_{\text{lim}} = P(1)\vec{a}_{1n} + \sum_{i=2}^{m} \hat{P}_{ij}\vec{a}_j \tag{8}$$

where $P(1)$ is the probability of damage formation under normal exploitation; $\vec{a}_{1n}$ is the vector of limited damages under regular exploitation. $P_{ij}a_j$ coordinate vector value in sum is equal the damage value of each j kind under realization of i kind accident. Obviously that at normal mountain lake we have the state $P(1) = 0$. At extreme cases of lake outburst the possible economic damage may be assessed as total material loss, caused by break-through water wave, resulting to numerous destructions of different agricultural and industrial constructions, pollution of vast territories and others. The assessment of this wave parameter and the size of the flooding territories have to be done for each selected lake based on the method described above.

Here, for example, we pay some attention to the mountain lakes of the Kyrgyzstan Republic in Central Asia. It is a region of great ecological risk, resulting from numerous natural catastrophes and cataclysms and intense industrial activity, stimulated by man-made accidents. In Kyrgyzstan the numerous U, S_b, A_u and H_g tailing storages are often located near: (1) the largest and main republic Naryn river with five power hydro electric stations (HES) and their huge reservoirs; (2) the large lakes in Tien-Shan mountains.The main natural and man-made water objects are presented in Fig. 1. The total or partial destruction of their natural and artificial

Fig. 1. Map of Kyrgyzstan Republic

dams from the above mentioned disasters or directed water terrorist acts may cause huge water waves and a great irreversible negative response on environment and populations with wide scale global hazard pollution of vast and transboundary territories (in Uzbekistan, Tajikistan and others), including water basins of Syrdarya and Amurdarya main rivers in Central Asia[1,2] (see Fig. 1). The unique huge mountain lake Petrov is located at the moraine and glacial complex of Peirov large glacier, flowing from the western slope of Ak-Shyirak mountain range, shown in Fig. 2.

The Kumkor river, tributary of the Naryn river, flows out the lake. The maximum registered Kumkor river flow is $66.2\text{m}^3/\text{s}$, its usual average value varies from $2,9$ to $25,2\text{m}^3/\text{s}$ from June to September. The Petrov lake parameters are the following: 3734 m height at sea level; $60,3$ million m^3 water volume; the ice dam; $15-24$m lake depth range; $19-2.1$km bottom width range The lake feeds on melting ice from the glacier. The total glacier area of is 69.8km^2.

The recent detailed geophysical and engineer-geologic study of the Petrov lake, including uranium isotope indicator for water dynamic, shows a dangerous growth of the lake outburst as a result of its constant water volume increase and current underground thermo-caster processes.[10,12] This outburst will cause destruction and hazard pollution of the Naryn river head with dangerous consequences for the operative mountain gold mines of the

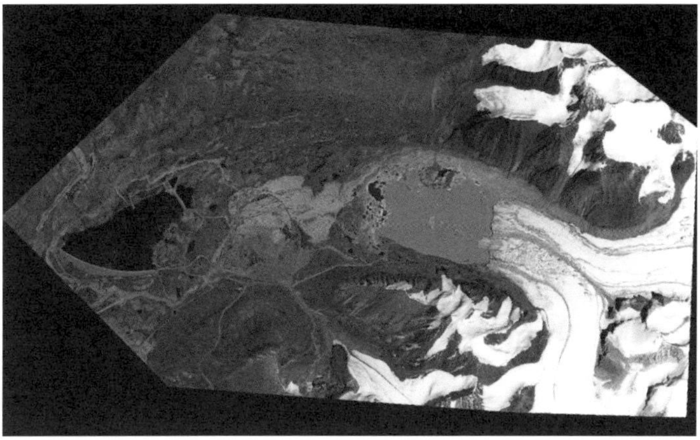

Fig. 2. Bulk of ice block at the subsidence area of grotto tunnel inside the glacier

Kumtor Operating Company. Also the outburst wave may stimulate the dam destruction at the huge water reservoirs of the five Naryn high power operated HES, that supply energy for the whole Republic. The Torkogul HES reservoir (Fig. 1) alone has accumulated 20 billion m^3 water with the height of its dam$\sim$ 200 m. Its dam destruction will have catastrophic consequences of global scale for all Central Asia.

5. Conclusion

Under consideration and analysis of high mountain lake outbursts we have dealt with a large scale open non-equilibrium system, exposed to numerous external natural and man-made factors.[13] The effect of most of these factors to the sytsem is not yet known. As a result the system behavior has a chaotic character. Direct modeling and correct qualitative description of this system is impossible at this point. That is why we use here only the first rough probability approximation for the qualitative theoretical-experimental description of the very complex phenomenon of lake outbursts by its presentation with the main elemental stages. We hope that the main environmental factors and own current parameters of lake and its dams, affected on lake outburst, were experimentally detected and described with their influence on the corresponding risk values. The presented results are in agreement with experimental ones. Our following steps will be directed to develop and obtain the correct approximation of functions, given by the earlier formulas. To assess the probabilities of possible terrorism acts with the use of explosives, directed to catastrophic lake outbursts, is impossible and absent in our consideration. Today some advanced methods for detection of explosives, for example, in dams and river power hydro-contractions,[12] have to be widely used for prediction/prevention of lake catastrophes.

As for the forecast and assessment of possible economic and ecological damages, caused by catastrophes and cataclysms, the most difficult problem is the assessment of elements for the P_{ij} matrix of risks and damages in (8). One of the main reasons here is connected with insufficient representative statistics on such dangerous objects or often its full absence. Some additional thematic results, that include photos and pictures of different lakes, their dams, glaciers and others will be presented in my communication.

The analysis of all calculation results will be taken into account for the classification of these lakes on their danger degrees, management of their risks, development of counter measures for prediction/prevention of the lake catastrophes, some terrorist acts, management of risks, reducing and softening of their negative damages. It will promote to substation and safety development of vast regional and transboundary territories of many countries. This approach has a universal character and may be used for many dangerous natural and man- made objects and also in development of common systems for emergency prevention/elimination.

References

1. A. N. Valyaev, S. V. Kazakov and A. A. H. D. Passell, Assessments of risks and possible ecological and economic damages from large-scale natural and man-induced catastrophes in ecology-hazard regions of central asia and the caucasus, in *Proc. of NATO Advanced Research Workshop: "Prevention, Detection and Response to Nuclear and Radiological Threat "*, ed. S. Apikyan, NATO Science Series (Springer, Netherlands, May 2008).
2. A. N. Valyaev, S. V. Kazakov and A. A. H. D. Passell, Development of monitoring system for studying of radionuclide and chemical contamination level in trans boundary river's basins of caspian and kara seas at russian federation territory., in *Proc. of NATO Advanced Research Workshop: "Nuclear Risk in Central Asia "*, eds. B. Salbu and L. Skipperud, NATO Science Series (Springer, Netherlands, 2008).
3. S. V. Kazakov, S. S. Utkin, I. I. Linge and A. N. Valyaev, Categorization of aqueous media and water bodies by contamination radioactive levels, in *Proc. of International Conference "Radioactivity after Nuclear Explosions and Accidents "*, (St. Peterburg, GIDROMETIZDAT, 2006).
4. A. N. Valyaev, S. V. Kazakov and A. A. H. D. Passell, Assessments of risks and possible ecological and economic damages from large-scale natural and man-induced catastrophes in ecology-hazard regions of central asia and the caucasus., in *Proc. of NATO Advanced Research Workshop: "Nuclear Risk in Central Asia "*, eds. B. Salbu and L. Skipperud, NATO Science Series

(Springer, Netherlands, 2008).

5. H. Vandeenhove and I. Torgaev, Mailuu-suu tailing problems and options for remendation, in *ISEM 03: The 9-th Intern/ Conf. on Rdioactive Waste Management and Inviromental Remendation*, (Oxford, England, 2003).
6. A. N. Valyaev and S. V. Kazakov, Some actual problems and their possible decisions with using of space monitoring, in *Proc. the First Intern. Conference "Earth from Space- the Most effective Solutions"*, (Moscow, 2003).
7. A. N. Valyaev, S. V. Kazakov and O. V. Stepanets, Using of space technologies for monitoring of large river's basins and prediction of large-scale natural and man-induced catastrophes in ecology-hazard regions of central asia and caucasus, in *Proc. the Second Intern. Conference "Earth from Space- the Most Effective Solutions" Section "Space Monitoring in Problems of Management of Territories'*, (Moscow, 2005).
8. T. V. Tuzova, V. V. Romanov, L. S. Vlasova, S. A. Erohin, A. A. Jerdev and V. I. Shatravin, *Water sources* **1**(2), 236 (1994).
9. T. V. Tuzova, *Investigations of waters of the Issykkyl basin with the use of uranium isotopic method. Study of the Issyk-Kyl lake hydrodynamics with the use of isotopic methods*, no. 132-140 (Institute of water problems and hydropower, 2006).
10. S. A. Erohin, *Outburst lake Petrov. Study of Issyk-Kyl lake hydrodynamics with the use of isotopic methods.* (Institute of water problems and hydropower, 2006).
11. L. K. Malik, N. I. Koronkevich and E. A. Barabanova, *Safety problems under the emergencies, Moscow, VINITI* **2**, 67 (1998).
12. A. N. Valyaev and V. A. Yanushkevich, Using of acoustic techniques for detection of explosives in gas, liquid and solid mediums, in *Proc. of NATO Advanced Research Workshop*, , , NATO Science Series II "Detection of Bulk Explosives: Advanced Techniques against Terrorism" Chemistry Vol. 138 (Kluwer Academic Publishers, Netherlands, 2003).
13. A. N. Valyaev, S. A. Erochin and T. V. Tusova, Processes under outbursts of mountain lakes and model for risk assessment, in *Proc. of International Conference: "Chaotic Modeling and Simulation, CHAOS2008"*, (2008).

State Space Reconstruction from Multiple Time Series

I. Vlachos* and D. Kugiumtzis

Department of Mathematical, Physical and Computation Sciences
Faculty of Engineering Aristotle University of Thessaloniki
Thessaloniki, 54124, Greece
*Email: *ivlaxos@gen.auth.gr*
web.auth.gr/gen

In the non-linear analysis of scalar time series the common practice is to reconstruct the state space using time-delay embedding. When there are more than one observed quantities, one can reconstruct the state space using a time-delay embedding scheme specifying embedding parameters for each quantity. In this study we investigate the state space reconstruction from multiple time series derived from continuous systems and propose a method for building the embedding vector progressively using information measure criteria. The proposed method is compared to other methods with simulations on known chaotic systems, such as individual and coupled Lorenz and Rössler systems. Our analysis showed that multivariate attractor reconstruction preserves better the dynamics of a system and our proposed method gives a parsimonious alternative to the simple extension of the univariate case.

Keywords: Non-linear analysis; multivariate time series; state space reconstruction.

1. Introduction

Since its publication, Takens' Embedding Theorem[1,2] is reputed as the most significant tool in non-linear time series analysis and has been used in many different settings ranging from system characterization and approximation of invariant quantities to prediction and noise-filtering.[3] The Embedding Theorem implies that although the true dynamics of a system may not be known, equivalent dynamics can be obtained using time delays of a single time series, seen as the one-dimensional projection of the system trajectory.

Most applications of the Embedding Theorem deal with univariate time series, but often measurements of more than one quantities related to the same dynamical system are available. One of the first uses of multivariate embedding was in the context of spatially distributed systems where embedding vectors were formed from simultaneous measurements of a variable at different locations.[4,5] Multivariate embedding was used among others for noise reduction[6] and surrogate data generation.[7] Another particular implementation was the prediction of a time series using local models on a state space reconstructed from a different time series of the same system.[8] In a later work, embedding was extended to all of the observed time series,[9] and recently non-uniform multivariate embedding using different delays was briefly addressed in [10,11]. All of these works are basically extensions of methods used in the scalar case with slight modifications without taking into account particular characteristics of multivariate embedding, such as possible high dependence of individual components obtained from the different time series.

In this work, we contemplate on the particularities of multivariate embedding and their effect on the state space reconstruction and suggest a new embedding scheme to treat them. We study only time series generated from continuous dynamical systems where complications mostly occur. Based on Fraser's idea of using information criteria to select the embedding parameters in the univariate case[12,13] we propose a method for progressive building of the embedding vectors allowing for different variables and delays. We compare our proposed method with the univariate reconstruction on each single time series and the direct extension to all the time series with respect to the distortion induced on the original attractor under the embedding processes. Simulations on known chaotic systems are used to evaluate the embedding techniques.

In Sec. 2, the embedding for univariate time series is briefly discussed. In Sec. 3, the discussion is extended to multivariate time series and the proposed embedding technique is presented, and in Sec. 4, the tools for comparing the reconstructions are given. In Sec. 5, the results of simulations are presented and, finally, in Sec. 6, the results are discussed and conclusions are given.

2. Univariate Embedding

A continuous dynamical system generates a trajectory in a D-dimensional manifold Γ and its evolution can be described by D ordinary differential equations

$$\dot{\mathbf{y}} = f(\mathbf{y}),$$

where $\mathbf{y} \in \Gamma$ is the state vector at time t.

The observed scalar time series $\{x_n\}_{n=1}^N$ of length N is the projection of the state vector of the system at discrete times $\{\mathbf{y}_n\}_{n=1}^N$ given by a measurement function $h : \Gamma \mapsto \mathbb{R}$ as $x_n = h(\mathbf{y}_n)$. Besides the apparent loss of information of the system dynamics by the projection, the system dynamics may be recovered through suitable state space reconstruction from the single time series.

Assume that the dynamical system "lives" on an attractor $A \subset \Gamma$. According to Taken's embedding theorem a trajectory formed by points $\mathbf{x}_n$ of the delay map

$$\mathbf{x}_n = (h(\mathbf{y}_n), h(\mathbf{y}_{n-\tau}), ..., h(\mathbf{y}_{n-(m-1)\tau})) = (x_n, x_{n-\tau}, ..., x_{n-(m-1)\tau}), \quad (1)$$

under certain genericity assumptions, is an one-to-one image of the original trajectory of $\mathbf{y}_n$ provided that m is large enough.

The reconstructed attractor $\tilde{A}$ obtained by the time-delay vectors is topologically equivalent to A as long as $m \geq 2d + 1$ where d is the box-counting dimension of A, and invariant quantities of the original system (dimensions, Lyapunov exponents, etc) can be estimated from the reconstructed dynamics.

The embedding process is visualized in the following diagram where $\mathbf{F}$ is the original map at discrete steps, e is the embedding procedure creating the delay vectors from the time series and $\mathbf{G}$ is the reconstructed dynamical system on $\tilde{A}$.

$$\begin{array}{ccc}
\mathbf{y}_n \in A \subset \Gamma & \overset{\mathbf{F}}{\longmapsto} & \mathbf{y}_{n+1} \in A \subset \Gamma \\
h\downarrow & & h\downarrow \\
x_n \in \mathbb{R} & & x_{n+1} \in \mathbb{R} \\
e\downarrow & & e\downarrow \\
\mathbf{x}_n \in \tilde{A} \subset \mathbb{R}^m & \overset{\mathbf{G}}{\mapsto} & \mathbf{x}_{n+1} \in \tilde{A} \subset \mathbb{R}^m
\end{array}$$

The two parameters of the delay embedding in (1) are the embedding dimension m, i.e. the number of components in $\mathbf{x}_n$ and the delay time τ. Among the approaches for the selection of m the most popular is the *false nearest neighbors* (FNN)[14] while τ is usually chosen as the one that gives the first minimum of the time-delayed mutual information (MI).[15]

3. Multivariate Embedding

Given that there are p time series $\{x_{i,n}\}_{n=1}^N$, $i = 1, \ldots, p$, generated through p measurement functions $h_i(\mathbf{y}_n)$ from the same system, the simplest extension of the reconstructed state space vector in (1) is of the form

$$\mathbf{x}_n = (x_{1,n}, x_{1,n-\tau}, \ldots, x_{1,n-(m-1)\tau}, x_{2,n}, \ldots, x_{p,n-(m-1)\tau}). \quad (2)$$

The corresponding diagram for the multivariate embedding process is given below, where the dimension of the reconstructed space is M

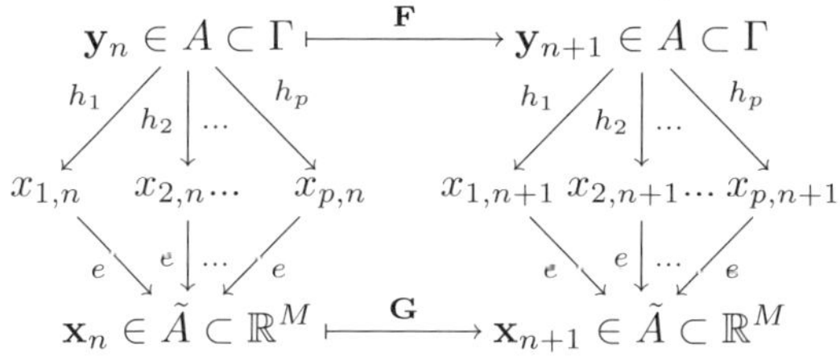

A question that arises immediately is if all the information contained in the reconstructed vectors is useful. Some time series may be contemporary and/or lagged dependent on others and then the embedding vectors carry redundant information. The inclusion of such components just complicates the state space reconstruction and this may have implications in the analysis, e.g. degradation of the estimates of measures evaluated on the reconstructed attractor.

Judd and Mees[16] introduced the non-uniform embedding scheme for univariate time series using unequal lags, so that $\mathbf{x}_n = (x_{n-l_1}, x_{n-l_2}, \ldots, x_{n-l_m})$. The use of unequal lags may reduce better the redundant information than the embedding with fixed lag. Extending this to multivariate embedding we get

$$\mathbf{x}_n = (x_{1,n-l_{11}}, x_{1,n-l_{12}}, \ldots, x_{1,n-l_{1m_1}}, x_{2,n-l_{21}}, \ldots, x_{p,n-l_{pm_p}}).$$

Inspection of all possible combinations of components for determination of the optimum embedding (with any criterion we may choose) can be computationally intractable when the embedding dimension and number of time series are large. For a moderate example when $p = 3$, $m = 3$ and the maximum lag for all time series is 10 we have 4050 cases, while when we increase m to 4 we have 27405 cases. It is evident that for most practical purposes we need a progressive method to build the embedding vectors.

Assume that we have somehow chosen the maximum lags L_i for each time series $i = 1, \ldots, p$. Then we have a set of $L_1 \times L_2 \times \ldots \times L_p$ candidate components to choose from. Using ideas developed for the state space reconstruction from univariate time series, the reconstructed vector must satisfy two properties, first its components must be least dependent to each other and secondly it must be able to explain best the future states of the system for one, two or several steps ahead. In the following we refer to only a single, immediate future state.

Starting from an empty vector $\mathbf{b}_0$, at each step j that we have already selected the $j-1$ components $\mathbf{b}_{j-1} = (x_1^\star, x_2^\star, \ldots, x_{j-1}^\star)$ we add the lag

$x_j^\star \in \{x_{1,n}, x_{1,n-1}, \ldots, x_{1,n-L_1}, x_{2,n}, \ldots, x_{p,n-L_p}\} \setminus \{x_1^\star, \ldots, x_{j-1}^\star\}$ that fulfills the following maximization criterion

$$\max_{x_j^\star} \left\{ \frac{1}{p} \sum_{i=1}^{p} \mathrm{I}\big(x_{i,n+1}; (x_j^\star, \mathbf{b}_{j-1})\big) - \mathrm{I}\big(x_j^\star; \mathbf{b}_{j-1}\big) \right\}. \tag{3}$$

The first mutual information term in the sum gets large when the new embedding vector (with the addition of $x_j^\star$) explains best the dynamics of the system, while the second term is small when the candidate component is independent to the already included components, so maximization of (3) leads to embedding vectors that satisfy the two desired properties. Note that in essence the components of the delay vectors can never be fully independent since the two mutual information terms are related and the maximization gives actually the best tradeoff between independency of the components and explaining the future of the system.

The mutual information in (3) regards vector variables and thus its estimation requires more data than for the standard case of scalar variables. Data demand increases dramatically with the dimension, at the level of millions of data points for large dimension. Thus we approximate multidimensional mutual information in (3) with average pair-wise mutual information as

$$\mathrm{I}\big(x_j^\star; \mathbf{b}_{j-1}\big) \approx \frac{1}{j-1} \sum_{l=1}^{j-1} \mathrm{I}\big(x_j^\star; x_l^\star\big)$$

$$\mathrm{I}\big(x_{i,n+1}; (x_j^\star, \mathbf{b}_{j-1})\big) \approx \frac{1}{j-1} \sum_{l=1}^{j-1} \mathrm{I}\big(x_{i,n+1}; (x_j^\star, x_l^\star)\big).$$

We use the standard mutual information estimate of equiprobable binning, which was found to perform satisfactorily among other estimates.[17]

For the determination of L_i we use the *Minimum Description Length*[18] selection criterion on each i time series separately, as was used by Judd and Mees.[16] As a stopping criterion for the scheme of progressive vector building we use the idea of FNN. The expansion of $\mathbf{b}_{j-1}$ terminates when the percentage of false neighbors going from $\mathbf{b}_{j-1}$ to $\mathbf{b}_j$ with the selected component added at step j is smaller than a properly chosen threshold. The final reconstructed vector is then $\mathbf{b}_{j-1}$.

4. Comparing Reconstructions

The reconstructed attractor is a distorted “image” of the original one, and the distortion may be with respect to its shape or its dynamics. The better the reconstruction, the less the distortion of the attractor. Thus we base the comparison of different reconstructions on two distortion measures, one for the shape and one for the dynamics.

Supposing that A is the original attractor and $\tilde{A}$ the reconstructed attractor, we measure the shape distortion with *Mutual False Nearest Neighbors*.[19] Let $\mathbf{y}_n$ be a point on A with $\mathbf{y}_{n_k}$ its nearest neighbor and respectively $\mathbf{x}_n$ the corresponding point on $\tilde{A}$ with nearest neighbor $\mathbf{x}_{n_r}$. The shape distortion is given by

$$Sd = \left\langle \frac{\|\mathbf{y}_n - \mathbf{y}_{n_k}\|}{\|\mathbf{x}_n - \mathbf{x}_{n_k}\|} \frac{\|\mathbf{x}_n - \mathbf{x}_{n_r}\|}{\|\mathbf{y}_n - \mathbf{y}_{n_r}\|} \right\rangle,$$

where $< . >$ is the average over all points and $\|.\|$ is the Euclidean distance.

To measure the distortion of the dynamics we use the *Normalized Root Mean Squared Error* of the one-step ahead nearest neighbor prediction on the original attractor. If $\mathbf{x}_n$ is a point on the reconstructed attractor with nearest neighbor $\mathbf{x}_{n_r}$, then the dynamics distortion is given by

$$Dd = \sqrt{\frac{\sum\limits_n \|\mathbf{y}_{n+1} - \mathbf{y}_{n_r+1}\|^2}{\sum\limits_n \|\mathbf{y}_{n+1} - \bar{\mathbf{y}}\|^2}},$$

where the summation is over all points and $\bar{\mathbf{y}}$ is the center of mass of the points in A. The larger Sd is and the smaller Dd is, the less distortion there is on the shape and on the dynamics of the attractor respectively. Sd is a stricter criterion. The dynamics may be preserved though the reconstructed attractor is a stretched or squeezed (or both) version of the original attractor. On the other hand, since the shape of the attractor is formed by the dynamics, the preservation of the shape guarantees the preservation of the dynamics.

An example for the dependence of the measures on the embedding is given in Fig. 1 for the reconstruction of the Lorenz attractor with the x and z variables using non-uniform multivariate embedding with $L_i = 10, i = 1, 2$ and for embedding dimensions 3 and 4. All possible combinations where taken and the measure values where sorted for clarity. We notice that there are more than one "optimal" (with respect to the dynamics distortion) embeddings for a given embedding dimension m that unfolds the attractor and there are even more "good" embeddings for larger m. For $m = 3$ we have approximately 1200 possible embeddings and 40 with Dd less than 0.03 that become 5000 and 400, respectively when m increases to 4. This fact exhibits the problem of identification that appears when dealing with multiple time series.

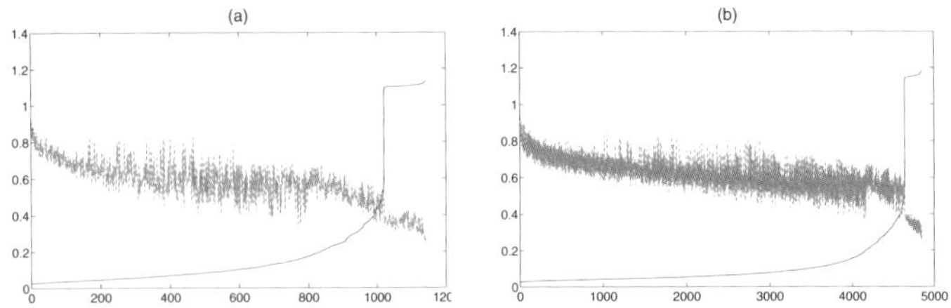

Fig. 1. Dynamics and shape distortion of reconstruction with Lorenz (x,z) variables for embedding dimension 3 (a) and 4 (b). The values of dynamics distortion are sorted to increasing order (black line) and the values of shape distortion are plotted with regards to the sorting of the dynamics distortion measure (grey line).

5. Simulations and Results

We compare by means of simulations the proposed embedding scheme to two other methods. These are the uniform univariate embedding (UU), i.e. delay embedding on a single time series disregarding the presence of other time series, and the extension of the uniform embedding to the multivariate case (UM), i.e. for p time series the dimension of the embedding vectors is $M = mp$. Uniform embedding suggests that a fixed delay τ is used, which is selected by the time-delayed mutual information criterion. Note that τ is separately computed for each time series and their average is used in UM. The embedding dimension M is selected by the FNN procedure with threshold 0.01. The FNN procedure is used in the same way as in the univariate case, but at each step the embedding vectors are augmented by the addition of the p components corresponding to the lagged multivariate time series. For the proposed non-uniform multivariate embedding (N-UM), the same threshold is used for the stopping criterion.

Noise-free bivariate time series as well as time series contaminated with 5% normal observational noise are generated from chaotic systems. The data are scaled to be in the same range [0,1] to compensate for different variable ranges. For each system we generate 100 time series of length $N = 8192$. For each realization we obtain the state space reconstruction with the three methods and compute Sd and Dd.

The bivariate time series are from the x and z variables of the Lorenz system

$$\dot{x} = 10(y - x), \quad \dot{y} = 28x - y - xz, \quad \dot{z} = -8/3z + xy,$$

the x and y variables of the Rössler system

$$\dot{x} = -y - z, \quad \dot{y} = x + 0.15y, \quad \dot{z} = 0.2 + xz - 10z,$$

and the y_1 and y_2 variables of the coupled Rössler–Lorenz system at two regimes: independent ($C = 0$) and strong coupling ($C = 4$):

$$\dot{x}_1 = 6(-y_1 - z_1) \quad \dot{y}_1 = 6(x_1 + 2y_1) \quad \dot{z}_1 = 6(0.2 + x_1 z_1 - 5.7 z_1)$$
$$\dot{x}_2 = 10(y_2 - x_2) \quad \dot{y}_2 = 28x_2 - y_2 - x_2 z_2 + C y_1^2 \quad \dot{z}_2 = -8/3 z_2 + x_2 y_2$$

For each bivariate time series, we consider the uniform and non-uniform multivariate embeddings, UM and N-UM, as well as the uniform univariate embedding from the first and second variable, UU1 and UU2, respectively.

In Table 1 we give the results for the shape distortion measure for each reconstruction scheme and each system, where we highlight the best reconstruction scheme for each system. We see that both multivariate embedding procedures preserve better the shape of the attractor in comparison to the univariate embeddings, as expected. Uniform and non-uniform multivariate embeddings give equivalent values for the shape distortion measure with uniform being slightly better for most of the systems.

Table 1. Average of the shape distortion measure

Shape Distortion (Sd)	UU1	UU2	UM	N-UM
Lorenz (x, z)	0.647	0.357	0.805	**0.878**
Lorenz (x, z)+5% noise	0.154	0.102	**0.294**	0.278
Rössler (x, y)	0.797	0.813	**0.899**	0.869
Rössler (x, y)+ 5% noise	0.196	0.202	**0.382**	0.250
Coupled Rössler Lorenz $C = 0$ (y_1, y_2)	0.046	0.036	0.528	**0.586**
Coupled Rössler Lorenz $C = 0$ +5% noise	0.062	0.046	**0.474**	0.447
Coupled Rössler Lorenz $C = 4$	0.667	0.713	**0.854**	0.832
Coupled Rössler Lorenz $C = 4$ +5% noise	0.301	0.377	**0.470**	0.459

Table 2. Average of the dynamics distortion measure

Dynamics Distortion (Dd)	UU1	UU2	UM	N-UM
Lorenz (x, z)	0.036	1.144	0.031	**0.030**
Lorenz (x, z)+5% noise	0.131	1.146	0.106	**0.099**
Rössler (x, y)	0.036	0.025	**0.024**	**0.024**
Rössler (x, y)+ 5% noise	0.115	0.111	**0.106**	0.107
Coupled Rössler Lorenz $C = 0$ (y_1, y_2)	0.950	1.048	0.246	**0.213**
Coupled Rössler Lorenz $C = 0$ +5% noise	0.972	1.038	**0.270**	0.274
Coupled Rössler Lorenz $C = 4$	0.092	0.381	**0.044**	0.051
Coupled Rössler Lorenz $C = 4$ +5% noise	0.166	0.419	**0.116**	0.120

In Table 2, the results on the dynamics distortion measure are presented in the same way. Again multivariate embedding procedures preserve better the dynamics of the original system, and uniform and non-uniform embeddings give similar results. For systems where the observed variables behave differently, i.e., Lorenz and uncoupled Rössler–Lorenz system, non-uniform

Table 3. Average Embedding dimension

Average embedding dimension	UU1	UU2	UM	N-UM
Lorenz (x, z)	3	4.68	(2 , 2)	(2.03 , 1.44)
Lorenz (x, z)+5% noise	4.35	7.09	(3 , 3)	(2.07 , 3.44)
Rössler (x, y)	3	3	(2 , 2)	(2.4 , 1.12)
Rössler (x, y)+ 5% noise	4	4	(2 , 2)	(2 , 2)
Coupled Rössler Lorenz $C = 0$ (y_1, y_2)	3	3.6	(3.12 , 3.12)	(3.2 , 2)
Coupled Rössler Lorenz $C = 0$ +5% noise	4	4.7	(3.14 , 3.14)	(3.9 , 2)
Coupled Rössler Lorenz $C = 4$	3	3	(2 , 2)	(2.9 , 1)
Coupled Rössler Lorenz $C = 4$ +5% noise	4	4.9	(2 , 2)	(3 , 1)

embedding performs marginally better, while uniform embedding performs marginally better when the variables have similar behavior, i.e., Rössler x and y and strongly coupled Rössler–Lorenz.

Finally, in Table 3 we see the average embedding dimension obtained from the 100 simulations, i.e. the average number of components included in the embedding vector (a pair of values for the methods on the bivariate time series). The non-uniform multivariate embedding gives reconstruction vectors with smaller or equal dimension to these of the uniform multivariate embedding. Univariate embeddings mostly regard smaller dimensions of the embedding vectors because the full dynamics of the system may not be well represented in the observed variable. With reference to Table 2, N-UM has less dynamic distortion than UM whenever N-UM embedding dimension is less than that of UM. In general N-UM turns out to give reconstructions at state space of smaller dimensions without distorting the dynamics any more than UM does. This is particularly useful in applications where high-dimensionality may cause problems.

6. Conclusions

It is evident that multivariate embedding significantly improves attractor reconstruction when compared to univariate embedding. Both multivariate approaches, the standard uniform approach and the non-uniform approach we introduce here, attain equivalently good state space reconstruction as evaluated by the shape and dynamics distortion measures. As shown by the average embedding dimensions, the proposed method is more parsimonious and gives embedding at lower dimension. Due to data size limitation, the implementation of the proposed method requires simplifications that result in a sub-optimal solution to the embedding problem as set by the conditions for component-wise independence and best correlation to future states. Further investigation is needed towards optimal solution that would allow a more fair comparison to the standard uniform multivariate embedding, which matches directly the criteria for optimum embedding (minimum mutual information for the delay parameter and insignificant percentage of false nearest neighbors for the embedding dimension).

Acknowledgments

This paper is part of the 03ED748 research project, implemented within the framework of the "Reinforcement Programme of Human Research Manpower" (PENED) and co-financed at 90% by National and Community Funds (25% from the Greek Ministry of Development–General Secretariat of Research and Technology and 75% from E.U.–European Social Fund) and at 10% by Rikshospitalet, Norway.

References

1. F. Takens, *Lecture Notes in Mathematics* **898**, 365 (1981).
2. T. Sauer, J. A. Yorke and M. Casdagli, *Journal of Statistical Physics* **65**(3), 579 (1991).
3. H. Kantz and T. Schreiber, *Nonlinear Time Series Analysis* (Cambridge University Press, 1997).
4. J. Guckenheimer and G. Buzyna, *Physical Review Letters* **51**(16), 1438 (1983).
5. M. Paluš, I. Dvořak and I. David, *Physica A* **185**, 433 (1992).
6. R. Hegger and T. Schreiber, *Physics Letters A* **170**, 305 (1992).
7. D. Prichard and J. Theiler, *Physical Review Letters* **73**(7), 951 (1994).
8. H. D. I. Abarbanel, T. A. Carroll, L. M. Pecora, J. J. Sidorowich and L. S. Tsimring, *Physical Review E* **49**(3), 1840 (1994).
9. L. Cao, A. Mees and K. Judd, *Physica D* **121**(1-2), 75 (1998).
10. S. P. Garcia and J. S. Almeida, *Physical Review E* **72**(2), p. 027205 (2005).
11. L. M. Pecora, L. Moniz, J. Nichols and T. L. Carroll, *Chaos: An Interdisciplinary Journal of Nonlinear Science* **17**(1), p. 013110 (2007).
12. A. M. Fraser, *Physica D* **34**(3), 391 (1989).
13. A. M. Fraser, *IEEE Transactions on Information Theory* **35**(2), 245 (1989).
14. M. B. Kennel, R. Brown and H. D. I. Abarbanel, *Physical Review A* **45**(6), 3403 (1992).
15. A. M. Fraser and H. L. Swinney, *Physical Review A* **33**(2), 1134 (1986).
16. K. Judd and A. Mees, *Physica D* **120**(3-4), 273 (1998).
17. A. Papana and D. Kugiumtzis, *Nonlinear Phenomena in Complex Systems* **11**(2), 225 (2008).
18. J. Rissanen, *Automatica* **14**, 465 (1978).
19. N. F. Rulkov, M. M. Sushchik, L. S. Tsimring and H. D. I. Abarbanel, *Physical Review E* **51**, 980 (1995).

Recurrence Quantification Analysis as a Tool for Investigation of Short-Term Traffic Flow: Application to Urban Freeways

E. I. Vlahogianni*, M. G. Karlaftis and J. C. Golias
National Technical University of Athens
Athens, Greece
*Email: *elenivl@central.ntua.gr*

We discuss recent findings on the statistical properties of short-term traffic flow in urban freeway systems. In, particular, the complexity, nonlinearity and nonstationarity of freeway travel speed patterns are studied using the recurrence plots and the recurrence quantification analysis. Results indicate a variable statistical behavior of the speed's temporal evolution, especially during the morning and peak periods of both the weekdays and weekend. Moreover, a clear deterministic evolution of speed patterns with variable nonlinear and laminar characteristics is observed during congested conditions.

Keywords: Traffic flow; travel speed; urban freeways; recurrence plots; recurrence quantification analysis.

1. Introduction

In recent years traffic engineering practice has been marked by the abundance of available data collected in real-time enabling the physicists and statisticians to analyze traffic as a dynamic system. Traffic flow literature emphasizes the significance of interdisciplinary work towards the better understanding of the complexity of traffic flow evolution, especially when studied in the short-run.

It is frequently observed that flow exhibits an irregular (non-periodic) behavior when studied in short time intervals; this irregularity has been reported by several traffic studies.[1–3] This irregularity may be a reflection of changing traffic conditions and, therefore, must be predicted in order to acquire an accurate picture regarding the prevailing traffic conditions, especially in the case of onset of congestion, as well as the formation and dissipation of queues.[2,3]

In the effort to clarify the freeway traffic flow dynamics, the focus has been placed on the speed variations as well as the bivariate speed-volume relationship, because they form consistent congestion and mobility indicators in intelligent transportation systems and provide valuable information regarding the onset of congestion and incidents.[4] The bulk of literature dedicated to freeway traffic analysis has pointed out to several observed complex phenomena that relate the temporal evolution of speed to the formation of queues at bottlenecks.[5] Other results have indicated strong complexity in the temporal evolution of freeway traffic that occur spontaneously and in a deterministic manner.[6] This complexity observed may increase due to the vehicle interactions taking place in at specific freeway sections such as freeway merges and diverges, as well as due to weaving, lane drop and so on.[7]

Regardless of the spatial extend of traffic phenomena, as well as the causes of formation and dissipation of congested conditions, there is a need to reveal traffic's temporal evolution, as well as its relation to the past events in order to improve on the predictability of traffic variables during the occurrence of complex or non-recurrent traffic phenomena. A certain *a posteriori* knowledge of traffic dynamics must be extracted from the available data (e.g. traffic flow outputs) that could give answers to issues such as the non-stationarity and the non-linearity of the traffic processes. Theoretically, the reconstruction of a topologically equivalent picture (for example a set of rules that describe the system's behavior) of the original multidimensional system behavior by using the time series of a single observable variable, was first tackled by Taken.[8] The underlying concept is that the effects of all other critical, but unobservable, variables is already reflected in the series of the observed output variable and any rules that could best describe the original system's behavior can be extracted and reconstructed from its output.

The present paper implements nonlinear tools for the analysis of trafffic flow dynamics. Freeway traffic patterns are visualized based on the Recurrence and Cross-recurrence plots. Moreover, a quantitative analysis of recurrences is also conducted to investicate the statistical characteristics of short-term traffic flow evolution.

2. Recurrence Plots and Recurrence Analysis of Traffic Flow Dynamics

Recurrence is a fundamental attribute of the nonlinear deterministic systems. The recurrence of a state x the time i in a different time j is given by the equation:

$$R_{i,j}^{m,\varepsilon_i} = \varepsilon_i - \|\vec{x}_i - \vec{x}_j\|, \qquad i,j = 1,\ldots,N \tag{1}$$

where N is the number of states x_i in the time window of study, m is the embedding dimension for developing the scalar measurement x_i to an N-dimensional vector space (State-Space) and ε_i is the threshold (Euclidean distance) for the distances $r_{i,j}$. Equation (1) when plotted provides the Recurrence (RP). The usefulness of such plots is in distinguishing several patterns that point out the dynamic properties of the system under study; these properties refer to its deterministic or nonlinear evolution, its laminar behavior and so on. In general, four different structural patterns can be traced in a RP with $m = 1$, such as the parallels to 45° diagonal or to 135° diagonal, the block-like structures and the square-like structures (sojourn points).[9]

In real-world, systems dynamics are complex and difficult to visualize. For this the quantitative analysis of RPs was introduced by defining five statistical measures:[10] the (%) *recurrence* (%REC) (the percentage of points that have Euclidean Distance lower than a user-defined threshold (**e**), the (%) *determinism* (%DET) (the percentage of recurrent points that form lines parallel to the main diagonal), the *entropy* (ENT) (the Shannon entropy of line segment distribution), the *maxline* or $l/line_{\max}$ (DIV) (reciprocal of the longest diagonal line segment) and the *trend* (TREND) (measure of the positioning of recurrent points away from the central diagonal). This analysis is known as Recurrence Quantification Analysis (RQA).

RQA is simply a study of the density of recurrent points in a specific time window. From the states that are recurrent in the State-Space, there exist some states that are spatially close and are settled in segments of orbit that are parallel in different times (deterministic structures) in the time window, while others are isolated (stochastic structures).

Moreover, it is of importance the time that such a described behavior occurs in the window of study or else the evolution of its structure in time; states moving closely for a long time are sign of cyclic behavior, while the opposite case indicates strong nonlinearity (chaos).

Marvan and Kurths extended the RQA by introducing the LAM measure that is the percent of recurrent points in the recurrence matrix that form vertical lines.[11] The mathematical formulation of the above measures can be found in.[12] RQA can also be extended to joint consideration of the patterns of two variables; this analysis is known as *Cross-Recurrent Quantitative Analysis* (CRQA).[13]

3. Analysis of Urban Freeway Traffic Flow Dynamics using Recurrences

The available data comes from a set of highly congested urban freeway sections located on the greater area of a major interchange that connects the center of Athens with the Southern the Western and the Eastern regions of Athens (Figure 1). Volume and travel speed are collected every 90 seconds by both loop detectors and camera in 18 distinct control points. For the specific paper, data between July 2006 and June 2007 are available.

Figure 2 shows the time-series of travel speed (km/h) for selected weekdays and weekend during the busiest week of the year. As can be observed the short-term temporal evolution of speed is different in each day of the week. Speed drops that could signify a formation of queue in a downstream section occur in specific time periods — for example the morning peak 6 to 9 am and the afternoon peak 3 to 8pm — but with a variable strength and duration.

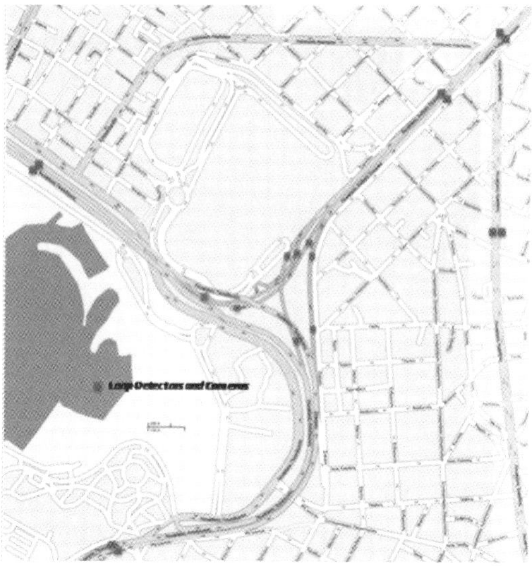

Fig. 1. Graphical representation of the freeway sections under study and the controlled locations. Points signify the locations of loop detectors and cameras.

This variability in the pattern-based evolution of traffic flow can be also observed in the speed-volume relationships during the same week, as seen in Figures 3. In these relationships, two regions are distinguished: the free-flow region, that is, (a) free-flow (unqueued) region a region of high speeds and densely populated traffic states with volume values ranging from zero to effective capacity and (b) the congested (queued) region.

From the congested region the speed drop section (steady values of volume) is more apparent form the section where speed and volume significantly decreases. From the observed speed-volume relationships, it is evident that although the same regions can be traced, the temporal transitional conditions seem to be complex and dependent on the temporal evolution of travel speed.

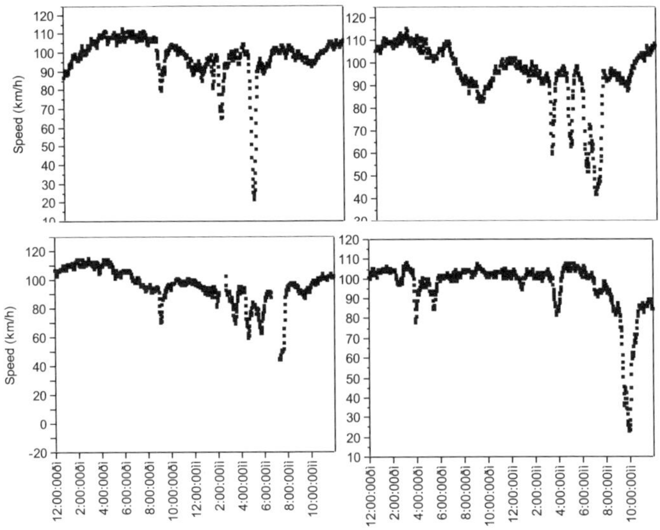

Fig. 2. Time-series of speed and volume for the busiest weekdays and weekend of the year in the study area (from left to right and top to bottom: Monday, Wednesday, Saturday, Sunday).

In order to visualize the variable flow dynamics, the morning and peak time periods for distinct days is further studied. Figure 5 demonstrates the RPs for the morning and afternoon peak of three different days. As can be observed, traffic dynamics during the morning peak are more complex and variable when compared to the afternoon peak. However, abrupt changes to dynamics (dark color bands) may occur during both periods and alter the characteristics of traffic flow.

Based on the RQA measures, this variability is quantified and graphically presented in Figure 6. RQA analysis is conducted in a sliding time window of 1 hour updated every 1 interval (90 sec). The embedding pa-

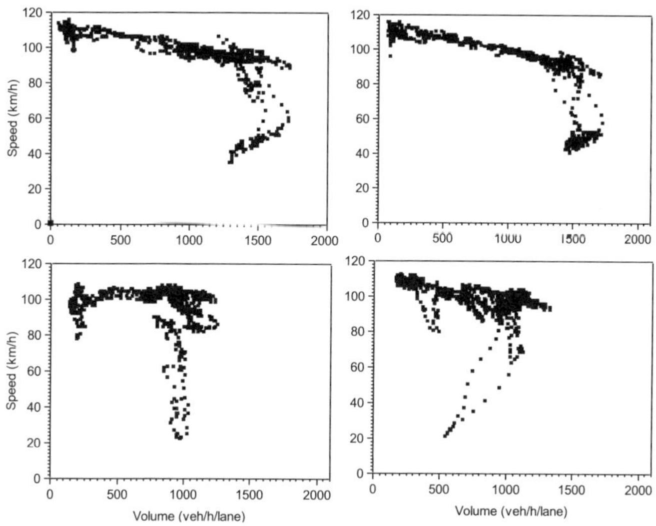

Fig. 3. Speed-volume relationship for the busiest weekdays and weekend of the year (from left to right and top to bottom: Monday, Wednesday, Saturday, Sunday).

rameters of travel speed data are estimated using the mutual information criterion (for the estimation of the embedding delay)[14] and the false nearest neighbor (for the embedding dimension).[15] Both during morning and evening peak, speed patterns exhibit a variable statistical behavior with respect to the deterministic and nonlinear characteristics. The basic outcome is that it seems that there is no general pattern in the manner speed evolves in time. Peaks in the RQA measures reveal transitional movements of traffic flow. However, both deterministic and nonlinear features seem to be variable revealing the strong non-stationary character of travel speed.

Figure 7 shows a specific congested area in a speed-volume graph utilized to better understand the transitional dynamics during the formation of congestion.

As can be seen, strong determinism lays during the formation of congested traffic flow conditions. Interestingly, the nonlinear behavior, the complexity in structure, as well as the laminarity seem to fluctuate during congestion. It is evident that peaks in $L_{\max}$ correspond to transitions between strong deterministic nonlinear behavior to more periodic traffic states. Moreover, the peaks in %LAM demonstrate that during congested conditions, traffic flow may have laminar behavior, indicating slow changing dynamic and states where effect of shocks may slowly fade in time.

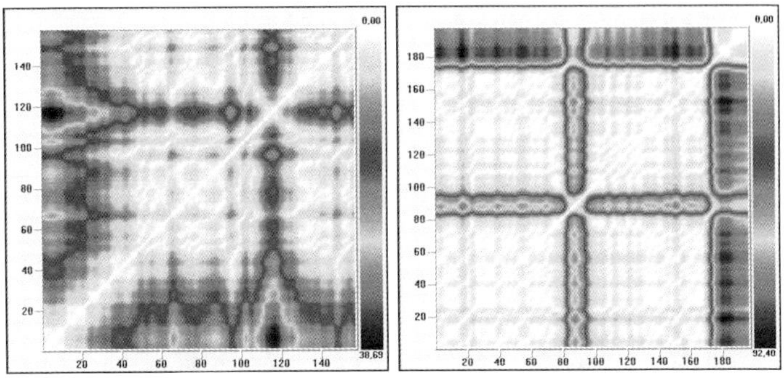

Fig. 4. Wednesday (morning peak) Wednesday (afternoon peak)

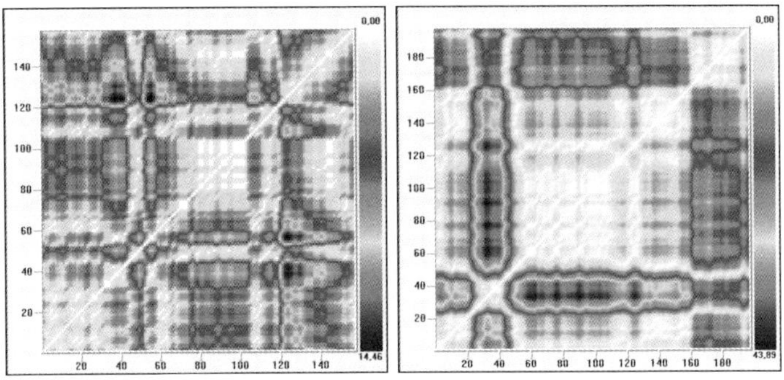

Fig. 5. Saturday (morning peak) Saturday (afternoon peak)

Recurrence Plots during morning and afternoon peak for Wednesday and Saturday ($m = 5$, $\varepsilon = 0.3$)

4. Conclusions

We implemented advanced nonlinear tools in order to visualize and statistically quantify the statistical properties of travel speed pattern evolution in urban freeways. Speed patterns as reflected in this study are complex and variable; their temporal evolution significantly differs between the morning and afternoon peak.

This interdisciplinary study provides evidence that irregularities in traffic flow can be revealed and treated independently of the traffic setting (urban arterial or freeway, signalization dense or not, etc.) but only on the basis of prevailing traffic conditions. The methodology applied provides the ability to study traffic's short-term dynamics and its statistical characteristics without prior assumptions on the non-stantionary and nonlinear

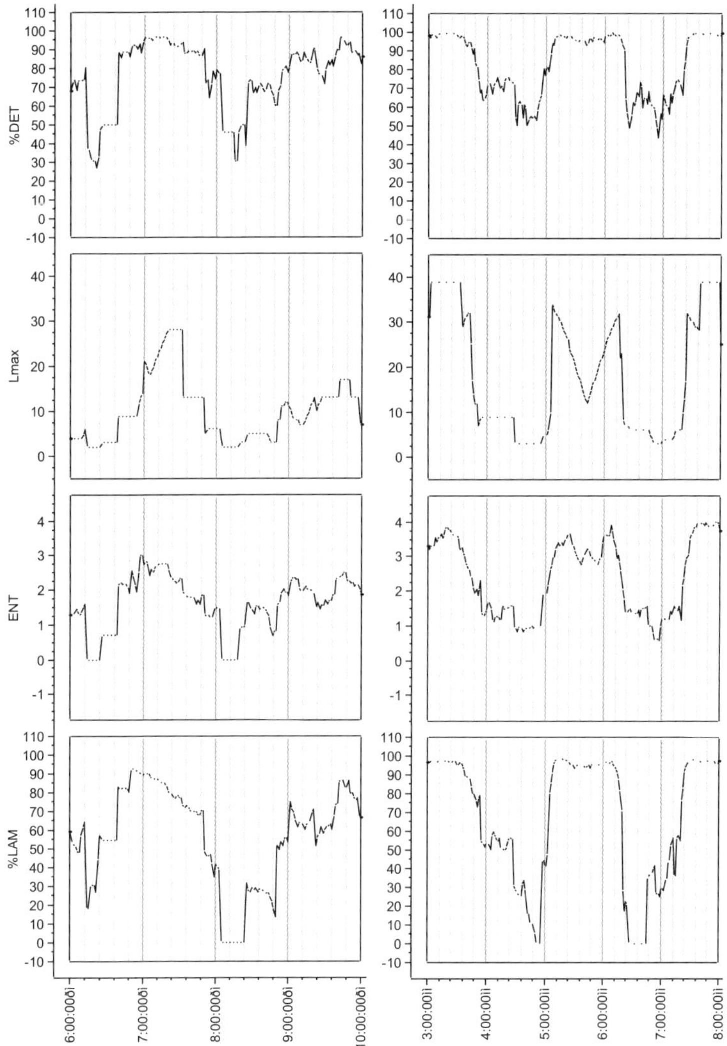

Fig. 6. RQA variables for the morning and afternoon peak of the busiest Wednesday.

features of the data. The identification and analysis of the dynamics of the available traffic flow data could lead to the better understanding of the frequently observed boundary conditions caused by phenomena of congestion and non-recurrent incidents and enhance modeling.

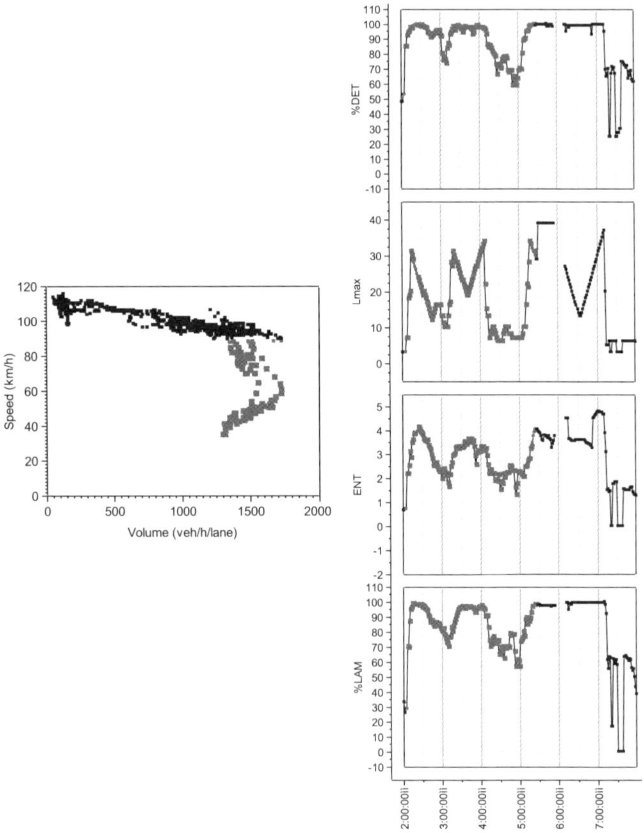

Fig. 7. The statistical characteristics of the pattern based speed evolution during congestion formation (grey corresponds to traffic during congested conditions).

References

1. T. Nagatani, *Physica A: Statistical Mechanics and its Applications* **348**, 561 (2005).
2. A. Stathopoulos and M. G. Karlaftis, *Transportation Research Part C* **11**(2), 121 (2003).
3. E. I. Vlahogianni, C. L. J. Webber, N. Geroliminis and A. Skabardonis, *Transportation Research Part C* **15**(6), 392 (2007).
4. F. L. Hall, V. F. Hurdle and J. H. Banks, *Synthesis of Recent Work on the Nature of Speed-Flow and Flow-Occupancy (Or Density) Relationships on Freeways*, Transportation Research Record 1365, TRB (National Research Council, Washington, DC, 1992).
5. M. J. Cassidy and R. L. Bertini, *Transportation Research Part B* **33**, 25 (1999).

6. B. S. Kerner and H. Rehborn, *Physical Review Letters* **79**(20), 4030 (1997).
7. C. Daganzo, M. Cassidy and R. Bertini, *Transportation Research Part A* **33**, 365 (1999).
8. F. Taken, Detecting strange attractors in turbulence, in *Dynamical Systems and Turbulance*, eds. D. A. Rand and L. S. Young, Lecture Notes in Mathematics, Vol. 898 (Springer-Verlag, Berlin, 1981) pp. 366–381.
9. J. Gao and H. Cai, *Physics Letters A* **270**, 75 (2000).
10. J. C. L. Webber and J. P. Zbilut, *Appl. Physiol.* **76**, p. 76 965 (1994).
11. N. Marwan and J. Kurths, *Physics Letters A* **302**(5–6), 299 (2002).
12. N. Marwan, C. M. Romano, M. Theiler and J. Kurths, *Physics Reports* **438**, 237 (2007).
13. J. Zbilut, A. Giuliant and C. Webber, *Physics Letters A* **237**, p. 131 (1998).
14. A. M. Fraser and H. L. Swinney, *Physical Review A* **33**(2), 1134 (1986).
15. M. B. Kennel, R. Brown and H. D. I. Abarbanel, *Phys. Rev. A* **45**, p. 3403 (1992).

On the Reliability of Nonlinear Modeling using Enhanced Genetic Programming Techniques

S. M. Winkler*, M. Affenzeller and S. Wagner

Heuristic and Evolutionary Algorithms Laboratory
Upper Austria University of Applied Sciences
4232 Hagenberg, Austria
*Email: *stephan.winkler@heuristiclab.com*
www.heuristiclab.com

The use of genetic programming (GP) in nonlinear system identification enables the automated search for mathematical models that are evolved by an evolutionary process using the principles of selection, crossover and mutation. Due to the stochastic element that is intrinsic to any evolutionary process, GP cannot guarantee the generation of similar or even equal models in each GP process execution; still, if there is a physical model underlying to the data that are analyzed, then GP is expected to find these structures and produce somehow similar results. In this paper we define a function for measuring the syntactic similarity of mathematical models represented as structure trees; using this similarity function we compare the results produced by GP techniques for a data set representing measurement data of a BMW Diesel engine.

Keywords: Nonlinear system identification; genetic programming; modeling of mechatronical systems; results similarity estimation.

1. Nonlinear Evolutionary System Identification

Genetic programming (GP, [1]) was first explored in depth in 1992 by John R. Koza who pointed out that virtually all problems in artificial intelligence, machine learning, adaptive systems, and automated learning can be recast as a search for computer programs, and that genetic programming provides a way to successfully conduct the search in the space of computer programs. Similar to GAs, GP works by imitating aspects of natural evolution to generate a solution that maximizes (or minimizes) some fitness function; a population of solution candidates evolves through many generations towards a solution using evolutionary operators and a "survival of the fittest" selection scheme. Whereas GAs are intended to find an array of characters or integers representing the solution of a given problem, the goal of a GP process is to produce a computer program (or, as in our case, a formula) solving the optimization problem at hand. Typically, a GP population contains a few hundred or thousand individuals and evolves through the action of operators known as crossover, mutation and selection; as in every evolutionary process, new individuals (in GP's case, new programs) are created, tested, and the fitter ones in the population succeed in creating children of their own, whereas unfit ones are removed from the population. Theory and application scenarios have been widely discussed[1–3].

One of the ways how GP can be used in data based modeling is its application in systems analysis: A given system is to be analyzed and its behavior modeled by a mathematical model; this process is (especially in the context of modeling dynamic physical systems) called system identification[4]. The main goal here is to determine the relationship of a dependent (target) variable to a set of specified independent (input) variables. Applying this procedure we assume that a model can be created with which it will also be possible to predict correct outputs for other data examples.

2. Comparing Results of Evolutionary Modeling

Even though GP is widely considered a method that is able to identify models for several kinds of linear as well as nonlinear systems[1–3,5,6], the following problem is repeatedly discussed: Due to the stochastic element, that is intrinsic to any evolutionary process, and the extremely wide range of possible (arbitrarily complex) models that can be formed using function and terminal definitions, genetic programming cannot guarantee that the results of GP processes applied to a given data set will always be syntactically similar to each other. Still, if there is a physical model underlying the data that are analyzed, then GP is expected to find these structures and produce somehow similar results. For analyzing this aspect, we shall use a function described in this section that can be used for estimating the similarity of nonlinear models represented as structure trees.

Structure trees representing formulas have to store several aspects that are specific to mathematical aspects including for example coefficients, time-offsets, variable indices, or information about function types referenced by function nodes. This is why we have designed and implemented a method that systematically collects all pairs of ancestor and descendant nodes and information about the properties of these nodes. Additionally, for each pair we also document the distance (with respect to the level in the model tree) and the index of the ancestor's child tree containing the descendant node. The similarity of two models is then calculated by comparing all pairs of ancestors and descendants in one model to all pairs of the other model and averaging the similarity of the respective best matches. Figure 1 shows a simple formula and all pairs of ancestors and descendants included in the structure tree representing it; the input indices as well as the level differences ("level deltas") are also given. Please note: The pairs given on the right side of Figure 1 are shown intentionally as they symbolize the pairs of nodes with level difference 0, i.e. nodes combined with themselves.

We define a genetic item gi as a 6-tuple storing the following information about the ancestor node a and descendant node d: $type_a$ (the type of a), $type_d$ (the type of d), δl (the level delta), $index$ (the index of the child branch of a that includes d), np_a (the node parameters characterizing a), and np_d (the node parameters characterizing d), where the parameters characterizing nodes are represented by tuples containing the following information: var (the variant (of functions)), $coeff$ (the coefficient (of terminals)), to (the time offset (of terminals)), and vi (the variable index (of terminals)). When it comes to calculating the similarity of two genetic items gi_1 and gi_2 as $sim(gi_1, gi_2)$, the most important aspect are the types of the definitions referenced by the nodes; if these are not equal, then the similarity is 0 regardless of all other parameters:

$$\forall_{gi_1,gi_2} : gi_1.type_a \neq gi_2.type_a \Rightarrow sim(gi_1, gi_2) = 0 \tag{1}$$

$$\forall_{gi_1,gi_2} : gi_1.type_d \neq gi_2.type_d \Rightarrow sim(gi_1, gi_2) = 0 \tag{2}$$

If the types of the nodes correspond correctly, then the similarity of gi_1 and gi_2 is calculated using the similarity contributors $s_1 \ldots s_{10}$ of the parameters of gi_1 and gi_2 weighted with coefficients $c_1 \ldots c_{10}$.

The differences regarding input index, variant and variable index are not in any way scaled or relativized, their similarity contribution is 1 in the case of equal parameters for both genetic items and 0 otherwise. The differences regarding level difference, coefficient and time offset, on the contrary, are indeed scaled: The level difference is divided by the maximum tree height $height_{max}$, the difference of coefficients is divided by the range of the referenced terminal definition (in case of uniformly distributed coefficients) or divided by the standard deviation σ (in case coefficients are normally distributed), and the difference of the time offsets is divided by the maximum time offset allowed $offset_{max}$.

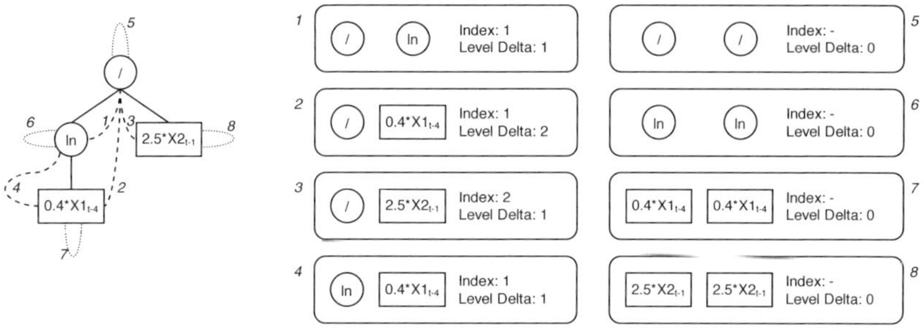

Fig. 1. Simple formula structure and all included pairs of ancestors and descendants.

$$\forall(gi_1, gi_2 : gi_1.type_a = gi_2.type_a \ \& \ gi_1.type_d = gi_2.type_d) :$$

$$s_1 = 1 - \frac{|gi_1.\delta l - gi_2.\delta l|}{height_{max}}, s_2 = \begin{cases} gi_1.index \neq gi_2.index & : \ 0 \\ gi_1.index = gi_2.index & : \ 1 \end{cases} \tag{3}$$

$$s_3 = \begin{cases} gi_1.np_a.var \neq gi_2.np_a.var & : \ 0 \\ gi_1.np_a.var = gi_2.np_a.var & : \ 1 \end{cases} \tag{4}$$

$$s_4 = \begin{cases} gi_1.np_d.var \neq gi_2.np_d.var & : \ 0 \\ gi_1.np_d.var = gi_2.np_d.var & : \ 1 \end{cases} \tag{5}$$

$$\delta c_a = |gi_1.np_a.coeff - gi_2.np_a.coeff| \tag{6}$$

$$s_5 = 1 - \begin{cases} isUniform(gi_1.type_a) & : \ \frac{\delta c_a}{gi_1.type_a.max - gi_1.type_a.min} \\ isGaussian(gi_1.type_a) & : \ \frac{\delta c_a}{gi_1.type_a.\sigma * 4} \end{cases} \tag{7}$$

$$\delta c_d = |gi_1.np_d.coeff - gi_2.np_d.coeff| \tag{8}$$

$$s_6 = 1 - \begin{cases} isUniform(gi_1.type_d) & : \ \frac{\delta c_d}{gi_1.type_d.max - gi_1.type_d.min} \\ isGaussian(gi_1.type_d) & : \ \frac{\delta c_d}{gi_1.type_d.\sigma * 4} \end{cases} \tag{9}$$

$$s_7 = 1 - \frac{|gi_1.np_a.to - gi_2.np_a.to|}{offset_{max}} \tag{10}$$

$$s_8 = 1 - \frac{|gi_1.np_d.to - gi_2.np_d.to|}{offset_{max}} \tag{11}$$

$$s_9 = \begin{cases} gi_1.np_a.vi \neq gi_2.np_a.vi & : \ 0 \\ gi_1.np_a.vi = gi_2.np_a.vi & : \ 1 \end{cases} \tag{12}$$

$$s_{10} = \begin{cases} gi_1.np_d.vi \neq gi_2.np_d.vi & : \ 0 \\ gi_1.np_d.vi = gi_2.np_d.vi & : \ 1 \end{cases} \tag{13}$$

Finally, we can now calculate the structural similarity of gi_1 and gi_2, $sim(gi_1, gi_2)$: We first calculate a punishment factor p_i for each s_i (using weighting factors c_i, $0 \leq c_i \leq 1$ for $i \in [1, 10]$) as $p_i = (1 - s_i) \cdot c_i$ and then get the temporary similarity result as

$$sim_{tmp}(gi_1, gi_2) = \prod_{i=1}^{10}(1 - p_i). \tag{14}$$

In the worst case scenario we get $s_i = 0$ for all $i \in [1, 10]$ and therefore the worst possible sim_{tmp} is

$$sim_{worst} = \prod_{i=1}^{10}(1 - ((1 - s_i) \cdot c_i)) = \prod_{i=1}^{10}(1 - c_i). \tag{15}$$

As sim_{worst} is surely greater than 0 we linearly scale the results to the interval $[0, 1]$:

$$sim(gi_1, gi_2) = \frac{sim_{tmp}(gi_1, gi_2) - sim_{worst}}{1 - sim_{worst}}. \tag{16}$$

By setting a weighting coefficient c_j to a rather high value (i.e., near or even equal to 1.0) the total similarity will become very small for pairs of genetic items that do not correspond with respect to this specific aspect, even if all other aspects would lead to a high similarity result.

Based on this similarity measurement it is easy to formulate a similarity function that measures the similarity of two model structures: For comparing models m_1 and m_2 we collect all pairs of ancestors and descendants (up to a given maximum level difference) in m_1 and m_2 and look for the best matches in the respective opposite model's pool of genetic items, i.e. pairs of ancestor and descendant nodes. For each genetic item gi_j in the structure tree of m_1 we elicit that genetic item gi_k in the model structure m_2 with the highest similarity to gi_j; the similarity values $\mathbf{s}$ are collected for all genetic items contained in m_1 and their mean value finally gives us a measurement for the structure based similarity of the models m_1 and m_2, $sim(m_1, m_2)$.

Obviously, this similarity measure sim is not symmetric, which means that $sim(m_1, m_2)$ does not necessarily return the same result as $sim(m_2, m_1)$ for any pair of models m_1 and m_2 (simply because genetic items of any formula might be included in other ones, but not necessarily vice versa).

3. Empirical Tests

The NO_x data set contains the measurements taken from a 2 liter 4 cylinder BMW diesel engine at a dynamical test bench (simulated vehicle: BMW 320d Sedan). Several emissions (including NO_x, CO and CO_2) as well as several other engine parameters have been recorded over approximately 30 minutes and downsampled to 20 Hz; 40 signals have been recorded, but only 9 variables have been considered for identification. We have used the NO_x data set and executed 10 separate GP runs for identifying models for the given NO_x target values; this procedure has been repeated 5 times. By comparing the results we get an impression of the similarity of the models produced by GP. The population size has been set to 1,000, the mutation rate to 0.15, the maximum tree size to 50, and strict offspring selection[7] has been used (which means that both success ratio and comparison factor were set to 1.0) as well as a combination of random and proportional parents selection. We have used the HeuristicLab framework[8] as underlying optimization framework.

The analysis of the results has been done in the following way: For each test run (including 10 separate GP processes) we have collected the models that were eventually returned by the modeling algorithms; all models have been syntactically compared to each other using the comparison function presented in Section 2 (all coefficients $c_1 \dots c_{10}$ were set to 0.2, only the coefficient c_1 weighting the level difference contribution s_1 was set to 0.8.). The results are summarized in Table 1: For each test run we give the average and standard deviation values of the similarities of the models; additionally, for each model we have collected the maximum similarity values $maxSim$ and calculated average and standard deviation for these values, too. Obviously, the results produced are not extremely similar to each other: Given the fact, that 1,000 randomly generated models for the NO_x data set show an average similarity of 0.1287 (std: 0.0255), an average similarity of 0.4149 (std: 0.1442) seems plausible, but unfortunately not really high; in fact, we had expected higher values.

One fact has to be considered here: Models produced by GP are likely to become maximally big and are thus prone to over-fitting. This is why pruning methods are frequently used for removing parts of the models produced in order to become smaller, more compact formulas; still, pruning should not decrease the quality of the models more than necessary (this trade-off always has to be kept in mind when it comes to pruning). We have used a pruning method based on evolution strategies (ES, [9]), as described recently[3], for reducing the complexity of the models that are returned as

Table 1. Similarities of models for the NO_x data set.

Test run	*mean(sim)*	*std(sim)*	*mean(maxSim)*	*std(maxSim)*
1	0.3991	0.1705	0.6038	0.1673
2	0.4279	0.1546	0.6518	0.0825
3	0.3966	0.1359	0.5761	0.1608
4	0.4009	0.1299	0.6780	0.1265
5	0.4498	0.1300	0.5555	0.0904
avg.	*0.4149*	*0.1442*	*0.6130*	*0.1255*

results of the GP processes by cutting out nodes and deleting subtrees of the given model structures; no quality deterioration was allowed. After doing so, the pruned results were once again compared to each other leading to the results summarized in Table 2.

Table 2. Similarities of pruned models for the NO_x data set.

Test run	*mean(sim)*	*std(sim)*	*mean(maxSim)*	*std(maxSim)*
1	0.6049	0.1995	0.7774	0.1853
2	0.5731	0.1479	0.7089	0.0750
3	0.6055	0.1217	0.7252	0.0757
4	0.6435	0.1352	0.7389	0.0940
5	0.6921	0.0926	0.7702	0.0752
avg.	*0.6238*	*0.1394*	*0.7441*	*0.1010*

Obviously, the mutual similarity of the pruned results is a lot higher than the original ones, on average even reaching 0.6238. This result is in fact very satisfying because it shows that independent GP processes have in this case really produced formulas with rather similar model structures; pruning reveals the really essential parts of the models, and these essential parts are more similar to each other than the original ones.

4. Conclusion

We have here presented a method for comparing models produced by evolutionary system identification; using this similarity estimation we have systematically compared the results obtained for the identification of NO_x emissions of a diesel engine. Comparing these results we see that especially the pruned models show higher similarities with each other. Thus, the results presented here can be seen as an indication that GP using enhanced selection concepts does not only work, but can also be considered reliable with respect to comparability and similarity of results if there really is a system whose behavior is represented in the given training data.

Acknowledgements

The work described in this paper was done within the Translational Research Project L284-N04 "GP-Based Techniques for the Design of Virtual Sensors" sponsored by the Austrian Science Fund (FWF). We thank members of the Institute for Design and Control of Mechatronical Systems at JKU Linz who provided and helped us with the NO_x data set.

References

1. J. R. Koza, *Genetic Programming: On the Programming of Computers by Means of Natural Selection* (The MIT Press, 1992).
2. W. B. Langdon and R. Poli, *Foundations of Genetic Programming* (Springer Verlag, Berlin Heidelberg New York, 2002).
3. S. Winkler, Evolutionary system identification - modern concepts and practical applications, PhD thesis, Institute for Formal Models and Verification, Johannes Kepler University Linz(2008).
4. L. Ljung, *System Identification – Theory For the User, 2nd edition* (PTR Prentice Hall, Upper Saddle River, N.J., 1999).
5. S. Winkler, H. Efendic, M. Affenzeller, L. D. Re and S. Wagner, *International Journal on Intelligent Systems Technologies and Applications* **2**(2/3), 255 (2006).
6. S. Winkler, M. Affenzeller and S. Wagner, *Journal of Mathematical Modelling and Algorithms* **6**(3), 455 (2007).
7. M. Affenzeller and S. Wagner, Offspring selection: A new self-adaptive selection scheme for genetic algorithms, in *Adaptive and Natural Computing Algorithms*, ed. B. R. et al.Springer Computer Science (Springer, 2005).
8. S. Wagner and M. Affenzeller, HeuristicLab: A generic and extensible optimization environment, in *Adaptive and Natural Computing Algorithms*, ed. B. R. et al.Springer Computer Science (Springer, 2005).
9. H.-P. Schwefel, Evolutionsstrategie und numerische optimierung, PhD thesis, Technische Universität Berlin(1975).

Plasma-Surfaces Interaction and Improvement of Plasma Conservation System by Application of a d. c. Electrical Potential

C. L. Xaplanteris

Plasma Physics Lab, IMS, NCSR "Demokritos", Athens, Greece and Hellenic Military Academy

E. Filippaki

Plasma Physics Lab, IMS, NCSR "Demokritos", Athens, Greece Email: lfilip@ims.demokritos.gr

The effect on surfaces that are in plasma environment is one of the problems of plasma chemistry. A variable electric field layer (called sheath) between plasma and surface is capable of changing ion concentration which results to the reduction of surface of the objects that exist in plasma environment (cleaning of archaeological objects). A proposed improvement of plasma restoration system with the application of external electric field is expected to have better results on the conservation of archaeological objects.

Keywords: Plasma-surface interaction; impurities; plasma sheath; conservation of iron objects; hydrogen plasma.

1. Introduction

The impurities' problem in the thermonuclear plasma into TOKAMAK reactors is well known.[1] The majority of the impurities are caused from the particles, which are released from restrained plasma solid surfaces or from surfaces, which in various ways touch the ionized gas.[2,3] There, at the contact place, the 'plasma sheath' is created. In this area, a negative potential is developed which reduces the number of electrons, while the number of ions is increased and finally the plasma has reductive behavior.[1,4,5] The plasma sheath with its negative potential is used for the cleaning and restoration of archaeological metallic objects. During the last few years, this activity has successfully been carried out at the plasma Physics Laboratory in NCSR 'Demokritos'.

In the present work some results from previous studies on plasma-solid surfaces interaction are presented. A theoretical elaboration adapted on our experimental conditions shows the existence of plasma sheath potential, which makes plasma drastic. From the same study, other physical quantities (electron density, ion density, ion velocity etc.), which affect the plasma behavior, are also shown.[6] In addition to theoretical results, an improvement of the plasma-cleaning device has been made, aiming at the faster cleaning of the objects, as well as their possible hardening. This improvement is achieved, when a negative potential (at sheath potential order) is placed on the metallic objects, so that the sheath negative potential increases with simultaneous increase of ion concentration, and the plasma becomes more effective. Similarly, by using positive external potential the sheath electric field is reduced and the plasma results are limited. Moreover, nitrogen plasma treatment at our laboratory creates a nitrogen layer on metallic objects, which leads to their hardening.

The next step could be the plasma-alkalis interaction, which is likely to restore the fading of treated objects.

2. Plasma Impurities

One serious problem, which is caused from plasma impurities, is the energy loss due to radiation from partially ionized atoms. In addition, large concentration of the impurities makes the plasma **heating** difficult. This is a problem noticed particularly at the beginning of plasma heating, because the impurities radiate much more before their ionization becomes very large. Furthermore, the impurities may cause plasma interruptions due to boundaries' cooling, and consequently the current's profile is modified.[1]

Figure 1 shows the impurities' percentage (logarithm axis), which gives radiation energy equal to 10% of all plasma thermodynamic power, versus the atomic number of atoms.

It must be noticed that impurities with low atomic number are not a serious problem as the impurities which contain heavy atoms. Obviously, impurities are created from material surfaces, which touch the plasma, due to various interactions between ionized gas and surfaces. It is possible for atoms from wall surfaces to be released into the plasma with procedures such as **sputtering, arcing and evaporation**. Sputtering is momentum transportation, as ions with big energy strike on the surfaces. Arcing may happen between plasma and surfaces, leaded from plasma sheath potential. Evaporation occurs when suitable energy heats the surface near to 'melting point'.

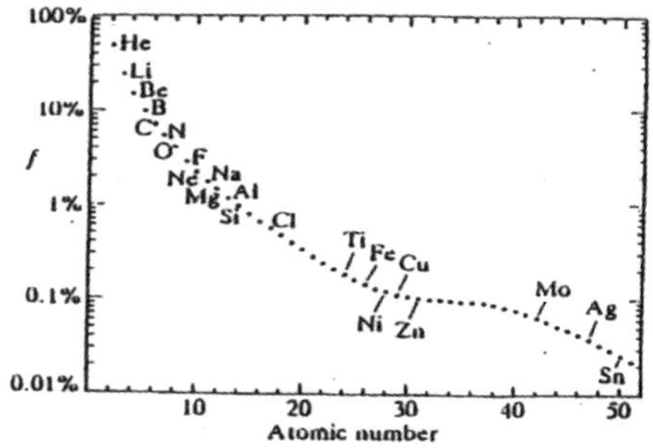

Fig. 1. Percentage of impurity versus the atomic number.

3. The Plasma Sheath

The atoms of a neutral gas, which is at touch with an absorbing surface, flow towards the surface with the **thermal velocity**. Only a numerical factor makes the difference between the thermal velocity and **acoustic velocity**. When we have plasma, the flow velocity towards the surface is again nearly to the acoustic one, but the understanding of this behavior is more complicated.

The electron thermal velocity is bigger than the ion one, by the square root of the mass ratio,

$$u_{th,e} = \sqrt{\frac{m_i}{m_e}} \cdot u_{th,i}$$

However, due to an electrical field, which is developed into plasma, the electron and ion flux becomes equal by decreasing the electron flux and increasing the ion one. The final flux is defined from electron and ion pressure combination, and is restrained by ion inertia. The initial electric field is located at a narrow sheath in touch with the surface whose thickness is some Debye lengths. One small electrical field, called pre-sheath, extends more deeply into the plasma.

The Poisson equation gives the electrical potential,

$$\nabla^2 \phi = \frac{e}{\varepsilon_0}(n_e - n_i) \quad \text{or} \quad \frac{d^2\phi}{dx^2} = \frac{e}{\varepsilon_0}(n_e - n_i) \tag{1}$$

where n_e and n_i are the electron and ion number density respectively. Considering that $\phi = 0$ at sheath edge, then the electrons distributed by Boltzmann are:

$$n_e = n_0 e^{\frac{e\phi(\chi)}{T_e}} \tag{2}$$

The ion density n_i may be obtained from energy balance,

$$\frac{1}{2}m_i u_i^2 = \frac{1}{2}m_i u_0^2 - e\phi$$

Where u_0 is the ion velocity when it enters into the sheath. From above we have,

$$u_i = \left(\frac{m_i u_i^2 - 2e\phi}{m_i}\right)^{\frac{1}{2}} \tag{3}$$

Due to the ion flux constancy it could be written, that $n_i u_i = n_0 u_0$, and a combination with (3) gives:

$$n_i = n_0 \frac{u_0}{u_i} \Rightarrow n_i = n_0 \left(\frac{\frac{1}{2}m_i u_0^2}{\frac{1}{2}m_i u_0^2 - e\phi}\right)^{\frac{1}{2}} \tag{4}$$

With substitution of the last into (1) we have,

$$\frac{d^2\phi}{d\chi^2} = \frac{n_0 e}{\varepsilon_0}\left[e^{\frac{e\phi}{T}} - \left(\frac{\frac{1}{2}m_i u_0^2}{\frac{1}{2}m_i u_0^2 - e\phi}\right)^{\frac{1}{2}}\right] \tag{5}$$

One interesting solution of equation (5) is in the range where the narrow sheath touches the pre-sheath (low electrical potential) At this point we have $\phi \prec\prec$ and then is valid,

$$e^{\frac{e\phi}{T_e}} = 1 + \frac{e\phi}{T_e}$$

With substitution the (5) is written,

$$\frac{d^2\phi}{dx^2} = \frac{n_0 e}{\varepsilon_0}\left[1 + \frac{e\phi}{T_e} - \left(1 + \frac{e\phi}{\frac{1}{2}m_i u_o^2 - e\phi}\right)\right] \tag{6}$$

As $e\phi \ll \frac{1}{m\,i} u_0^2$ is valid,

$$\frac{\frac{1}{2}m_i u_0^2}{\frac{1}{2}m_i u_0^2 - e\phi} = 1 + \frac{e\phi}{\frac{1}{2}m_i u_0^2 - e\phi} = 1 + \frac{2e\phi}{m_i u_0^2} \cong \left[1 + \frac{e\phi}{m_i u_0^2}\right]^2$$

And equation (6) becomes,

$$\frac{d^2\phi}{dx^2} = \frac{n_0 e^2 \phi}{\varepsilon_0 T_e}\left(1 - \frac{T_e}{m_i u_0^2}\right)$$

With $\lambda_D = \left(\frac{\varepsilon_0 T_e}{n_0 e^2}\right)^{1/2}$ (λ_D the Debye length), the last equation is written:

$$\frac{d^2\phi}{dx^2} = \frac{\phi}{\lambda_D^2}\left[1 - \frac{T_e/m_i}{u_0^2}\right] \tag{7}$$

When the potential ϕ changes slowly, the solution requires,

$$u_0 \cong \left[\frac{T_e}{m_i}\right]^{1/2}$$

If ion temperature is taken into consideration, and generalize the previous relation, then $u_o = c_s$, where into the model limits is valid: $c_s^2 \cong \frac{T_e+T_i}{m_i}$. Thus, the plasma comes in the sheath with the acoustic velocity.

With consideration that the total current on surface must be zero, the potential ϕ_0 at the sheath edge can be estimated. The ion current density into sheath and consequently forward the surface is,

$$j_i = n_0 e u_0 \quad \text{or} \quad j_i = n_0 e c_s \tag{8}$$

For electron flux from sheath to surface is: $\frac{1}{4} n_e c_s$, where n_e is the electron number density on surface, and $c_s = \sqrt{\frac{8T_e}{\pi m_e}}$. If we take into consideration (2), the electron current density is written:

$$j_e = -\frac{1}{4} n_0 e \left(\frac{8T_e}{\pi m_i}\right)^{1/2} e^{\frac{e\phi_0}{T_e}}$$

However it is valid that, $j_i + j_e = 0$, and taking into consideration (8) with the last one, we write:

$$n_0 e. \left(\frac{T_e+T_i}{m_i}\right)^{1/2} - \frac{1}{4} n_0 e \left(\frac{8T_e}{\pi m_e}\right)^{1/2} e^{\frac{e\phi_0}{T_e}} = 0$$

or

$$-\frac{e\phi_0}{T_e} = \frac{1}{2} \ln \frac{m_i/m_e}{2\pi\left(1+T_i/T_e\right)} \tag{9}$$

For deuterium plasma into the thermonuclear reactor we have, $-e\phi_0 \cong 2,8T_e$, whereas for hydrogen plasma as in our experiment, respectively it becomes, $-e\phi_0 = 2,5T_i$

Now returning to equation (7), and with $k^2 = \frac{1}{\lambda_D^2}\left(1 - \frac{T_e/m_i}{u_0^2}\right)$ equation (7) is written,

$$\frac{d^2\phi}{dx^2} - k^2\phi = 0 \tag{10}$$

One possible solution has the form, $\phi = \phi_0 e^{\rho\chi}$, where ρ is to be defined. Then, the eq. (10) with the proposed solution gives $\rho = \pm k$.

Finally, the solution is

$$\phi = \phi_0 e^{\pm kx} \tag{11}$$

with $k^2 = \frac{1}{\lambda_D^2}\left(1 - \frac{T_e/m_i}{u_0^2}\right)$ and $\phi_0 \cong -2,5\frac{T_e}{e}$

Approximately, it could be taken $k \cong \frac{1}{\lambda_D}$, and the solution (11) is represented graphically by Figure 2.

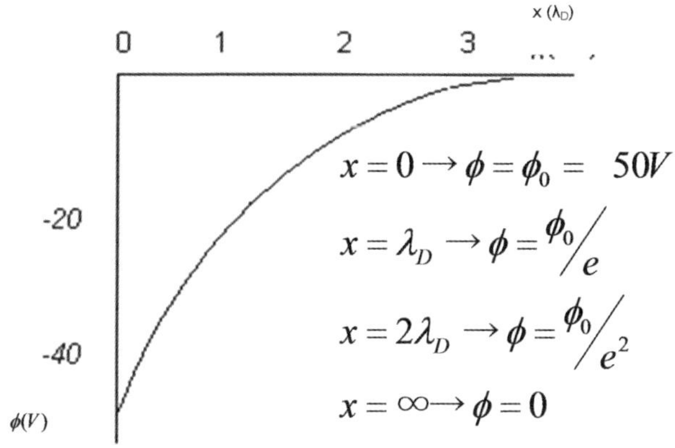

Fig. 2. Sheath potential versus distance from surface.

4. Experimental

4.1. *Preparation of samples-treatment of samples*

Two types of samples were used in this work: i) Steel coupons that were pre-corroded in a salt spray cabinet. A film of corrosion products was formed on the side subjected to the spray. ii) Selected corroded iron objects from the collection of the Holy Monastery of Simonopetra of Mount Athos. An RF glow discharge plasma device made by Vacotec S.A., similar to Veprek's prototype apparatus Veprek et al, 1988,[7] was used to treat the samples. Figure 3 shows the plasma reactor at "Demokritos" Plasma Laboratory, at full function during the cleaning procedure. The details of the plasma reactor have been published in [8].

A reducing plasma treatment of 100% H2 was applied on the samples at pressure of 1 to 1.2 Torr, and at a temperature range of 200°C to 220°C. Two different plasma procedures took place. The first procedure was as follows: steel coupons or excavated objects divided into 2 similar pieces were treated in plasma for 3 hours. One piece was treated with an external negative potential of 50V applied on it, while the other one was cleaned only by the action of hydrogen plasma. The second procedure was the same as the former with the difference that the external potential was positive (50V).

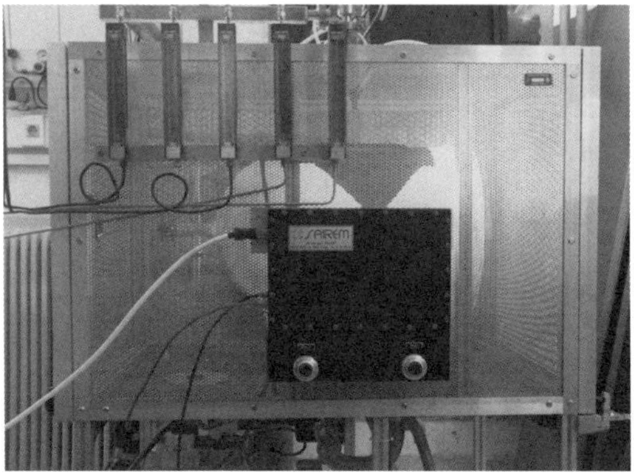

Fig. 3. RF plasma device during plasma treatment.

During the treatment process, the outer oxide layer became more porous and (therefore) less hard by transformations and reduction. The layer was removed mechanically (by means of a scalpel) until the original surface of the samples was revealed.

The identification of crystalline phases within the oxide layer before and after each stage of treatment of the samples was followed by means of X-ray diffraction method (XRD). A Siemens model D500 (Cuka radiation, 40kV, 35mA) automatic diffractometer was used for this work. 50-200mg of powder from the external oxide layer were removed mechanically and in the case of the steel coupons the surface oxide was sampled. Samples of the objects and coupons before treatment were also analyzed as reference samples for the plasma treatment effect.

4.2. *Results*

Results from XRD spectra from each category of samples are shown in Tables 1 and 2. In the case of the excavated objects silicates from the environment were evident in the spectra.

It was obvious that the main corrosion products in all steel coupons were the iron oxyhydroxides $Fe(OH)_3$ and akaganeite (β-FeOOH), while the presence of iron (Fe) is dominant. Concerning the objects, the hydroxides goethite (α-FeOOH) and lepidocrocite (γ-FeO(OH, Cl)) were the dominant corrosion products together with hematite (α-Fe_2O_3). It is also noted the presence of maghemite (γ-Fe_2O_3) as a corrosion product, which can be formed by the dehydration of lepidocrocite.[9]

The results obtained after the two plasma procedures differ between the two types of samples. As far as concerns the steel coupons treated with the additional positive potential, iron hydroxides are still identified followed by the presence of maghemite. In addition hematite is also present in the reference sample. On the other hand, spectra of the coupons that were treated with the external negative potential, consisted of magnetite and hematite. By contrast, spectra of the corroded objects of the two different plasma procedures consisted of magnetite (Fe_3O_4) and hematite.

Dehydration of iron hydroxides to the trivalent iron oxide Fe2O3 (hematite or maghemite) took place according to the reaction $2FeOOH \rightarrow Fe_2O_3 + H_2O$. In the reducing plasma environment, the reduction $F^{3+} \rightarrow Fe^{2+}$ also took place through the action of H atoms of the discharge. H atoms originate from the dissociation of molecular hydrogen in the gaseous phase of the discharge by the action of energetic electrons. Since negative potential is applied on the surface of treated coupons more H atoms are attracted so that plasma action is reinforced. By contrast when positive potential is applied on the surface of the treated coupons, concentration of H atoms is decreased and this results to the weakening of plasma action. Finally, hematite is reduced to magnetite according to the reduction $3Fe_2O_3{+}2H \rightarrow 2Fe_3O_4{+}H_2O$. Evaluating the above results, and comparing the results between plasma treatment with the additional positive potential to plasma treatment to the negative potential, we can conclude that additional negative potential applied on the surface of the coupons reinforces plasma so that chemical transformations take place faster. The above results were not obvious on the treated objects. The corrosion products identified on the excavated objects were hematite (α-Fe_2O_3), maghemite (γ-Fe_2O_3), goethite (α-FeOOH) and lepidocrocite (γ-FeO(OH, Cl)). After plasma treatment, regardless of applied potential (negative or positive), only hematite and magnetite were identified. Presence of hematite before treatment shows that the dehydration of hydroxides had already partly taken place. Consequently dehydration was completed during plasma treatment and reduction of hematite had begun. It is concluded that the extent of the corrosion layer, if treated in the same plasma conditions, affects on the reduction of the oxides. A shorter treatment on corroded objects might show the expected results (different influence between positive or negative potential) so that chemical transformations were not saturated.

5. Conclusions

According to predictions that originated from the above theory, when we apply negative potential on a corroded sample, faster and more evident

Table 1. Results of XRD in Steel coupons before and after plasma treatment.

Treatment	Sample	Iron Fe	Akaganeite β-FeO(OH,Cl)	Iron Hydroxide $Fe(OH)_3$	Hematite Fe_2O_3	Maghemite γ-Fe_2O_3	Magnetite Fe_3O_4
No treatment	D1	+	+				
Treatment with additional positive external potential (reference sample)	D1+	+	+	+	+	+	
Treatment with additional positive external potential	D1+V	+	+	+		+	
Treatment with additional negative external potential (reference sample)	D1-	+			+		+
Treatment with additional negative external potential	D1-V	+			+		+

Table 2. Results of XRD in Steel coupons before and after plasma treatment.

Treatment	Sample	Hematite Fe_2O_3	Maghemite γ-Fe_2O_3	Goethite α-FeOOH	Lepidocrocite γ-FeOOH	Magnetite Fe_3O_4
No treatment	S1	+	+	+	+	
Treatment with additional positive external potential (reference sample)	S1+	+				+
Treatment with additional positive external potential	S1+V	+				+
Treatment with additional negative external potential (reference sample)	S1-	+				+
Treatment with additional negative external potential	S1-V	+				+

results are expected. On the contrary, if the applied potential is positive results are expected to be weaker. In the present study, excavated objects and steel coupons artificially corroded have been treated in a reducing plasma environment with an additional external potential (negative and positive).

Difference in the transformation behavior of the corrosion layers on each category of samples has proved the expected interaction between plasma and sample, as mentioned in section 4, although a little difference may be noticed from predicted results. This difference may be caused by the following reasons:

(1) The initial corrosion of the sample, on steel coupons the influence of additional potential is obvious while on excavated objects is not.
(2) The elaboration time of treatment. When elaboration time is very short or very long the potential's influence is difficult to appear. Es-

pecially for long elaboration time, chemical transformations have been saturated. In addition it must be taken into account the relative values of the two d.c. potentials: the sheath's potential and the external one.

References

1. J. Wesson, *Tokamaks*, 2 edn. (Clarendon Press, Oxford, 1997).
2. D. Block, A. Piel, C. Schroder and T. Klingee, *Phys. Rev* **63**, p. 056401 (2001).
3. J. Vranjes, H. Sallem and S. Poedts, *Phys. Rev. E* **69**, p. 056404 (2002).
4. M. Salimullah, A. M. Rizwan, M. Nambu, H. Nitta and P. K. Shukla, *Phys, Rev. E* **70**, p. 026404 (2004).
5. M. Salimullah, I. Sandberg and P. K. Shukla, *Phys. Rev. E* **68**, p. 027403 (2003).
6. C. Xaplanteris, *Astrophysics and Space Science* **136**, p. 171 (1987).
7. S. Veprek, C. Eckmann and J. Elmer, *Plasma Chem. Plasma Processing* **8**(4), 445 (1988).
8. I. Kotzamanidi, P. Vassiliou, E. Sarris, A. Anastasiadis, L. Filippaki and S. E. Filippakis, *Anti-Corrosion Methods and Materials* **49**(4), 256 (2002).
9. J. D. Dana and E. S. Dana, *The system of mineralogy of James Dwight Dana and Edward Salisbury Dana, Elements, Sulfides, Sulfosalts, Oxides, vol I* (John Wiley & Sons, New York, NY, 1958).

AUTHOR INDEX

DATE DUE